Vancouver and Its Region

Vancouver and Its Region

edited by
Graeme Wynn and
Timothy Oke

UBCPress / Vancouver

ISBN 0-7748-0407-6 (hardcover)
ISBN 0-7748-0421-1 (paperback)

Canadian Cataloguing in Publication Data
Main entry under title:
Vancouver and its region

Includes bibliographical references and index.
ISBN 0-7748-0407-6 (bound). – ISBN
0-7748-0421-1 (pbk.)

   1. Vancouver (B.C.) – Geography. 2. Lower
Mainland (B.C.) – Geography. I. Wynn, Graeme,
1946-   II. Oke, T.R.
FC3847.3.V35 1992   917.11'33   C92-091114-5
F1089.5.V22V35 1992

Design: George Vaitkunas
Principal cartography: Eric Leinberger
Index: Annette Lorek

UBC Press
University of British Columbia
6344 Memorial Rd
Vancouver, BC V6T 1Z2
(604) 822-3259
Fax: (604) 822-6083

This book has been financially assisted by the
Province of British Columbia through the British
Columbia Heritage Trust and BC Lottery Revenues.

Publication of this book was also made possible by
ongoing support from The Canada Council, the
Province of British Columbia Cultural Services
Branch, and the Department of Communications
of the Government of Canada.

Contents

Preface and acknowledgments   *vii*

Introduction   *xi*
Graeme Wynn

1 Views of the metropolitan area   *1*
Alfred H. Siemens

2 The primordial environment   *17*
O. Slaymaker, M. Bovis, M. North, T.R. Oke, J. Ryder

3 The Lower Mainland, 1820–81   *38*
Cole Harris

4 The rise of Vancouver   *69*
Graeme Wynn

5 Primordial to prim order: A century of environmental change   *149*
T.R. Oke, M. North, O. Slaymaker

6 Vancouver, the province, and the Pacific Rim   *171*
Trevor J. Barnes, David W. Edgington,
Kenneth G. Denike, Terry G. McGee

7 Vancouver since the Second World War: An economic geography   *200*
Robert N. North, Walter G. Hardwick

8 Time to grow up? From urban village to world city, 1966–91   *234*
David Ley, Daniel Hiebert, Geraldine Pratt

9 The biophysical environment today   *267*
D.G. Steyn, M. Bovis, M. North, O. Slaymaker

Epilogue   *291*
Derek Gregory

Bibliographical notes   *299*
Credits   *317*
Contributors   *321*
Index   *323*

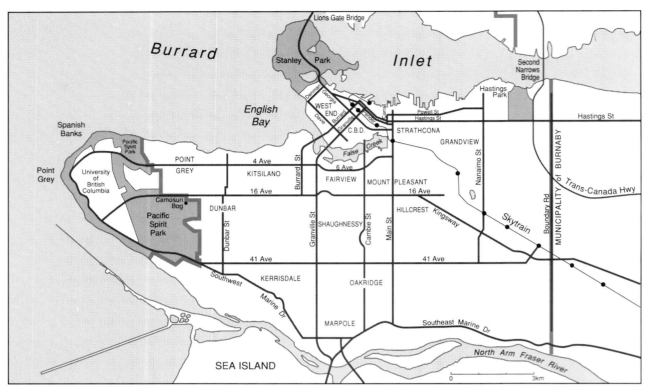

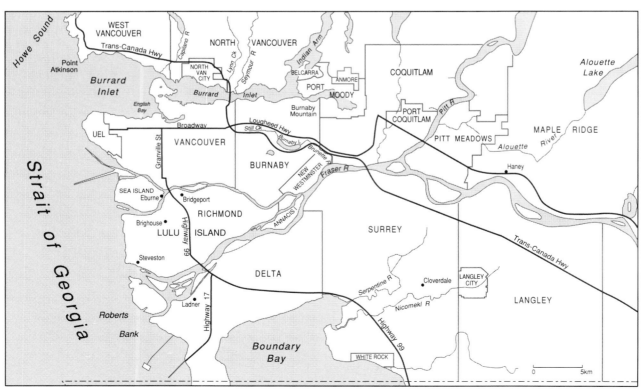

Vancouver and its region

Preface and acknowledgments

As a collaborative venture, drawing upon the expertise of a markedly diverse group of scholars to offer an accessible, integrated, and broadly inclusive account of a relatively small area, this book is unique. Remarkably, it is also entirely a product of the Department of Geography of the University of British Columbia, and these circumstances would seem to require a brief account of its genealogy.

Although the precise timing of the publication of *Vancouver and Its Region* is a direct result of the department's commitment to host the annual conference of the Canadian Association of Geographers in the spring of 1992, the idea for a book of this sort had rather hazier origins. As an historical geographer interested in understanding the ways in which, and why, people have transformed their environments over time, convinced that the broad compass of geographical scholarship offered a splendid window on this topic, and heartened by rising public interest in environmental questions, I had grown increasingly concerned, during the 1980s, about the fragmentation of geographical inquiry and the fact that many geographers seemed less interested in addressing the results of their work to a general audience than in communicating with small groups of like-minded experts through scholarly journals.

I was not, of course, alone in this unease. Other geographers, and many scholars in other disciplines, were similarly exercised. Anthropologists, and historians in particular, had been lamenting, for some time, the consequences of specialization, the tendency of their colleagues to know more and more about less and less, and the sense that their professions were becoming divorced from, and irrelevant to, the broader public. Calls for the return of stirring narrative and compelling synthesis were heard across a broad front. But however well-intentioned, such arguments are inherently contentious. In history, for example, feminist, labour, social, and cultural historians, among others, worry that the cry for more public history is really an argument for more political and military history and the return of powerful men (or at least their stories) to centre stage.

For all that, I remain convinced that synthesis need not be narrowly exclusive, and that by virtue of the diverse perspectives it brings to bear on the physical and human world, modern geography has much of importance to convey to late twentieth-century audiences. Further, I would argue that this message is likely to be most forceful when it is firmly tied to a particular place. There is no pretext for parochialism in this, because no part of the world can be understood in and of itself alone. But there is a profound conviction that places matter. How better to demonstrate this, I began to recognize, than through the collective wisdom of my departmental colleagues brought to bear on the place in which we all live and work.

Although few of us have made Vancouver or its region the main focus of our professional inquiries over the years, many of us have sought to understand facets of this place for ourselves and our students. Perhaps as a result, the proposal that we collaborate to produce this book was greeted with general enthusiasm although, in the end, our desire to produce a cohesive volume by May 1992 meant that a few colleagues with other especially pressing commitments or specialized research interests that did not fit the emerging shape of the volume were unable to participate in the project. To my considerable pleasure, Tim Oke agreed to join me in editorial harness with particular responsibility for the co-ordination of work on chapters 2, 5, and 9; and Olav Slaymaker, who had consistently attempted to maintain the discourse between physical and human geographers as Head of the Department of Geography since 1982, gave the enterprise his full support.

Thus we began. From the first, we strove to develop an integrated set of accessible essays that would, together, convey a sense of the broad range of fascinating and distinctive perspectives that Geography offers for the understanding of places while demonstrating the subject's capacity to put the increasingly fragmented pieces of modern scholarship together in a compelling and informative manner. To light us on our way, we noted relatively early in our efforts some observations offered by Stephen Jay Gould in a review of John Maynard Smith's *Games, Sex and Evolution* (published in *Nature*, 339, 8 June 1989). Here Gould distinguishes two strands of popular writing about the natural world. The Franciscan, he says, exalts nature in poetic terms and seeks empathy by appeal to the emotions; the Galilean – characteristically the product of writers who understand the process of science – emphasizes the loveliness of intellectual solutions, and uses clear, jargon-free prose to convey the intricacy, ambiguity, and beauty of science. Physical geographers, social scientists, humanists, we agreed that we would together follow Galileo's lead in our concerted effort to convey the intricacy and loveliness of our geographical perspective.

If we have succeeded, the credit lies firmly with the eighteen friends and colleagues who gave liberally of their research expertise, scholarly insight, time, and talent for this project. In particular, I thank those who co-ordinated work on the multi-authored chapters: to recognize their efforts we have placed their names at the head of the list of authors for these chapters; all other contributors appear, without regard for the extent of their contribution, in alphabetical order. Few of us, I suspect, realized quite what we were getting into when we began. None of my colleagues, I am sure, anticipated the vigour with which I wielded my editorial pencils. For their tolerance, endurance, and (apparent) good humour through the discussions and revisions of the shape of this book and its various parts, I thank each and every one of the authors whose contributions to these pages demonstrate why the UBC Geography Department has been such a vital and interesting place these last several years. Lew Robinson, the founding head of the department, was also of great assistance for twice reading the proof and saving us from several small errors of fact and interpretation. Further, I take the close co-operation and collaboration that lies behind a project of this magnitude, as well as the ready agreement of those involved to commit all proceeds from the sale of this book to a fund dedicated to enriching the intellectual environment of the department, as a significant reflection of the congenial, and collegial, spirit that marks our community.

In this, as in so many enterprises, editors and authors have benefitted considerably from the support and good work of others whose contributions also require acknowledgment. First, we thank the Royal Canadian Geographical Society for their generous support of this initiative through the award, in 1989, of a major research grant to the department in care of Tim Oke and myself. It is no exaggeration to say that the award made *this* book possible: it allowed the appointment of several graduate research assistants (whose efforts are acknowledged individually in the bibliographic end matter); it provided for the cartography that so enhances our story; and it enabled us to acquire the historical and contemporary photographs that illuminate the following pages.

Special mention is due the cartographic 'team' who worked effectively under considerable time pressure to allow us to meet our publication deadline. The cartography was begun by Angus Weller, but most of the maps and diagrams in this book were drawn by Eric Leinberger, who ably succeeded Angus, and by Maria Cavezza, who was employed for the 'stretch drive' toward the end of 1991. Paul Jance, the department's long-serving cartographic draftsman, also contributed a handful of figures. The skills and dedication of these people are much appreciated.

In tough times for Canadian publishers, Peter Milroy and his staff at UBC Press have done an exceptional job in producing a large and handsome book remarkably quickly. To Peter and his executive editor, Jean Wilson, we give thanks for their commitment to this project from its earliest stages, and for their faith that a manuscript would materialize as deadlines loomed. Associate editor Holly Keller-Brohman steered our words and figures through the production process quietly, competently, and astonishingly quickly; and George Vaitkunas brought an exceptional eye to the layout and design of a complex work.

Finally, I would like to take this opportunity to thank the Canada Council for honouring me with the award of a Killam Research Fellowship for 1988-90. Although it was primarily for work on a yet-unfinished study of the Canadian Maritime provinces, to which I shall now return, it allowed me an extended season of reading, reflection, and writing that played no small part in the genesis of this volume, and that has greatly influenced my ideas about the purpose and potential of geographical inquiry. Words are never enough to convey the sense of obligation and gratitude engendered by such benefits and the pleasures of engaging with spirited and able colleagues in ventures such as this.

Graeme Wynn

Introduction

Graeme Wynn

'The serenity of the climate, the innumerable pleasing landscapes and
the abundant fertility that unassisted nature puts forth,' wrote English
navigator George Vancouver in 1792, describing the area that would
later bear his name, 'requires only to be enriched by the industry of
man with villages, mansions, cottages and other buildings to render it
the most lovely country that can be imagined.' Two hundred years
later, the cottages and mansions envisaged by Captain Vancouver exist
in numbers beyond his dreams. Villages have grown into towns and
expanded into cities. More than 1.5 million people live in the 2,750
square kilometres of lowland flanking the Fraser River between
Abbotsford and the Gulf of Georgia. Yet the city of Vancouver and its
neighbouring municipalities are still often described in ways that echo
the expectations of the early European explorers of coastal British
Columbia. Nestled against the mountains, lapped by the ocean, possessed of one of the most equable climates in Canada, this area is
regularly characterized as 'lotus-land' or, even more extravagantly, as
paradise on earth. The image has become a cliché, an idea so familiar
that it is readily turned back on itself and rendered ironic. 'In Toronto,'
readers of the *Globe and Mail* were informed on 29 June 1991, 'people
lie about how many hours they work. In Vancouver they lie about how
few, because they have not only a career to pursue, but a lifestyle as
well, reveling in this wonderful environment.'

But images, like reputations, often outlive the circumstances of their
creation, and in the 1990s it seems that there are serpents in the pur-
ported Eden of southwestern British Columbia. Although Vancouverites
are not especially given to introspection about their city, the implications of growth have borne in upon them in recent years. George
Vancouver's benign confidence in the benefits of human industry has
given way, in the late twentieth century, to a host of concerns – social,
economic, political, and ecological – about the effects of human activity
on this place and its prospects. Task forces and think-tanks insist that
major economic changes will be necessary if the city is to 'stay in the
race' for prominence and prosperity. Many commentators see solutions

to the challenges ahead in sustainable development: growth that maintains the quality of life Vancouverites cherish. By this account, outlined in the *Vancouver Sun* of 15 November 1990, residents of the city will be able to have their cake and eat it too. Protected and healthy, the region's spectacular environment will be its 'greatest business opportunity' attracting 'some of the best brains in the world' to the area.

Be that as it may, the road ahead is far from clear. Problems, ranging from atmospheric pollution through homelessness to the rising incidence of social pathologies, make many wonder whether paradise has already been lost. Others contemplate recent trends of urban expansion and shudder at the future they seem to portend. So Vancouver artist and writer Michael Kluckner, mindful of the lyrics of Joni Mitchell's song 'Big Yellow Taxi' – in which asphalt turns paradise into a parking lot, and the only trees are found in museums – offers a critique of 'the binge of heritage destruction' in British Columbia since the Second World War in a book called *Paving Paradise*. Even more pithily, a local landscape architect informed a recent conference on land-use issues in the Greater Vancouver Regional District that the district's remaining green spaces will either be preserved or become an inauthentic 'Disneyland'; for her, the future of these areas is defined in simple binary terms: 'Mickey Mouse or Stanley Park.'

There has been no clearer, more public expression of the anxieties consequent upon these circumstances than a seven-part series of articles on the future of Vancouver and its region, published in the *Vancouver Sun* in mid-November 1990. Treating the area between Desolation Sound and the eastern limits of the Lower Fraser Valley under the general title 'Future Growth. Future Shock,' its pervasively optimistic tone was tinged with unease. Headlines declaring the area 'From Desolation to Hope – A Region of Riches,' pointing out that 'other cities have innovative solutions' to the problems of growth, and offering the reassuring intelligence that the 'communities' of the GVRD 'will shape [a] livable region' were matched by far less upbeat messages. Among them: 'Growing to Extremes,' 'Our Shrinking Farmland,' 'Gridlock,' 'Love Affair with Cars Becoming a Fatal Attraction,' and '"Ghettoization" of Neighbourhoods Feared.'

Reflecting on all of this, and upon the results of interviews and opinion surveys across the area, the principal authors of the *Sun* study suggest that the southwestern corner of mainland British Columbia is somehow 'the Peter Pan of urban regions,' a place that would like to set its face against growing up. But cities are not children's fantasies, and Vancouver is no island in a nameless sea. Increasingly linked to the global economy and far more easily reached than never-never land, the city and its region are likely to see considerable population growth and economic expansion in the immediate future. Current estimates suggest that another million people may come to live in the region in the next twenty years. If so, and 'if present land development trends continue,' warns a local planner, the lower mainland of British Columbia 'will end up like every other unattractive place on earth.'

How did the area upon which Captain Vancouver looked so optimistically come to the edge of this apparent abyss? That question lies at the heart of this book. But it defies a simple answer. Vancouver, like the province in which it is located, is no easy place to understand. As Cole Harris (one of the contributors to this volume) has observed elsewhere, 'writing on early B.C. is rather like watching water skimmers. The object of scrutiny is both near at hand and elusive, individuals appear and are lost in general movement, and activity is more obvious than pattern or purpose.' Others have despaired of grasping the essence of the contemporary city. 'There is no real centre to Vancouver' concluded one recent commentator (in the *Globe and Mail*, 29 June 1991), it is a place of 'pockets, strips, [and] urban moments,' each of which is but a fragment of an intricate urban kaleidoscope. Because most people are familiar with only a few pieces of this fabric, most views of the city elevate one or two facets of its character above others. Many, for example, think of Vancouver 'less as a city than as a resort,' but in doing so they ignore, or dismiss, striking economic and social disparities among its people. There is no room, in the glossy-brochure portrayals of grand hotels, fine restaurants, sailboats, and ski slopes, for images of 'the prostitutes and drunks holding up the walls of hotels' or the 'immigrant working class homes on the East side, shut off from the good life by the Prior Street viaduct.'

Vancouver and Its Region deliberately seeks a broader view. Written by geographers whose special interests and expertise lie, variously, within the realms of physical science, social science, and the humanities, it presents an integrated, accessible account of the physical, historical, economic, social, and environmental development of a relatively small but important area of British Columbia, Canada, and the Pacific basin. Although there are necessarily many strands to this story, and not all of them can be treated as fully as they deserve in a single volume, the central themes of these pages are relatively easily encapsulated. They describe the development of a spectacular physical setting shaped over many millennia but sculpted into approximately its present form only in the last few thousand years by glaciers and volcanism, and profoundly altered by urbanization in the last century. They tell of a complex, vigorous Native society, resting upon the exploitation of rich local resources well into the nineteenth century, disrupted and marginalized with astonishing rapidity by the impacts of the fur trade, a gold rush, the railroad, a massive influx of newcomers, and urban expansion. They trace the helter-skelter growth of a dynamic metropolis in little more than half a century. They chart the rise of that metropolis on provincial, Pacific, and world stages. And they explore the social, ethnic, and political tensions generated by these developments.

This is a story that seems worth the telling on at least three counts. First, it is inherently dramatic. The course of human affairs in this corner of the world turns on a striking transition. Here, an indigenous realm was remarkably quickly incorporated into – and largely subsumed by – the technological might and imperial impetus of a rapidly expanding European world. The process itself was far from unique, of course; in Europe, and on other, earlier New World frontiers, military, mercantile, and other extensive forms of authority had broken in upon the local worlds of people who lived by hunting, gathering, and fishing. But there change generally ran more slowly than it did in coastal British Columbia. It was the peculiar, and difficult, fate of the Musqueam and other Native peoples of the lower mainland to face, in the short span of a single lifetime, challenges that others had grappled with over centuries.

Moreover, the new, urban, European order that swept in upon the Native world of southwestern British Columbia possessed remarkable momentum. By comparison with what had gone before, the rise of Vancouver (and its counterparts in time – Seattle, Spokane, Los Angeles) proceeded at an astonishing pace. Town development in the late eighteenth and early nineteenth centuries was of 'utter insignificance' alongside the achievements of recent times, wrote Theodore Roosevelt in the fourth volume of *The Winning of the West*, published in 1905. To demonstrate his point, he urged his readers to consider Colorado and Kentucky. Although they had been settled approximately a century apart, their populations had increased at more or less the same rate; Denver, however, had expanded 'thirty or forty times as fast as Lexington had ever grown.' To Adna Ferrin Weber, the author of a major, statistical study of nineteenth-century urbanization, it seemed, in 1899, that 'the tendency towards [urban] concentration or agglomeration ... [was] all but universal in the western world.' Growing a thousand-fold in fifty years and including almost half the population of its province by the middle of the twentieth century, Vancouver was in many ways emblematic of this new age.

The second crucial aspect of this story is its profoundly geographical character. As much of this volume makes abundantly clear, the territory that we now know as the Lower Mainland of British Columbia has been dramatically, and irrevocably, transformed by human action. At one level, these changes are obvious. Towering skyscrapers have replaced lofty Douglas firs. The surveyor's geometric grid marches across the contours of nature. At another, they are more subtle. Now-familiar elements of the local flora and fauna, from starlings and grey squirrels to monkey-puzzle trees and daffodils, were quite unknown to inhabitants of the region two centuries ago. At yet a third level, they are hardly discernible. Ozone in the atmosphere, pesticides in rivers, cliff erosion attributable to changes in groundwater runoff – these are human impacts evident first and sometimes only through the lenses of science. Although the long view confirms that change has been the only constant through the millions of years in which the physical environment of southwestern British Columbia took shape, there can be no doubt that the pace of environmental change has quickened enormously in the last two hundred years. In many respects indeed, Vancouver and its region are new and distinctive places created by immigrant peoples and modern technologies in little more than half this span.

Significant as these environmental modifications have been, however, they are but part of the geographical story of this book. The appreciation and understanding of places – the very essence of the geographer's mandate – requires more than a knowledge of the physical environment (the topography, climate, and so on) and the ways in which human imprints have transformed it into something called (variously if never entirely satisfactorily) the human, cultural, or built environment. People not only build environments, they invest them with significance and meaning, and constitute them as places by their presence and behaviour. To know a place – to understand how and why it came to be, and to appreciate the vital inner dimensions of life within it – we need to know a good deal about its social, economic, and political characteristics. To recognize as much is to open discussion of a large suite of inter-related topics. Many of these – among them the imagined geographies of Vancouverites, the politics of alternative urban visions, the reworking of gender relationships, the changing ethno-cultural composition of the population, the intersection of fundamentally different traditions, and the impact of the new international division of labour on life and work in the city – are engaged in the pages that follow. Time and again, these emphases broaden the scope of this book far beyond common, but outmoded, conceptions of geography-as-the-height-of-mountains-and-an-inventory-of-regional-resources. By doing so they reflect a good deal of the vitality and interest of the modern discipline, while providing a broad, fresh, and yet distinctly geographical, perspective on the city and its adjacent territories.

The third important facet of this story is its relevance. This seems evident in at least two ways. With its resolute focus on Vancouver, this book addresses a question that has been of some concern to Vancouverites almost from the moment the city was incorporated: what is this place about? Established on the far western edge of the continent, an outlier of the Canadian state, initially linked to the rest of the country by little more than the winding ribbon of the Canadian Pacific railroad through the cordillera, and squeezed between the American border and the forbidding topography of the Coast range, this new and rapidly expanding settlement in the 'sea of sterile mountains' (as some of Vancouver's early Chinese residents characterized it), has long presented residents a challenge of identity and attachment.

Growing like topsy, and dominated, numerically, by people whose pasts were elsewhere, Vancouver has not been an easy place in which to feel 'at home.' Realtors and boosters have played upon this circumstance through the decades, and residents have responded in several ways to the challenge of defining and securing their place in the city. In doing so they have grappled repeatedly with the question that literary critic Northrop Frye saw confronting every Canadian: 'Where is here?' Nor has the search for answers grown easier with time. As the 'greed line' of development moves ever outward across the Fraser lowlands, and the speculative frenzy of a society that inclines to value land (and place) in purely economic terms spirals onward, tender roots of familiarity and attachment to particular landmarks and landscapes are severed by the accelerating cycle of change. The consequences of all of this are only slowly being recognized, but they have been explored recently by Tony Hiss in *The Experience of Place*, and his words are germane here. 'The danger,' he writes, is that 'whenever we make changes in our surroundings, we can all too easily shortchange ourselves by cutting ourselves off from the sights or sounds, the shapes or textures, or other [pieces of] information from a place that have helped mold our understanding and are now necessary for us to thrive. Overdevelopment and urban sprawl can damage our own lives as much as they damage our cities and countryside.' No string of words can make up for the destruction of what Hiss calls the 'long-lasting and familiar buildings' that we habitually use to 'organize and prompt our memories' (and thus to anchor ourselves in the city), but by providing a perspective on how this place came to be as it is, *Vancouver and Its Region* offers the means for Vancouverites to become more familiar with their surroundings, as well as a point of departure for readers who would answer Northrop Frye's enigmatic question.

At a second level, this book is relevant on a far wider canvas. For all its local detail, the account rendered in the following pages is in many ways the story of our world. In the years since George Vancouver sailed into the Strait of Georgia, much of the globe has been, and continues to be, transformed by the very forces that have made, and continue to re-make, Vancouver and its region. So the incorporation of southwestern British Columbia into the political and economic orbit of western Europe (and later eastern Canada) during the nineteenth century rested upon improvements in communication that were fundamental to the development of what historian Emmanuel Wallerstein has termed the 'World System.' As this system expanded, indigenous people on every continent were displaced and dispossessed in ways not very different, in the end, from those experienced by the Coast Salish of the Fraser Valley. As this system grew more elaborate, the cities, new and old, that were its pivots, extended tentacles of influence and control across their hinterlands in much the same manner as Vancouver became the centre of British Columbia's resource economy. Other towns engrossed their countrysides, just as Vancouver spilled across the Burrard Peninsula and onto the flatlands of the Fraser. Elsewhere, too, the patterns of environmental transformation that are so much a part of the story of

Vancouver quickened as the nineteenth century gave way to the twentieth. Migrations carried individuals to the ends of the earth and, as in
Vancouver, brought people from various homelands, who knew little of
each other, into always-challenging and often-threatening contact. To
defend their interests against the power of capital, workers around the
world rallied to the internationalist banner; early in the twentieth century, Vancouver strikers were described, symptomatically, as Bolsheviks.
Only a few decades later, the prosperity of British Columbia and
Vancouver rested, like that of the remainder of the developed world,
upon the mass-production of standardized goods by a largely unionized
labour force in the employ of enormous, integrated corporations. And,
in the last twenty years, many areas have had to adjust, as Vancouver
has, to the emergence of a global economy, a new international division
of labour, and a 'redefinition' of world cities as electronic crossroads of
information, money, and power. Vancouver, in short, is a product of the
massive economic, social, demographic, technological, and ecological
transformations that have made the modern world. To examine the past
and present of this place is to explore many of the roots and branches of
the contemporary condition, and to recognize the foundations upon
which the inevitably uncertain future must be built.

1

Views of the metropolitan area

photography by
Alfred H. Siemens

The photos in this section were taken
in the summer of 1990. Gordon Clark
did the darkroom printing and Walter
Hardwick provided advice on some of
the specifics of the city.

Looking northwestward over the entrance to Burrard
Inlet and English Bay. The lighthouse on Point
Atkinson is just visible: behind it is Passage Island in
Queen Charlotte Channel and behind that, Bowen
Island. The freighters are riding at anchor waiting
for berths in the harbour.

Residential high-rises in the West End and the taller
buildings of downtown behind them, as seen north-
eastward from Spanish Banks – with a hint of Simon
Fraser University on Burnaby Mountain in the
distance on the right

Eastward from over English Bay toward the West
End and downtown; Canada Place and the cranes of
the container port behind it; Hastings Street and residential Strathcona in the top right

The 'sails' of Canada Place and a changing foreshore. A line of gables marks the former Canadian Pacific Railway passenger terminal; the remaining tracks are now part of VanTerm's facilities but will be gone soon. Marathon Realty's Project 200 tower straddles the tracks; its Waterfront Hotel is under construction to the right. The SeaBus terminal is in the bottom left corner; a few truck trailers await barge transportation to Vancouver Island in the top right corner. Canada Place has cruise-ship docking facilities on its lower levels, convention and trade centres above that, and is surmounted by the Pan Pacific Hotel.

Downtown from across False Creek. The northern
margin of the creek was occupied by the railyards
of the Canadian Pacific Railway until recently; the
roundhouse has been preserved. These were the
Expo '86 grounds and are now the site of the projected Concord Pacific residential development.
Between them and downtown there is 'Yaletown,'
formerly a mixture of warehouses, industries, and
residences, now also being developed into an area
of upscale commerce and trendy apartments.

Between Main Street in the bottom right corner and
the approach to the Cambie Street Bridge in the top
left corner is what was once the residential district of
Mount Pleasant; houses are still apparent here and
there, but this is now mainly an area of wholesale
warehouses and services. False Creek is in the top
right corner; the land bordering it is owned by the
city and is to be developed for residential purposes.

Not very 'Super Natural!' The dome houses the
Omni-Max Theatre, initially part of Expo '86 and
now in 'Science World.' Beyond it loom the BC
Sugar Refinery and the cranes of the container port.
On the middle right margin is the arch of the former
Canadian National Railway terminal.

The eastern extremity of Burrard Inlet. Roche Point and the entrance to Indian Arm are on the left, the Shellburn Refinery in Barnet is on the right.

New housing in Port Coquitlam

The main channel of the Fraser River; Barnston Island
is immediately behind it and Mount Baker is in the
distance.

A forest in Langley, on the eastern limits of the
metropolitan area, as seen from the gondola of
a hot-air balloon. This provides a particular
perspective; the altitude is no more than a few
hundred feet, the speed is slow and the flight is
quiet, except when the pilot fires the burner.
One can hear dogs barking, and children shouting
as they cycle furiously along nearby roads, trying
to keep up.

A functioning dairy farm in Langley. There are several generations of barns, as is usual in the Lower Mainland. The cows are crowded around an entrance, waiting their turns in the 'milking parlour.' Machinery surrounds the buildings, including a manure spreader, seed drills, and the wagons that are pulled behind mowers to catch the green feed.

A Douglas fir (top centre) that has survived a lightning strike or two and a gallery of cottonwoods grown from seeds carried in the water of a drainage ditch. In the foreground is a row crop such as brussel sprouts.

The Trans-Canada Highway, looking westward
toward Vancouver, with the mountains on the north
shore of Burrard Inlet in the distance

2

The primordial environment

O. Slaymaker

M. Bovis

M. North

T.R. Oke

J. Ryder

Spectacular variety characterizes the biophysical setting of Vancouver. The rugged terrain, the exuberant vegetation, and the mild, moist climate vary over short distances. The North Shore Mountains, the rolling topography of the Fraser Lowland, and the island-studded Strait of Georgia create a dramatic physical setting (Figure 1).

Geology

An understanding of this setting requires a long view. Vancouver is located close to the western margin of the Canadian Cordillera, a belt of rugged mountains and dissected plateaus extending from the west coast of Vancouver Island to the eastern flanks of the Rocky Mountains in Alberta. Formed by regional compression and faulting of rocks at the 'leading edge' of the westward moving North American continental plate over the past 100 million years, the topography displays a strong northwest to southeast orientation (Figure 2).

Figure 1 (left)

Painting depicting the physical setting of Vancouver in 1792

Figure 2 (right)

LANDSAT image of the Vancouver region from space showing the major physiographic lineaments, glaciers, ice fields, and the Fraser River and delta

Within this general alignment, the North Shore Mountains are dominated by igneous rocks such as granite. These were intruded into a thick sequence of older sedimentary and volcanic rocks. Metamorphosed remnants of these host rocks now crop out along Howe Sound, between Vancouver and Squamish. Some of the youngest of the igneous rocks are relatively massive and form giant monoliths such as the Stawamus Chief, which towers above the town of Squamish. About 50 million years ago, sedimentation around the southern edge of the Coast Mountains was followed by minor episodes of basaltic volcanism. The resulting basaltic dykes helped to shape Siwash Rock and Prospect Point in Stanley Park as well as Queen Elizabeth Park.

In the past 10-15 million years, both the Western Cordillera and the Coast Mountains have been uplifted some 2-3 kilometres. In southwestern British Columbia and northern Washington State, this is attributed to the convergence of the Juan de Fuca and North American crustal plates (Figure 3). In essence, the oceanic Juan de Fuca plate is sliding beneath the continental North American plate at a rate of 4-5 centimetres per year, thus elevating the land mass. Because uplift was especially rapid between 5 and 2 million years ago, rivers such as the Fraser and Cheakamus cut deep canyons along their courses. In addition, older sedimentary rocks at the margins of the Coast Mountains were strongly tilted as uplift occurred and became sculpted into cuestas such as Burnaby Mountain and Grant Hill.

Within the last 2 million years, a further volcanic eruption occurred in the Garibaldi Volcanic Belt. This included the jagged peaks of Mount Garibaldi, Mount Cayley, and Meager Mountain, located 60, 80, and 150 kilometres north of Vancouver, respectively, as well as the Harrison Hot Springs geothermal area, 100 kilometres east of the city. Mount Garibaldi last erupted some 14,000 years ago. An eruption in the Meager Mountain area 2,400 years ago deposited volcanic dust and ash as far distant as western Alberta. Today, the towering outlines of Mount Garibaldi and Mount Baker within 100 kilometres of the city and frequent small seismic events under Vancouver Island and the Gulf of Georgia remind Vancouverites of the very active geologic environment in which they live. Building codes recognize the potential hazard of seismic activity, and there are periodic anxieties about the possibility of the 'Big One' – a regional megathrust earthquake on the scale of the 1964 Alaska quake, which caused 131 deaths and $500 million worth of property damage in Anchorage.

The Pleistocene glaciation

Approximately two million years ago, the Coast Mountains had been sculpted into a rugged mountain belt by the precursors of present-day rivers, although the average elevation of these mountains was probably lower than it is now. Continuous uplift and deteriorating climatic conditions eventually placed large areas well above the snowline. Ice

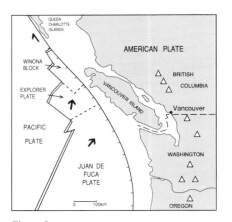

Figure 3

America-Pacific tectonic plate interaction. Arrows indicate plate motions relative to the American Plate. Triangles mark major Quaternary volcanoes.

accumulated, and the topography of the region was profoundly modified by glacial erosion and deposition. Together, these processes did much to create the interfingering of arms of the sea with mountainous peninsulas and islands that comprise the essence of Vancouver's natural setting (Figure 4).

Figure 4

The topography and bathymetry of the Vancouver region

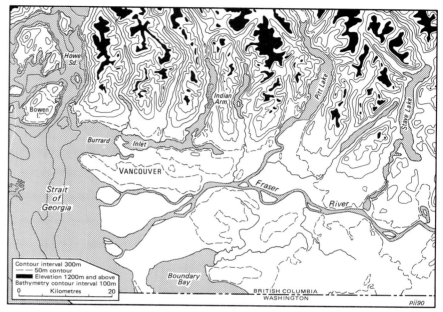

Evidence from ocean-floor sediments indicates that during the 2 million years of the Pleistocene Epoch there were at least a dozen 'ice ages' separated by warmer non-glacial intervals (such as the present). In the Vancouver area, there is evidence of only the last three glaciations, because older glacial sediments have been eroded. The deposits and landforms of the most recent phase of cold climate, which is known locally as the Fraser Glaciation, are extensive and well preserved. The progress of this glaciation can be charted in some detail using radiocarbon dates derived from wood, peat, and shells in the sediments and soils beneath and atop these landforms.

The Fraser Glaciation probably began rather innocuously, some 30,000 or more years ago, when increased snowfall led to the development of permanent snowbanks and small glaciers in the North Shore Mountains. Gradually, glaciers formed in the local cirques, such as those now occupied by Kennedy Lake on Goat Mountain and Enchantment Lake just north of The Lions. Further north, the larger glaciers in Garibaldi Park and the Tantalus Range began to expand. For about 10,000 years, the ice slowly accumulated, cirque glaciers lengthened into valley glaciers, and great branching ice streams eventually developed within the mountains.

In the Vancouver lowlands, the first effect of the approaching ice was a change in the behaviour of the rivers. Increased sediment loads in these glacial meltwater streams resulted in widespread deposition of stream sediments in the coastal lowlands. As the slow, inexorable advance of the ice continued, vast outwash plains developed. The accumulating sand and gravel filled in coastal inlets and lakes, and buried riverine wetlands, floodplains, and coastal lowlands. By the time the glaciers began to emerge from the mountains, an outwash system probably extended across the entire width of what is now the Strait of Georgia. An observer standing on this plain in the vicinity of Point Grey would have looked across a vast, sparsely vegetated, sandy plain crossed by the innumerable shifting channels of a major braided river. To the north, the advancing glacier in Howe Sound may well have been visible, and further to the west loomed the huge piedmont ice lobe (the main source of the braided river) that was advancing down the 'Strait' of Georgia toward Vancouver. In some parts of the Vancouver area, the outwash sands and gravels survived the ensuing glaciation. The thick sands that are exposed in the sea cliffs at Point Grey and Spanish Banks were deposited by the Strait of Georgia river. Thick gravels in the Coquitlam River valley also date from this time.

The ice may have hesitated briefly in its advance before it over-rode the Vancouver region. We know that the first piedmont lobe of the Coquitlam Valley glacier, which developed about 22,000 years ago, had melted 3,500 years later. But there is no evidence that the other encroaching glaciers behaved similarly.

Eventually, between 18,000 and 17,000 years ago, the Vancouver region succumbed to the glaciers. A great stream of ice, generated by increasingly heavy snowfalls in the Coast Mountains, flowed southward across the area, and the ice margin advanced rapidly southward through the Puget Lowland. By the time of the glacial maximum, 14,500 years ago, the southern tip of the vast Georgia-Puget ice lobe stood south of Seattle in the vicinity of Olympia. Meanwhile, the Vancouver region was buried by 1,500 metres of ice. Only the highest local peaks, such as Mt. Brunswick and The Lions, protruded as rocky islets in a vast, glacial sea.

At this time, virtually all of British Columbia was buried by the Cordilleran Ice Sheet, an enormous complex of inter-connected mountain ice sheets and coalescing piedmont glaciers. Although it would not have been apparent to an observer, the land had sunk by several hundreds of metres under the weight of ice. Sea level had also gone down by about 120 metres, because the ocean water was temporarily contained in the continental ice sheets.

A major climatic change must have occurred sometime between 15,000 and 14,000 years ago, because just as the Puget glacial lobe reached its southern limit, glacial melting began. By 13,000 years ago, the ice margin had receded northward, uncovering the Vancouver region and the Fraser Lowland, which were temporarily occupied by an icy sea that stood 200 metres higher than the present sea level. Muddy sediments that were carried in to the sea by meltwater streams, and rocky debris that was released by the melting icebergs, accumulated on the sea floor. The various molluscs and smaller creatures that inhabited this glaciomarine environment were the first living things to colonize the inhospitable landscape. Elsewhere there were even more dramatic events. An observer in the vicinity of Point Grey could not but have noticed a dark column of steam and ash on the northern horizon, where many of the volcanic features of Garibaldi Park were being created from the interaction of fire and ice. One final glacial event occurred in the eastern Fraser Lowland, where a piedmont lobe readvanced across the Sumas area about 11,300 years ago but melted shortly afterward.

For a few millennia after deglaciation, the Vancouver area gradually emerged from beneath the sea as the land rebounded from the weight of ice. All of the upland areas that presently lie below 200 metres elevation were, at some time, beaches within the zone of wave erosion. The uppermost glaciomarine sediments were reworked and transformed into beaches of sand and gravel, and marine terraces developed on some of the uplands, such as Point Grey. Delta formation at the mouths of streams resulted in staircases of gravel terraces, because no sooner had a delta surface begun to form at sea level, than it was elevated above sea level by the rising land. The delta terraces of the Capilano River are now particularly evident. Between the time of deglaciation 13,000 years ago, and 8,000 years ago when the sea had fallen to 12 metres below its present level, the land rose at an average rate of about 4 centimetres per year. Shortly after deglaciation, when the rebound was most rapid, the rate was about 11 centimetres per year. That the land was emerging would have been apparent to any inhabitants of the coastal region.

The legacy of the Fraser Glaciation and, to a lesser extent, some of the preceding glacial episodes, is written clearly in the local landscape. The erosional effects of the ice are most apparent in the mountains, where, to the delight of the summer hikers, the cirques and knife-edged ridges that date from the early phases of glaciation are now clothed with alpine parkland and meadows (Figure 5). In winter, the rounded summits of the lower mountains, which suffered greater erosion under the overriding ice sheet, are now the realm of the cross-country skier. And the glacial troughs and fjords that channelled the ice toward the lowlands provide the familiar scenic backdrop to the city.

These fjord and fjord-like finger lakes are one of the most striking physical features of the Vancouver region. Howe Sound, the largest inlet of the Vancouver area, is a fjord. It is 42 kilometres in length, and, although only 2 kilometres wide at its head near Squamish, it broadens rapidly to an island-filled sound that is 16 kilometres wide where it joins the Strait of Georgia (Figure 6). Like many of British Columbia's fjords, it exceeds 100 metres in depth in many places, though its outlet sill is only 70 metres deep. By contrast, Indian Arm is 20 kilometres long and averages only 1.25 kilometres in width, narrowing to 0.4 kilometres at its outlet where the sill depth is only 20 metres. Pitt Lake, 28

Figure 5

The Tantalus Range, Coast Mountains, showing glacial features and alpine parklands

Figure 6

Howe Sound, a typical Coast Mountain fjord

kilometres long and 2.5 kilometres in average width, is a fjord lake which discharges to the Fraser River via tidal Pitt River. On the other hand, Burrard Inlet and English Bay, which contain Vancouver's harbour and its approaches, are relatively shallow and are thought to be the drowned former course of the Alouette River. The shallowest of the coastal inlets are the Fraser estuary and Boundary Bay, both less than 10 metres in average depth and backed by the gently sloping surface of the Fraser Delta.

Much of the city itself is founded on glacial materials, and the varying properties of glacial till, glaciomarine muds, and outwash sands and gravels are incorporated by engineers into their plans for all types of construction. The outwash deposits are the prime source of the aggregate that is used for making concrete.

In the 10,000 years since the end of the Fraser Glaciation, smaller, but nonetheless significant, climatic changes have occurred. Analysis of the pollen grains that are preserved in bogs and lake sediments indicates that after deglaciation the Vancouver area enjoyed a relatively warm and dry climate. Mountain glaciers retreated beyond their present limits. Douglas fir and arbutus extended further up the mountainsides and the tree-line was probably higher. About 6,000 years ago, however, there was a gradual transition to a cooler and moister climate, and these conditions have persisted, more or less, to the present day.

Buried wood and soils in moraines that flank present-day alpine glaciers indicate that up to three glacier advances have occurred in the last 6,000 years, suggesting that phases of cooler or more snowy weather occurred within the most recent cool-moist interval. The first occurred between 6,000 and 5,000 years ago and the second between about 3,500 and 2,200 years ago. By far the greatest glacial advance began about

1,000 years ago and culminated during the eighteenth and nineteenth centuries with construction of the prominent moraines that flank the modern glaciers. This glacial advance, which is known as the Little Ice Age, was a worldwide event. In the European mountains, pasture-land, farms, and even villages were destroyed as the glaciers advanced further down their valleys than at any time in the last 10,000 years. Since the Little Ice Age, the glaciers have receded; some of the large glaciers in the Mount Waddington area have melted back several kilometres. At all glacier sites, this recession is so recent that the position of the Little Ice Age glacier terminus is still clearly indicated by vegetation trimlines or a partly vegetated morainal ridge (Figure 7).

Figure 7

Miller Creek, Coast Mountains, showing Little Ice Age moraine and former extent of glacier ice in the mid-nineteenth century

As the ice started to retreat from the Fraser Lowland about 13,000 years ago, plants rapidly moved in from the ice-free areas just south of the 49th parallel. Investigation of the pollens and fossils in sediments deposited in Marion Lake (northeast of Haney, BC), at an elevation of 405 metres, provides a record of the sequence of plant invasion that is probably representative of the entire upland area of the North Shore Mountains, currently part of the Coastal Western Hemlock biogeoclimatic zone.

The initial vegetation was dominated by lodgepole pine, buffalo berry, willow, and alder. This type of open forest and shrub community no longer occurs in the area, and its nearest present-day representation is at 1,200 metres in Manning Park, 200 kilometres east of Vancouver. About 10,500 years ago, Douglas fir (*Pseudotsuga menziesii*) appears in the pollen record in large quantities and the other conifers decrease. Western red cedar (*Thuja plicata*) was present in small numbers at this time as was western hemlock (*Tsuga heterophylla*). By 10,370 years ago, the shade-intolerant pine was replaced by more shade-tolerant conifers, fir (*Abies*), spruce (*Picea*), and mountain hemlock (*Tsuga mertensiana*). The increase in hemlock was initially slow but consistent; this may have been due to its preference for regenerating on podzolic soils or decaying coniferous wood. Podzolic soils usually develop in sandy materials under forest vegetation in areas where there is a cool, humid climate. As these soils developed over time, the conditions for the establishment and growth of hemlock improved. Cedar was rare until 6,600 years ago, but since then the forests of the Haney area have been dominated by hemlock and cedar. The sedimentary evidence since this date also shows that fossil mosses are identical with the present mosses of the area. The continuous presence of alder, pines, Douglas fir, willow, and other deciduous trees and shrubs indicate that disturbance of the vegetation was common throughout postglacial time. Charcoal fragments point to fire as the major cause of disturbance, as fires would destroy the closed forest and create open sites suitable for the wind-blown seeds of the fast growing, but shade-intolerant, deciduous trees and shrubs.

Approximately 6,000 years ago the vegetation cover of the region was similar to that encountered by nineteenth-century settlers. Forests of evergreen coniferous trees dominated all the undisturbed sloping land between the water's edge and the highest of the local mountain tops, such as Mount Strachan and the Lions.

Climate and hydrology

The climate of Vancouver makes the city one of the most desired residential locations in Canada. But it is not universally admired. Many who have lived in the Prairie provinces find the Vancouver winter damp, dismal, and soggy and long for blue skies, crisp air, and the squeaky crunch of snow underfoot. Nevertheless, the mild, sheltered climate of Vancouver creates one of the most benign cradles for living in Canada.

This kindly climate is attributable to the region's mid-latitude location and the presence of a mountain range to the west. The middle latitudes of the Northern Hemisphere are a zone of interaction between warm and often humid air masses to the south (formed over the subtropical ocean) and colder drier air to the north. In winter, on the west coast of North America, these air masses usually meet in the vicinity of Vancouver, but in summer the boundary between them generally lies to the north, bringing less variable conditions to the region.

These facts underlie the strongly seasonal and somewhat bimodal climate of Vancouver. The equator-to-pole temperature gradient and the rotation of the Earth produce a deep and predominantly westerly air flow. The strength of the flow depends on the steepness of the latitudinal temperature gradient, and is therefore strongest in winter. Low-pressure systems lead to condensation and considerable cloud and precipitation. The greatest precipitation is deposited on the windward slopes of the Beaufort Range on Vancouver Island, but the Coast Mountains of the Vancouver region also receive heavy orographic rain. Indeed, the increase of precipitation with elevation on these slopes is one of the most remarkable aspects of the local climate (Figure 8). Total

Figure 8

Mean annual precipitation (mm) for
Greater Vancouver

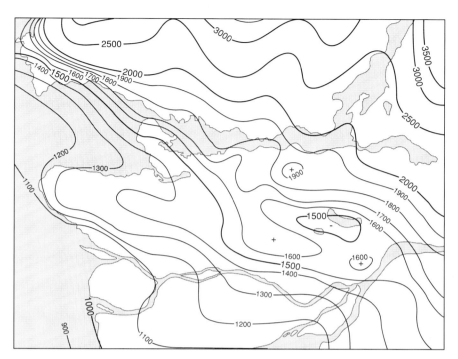

annual precipitation increases from 900 millimetres at sea level in Delta
to about 3,500 millimetres at 1,200 metres above sea level on the North
Shore slopes (Hollyburn, Seymour, and Grouse mountains) and an estimated 5,000 millimetres at 1,700 metres above sea level at The Lions.
Precipitation is not entirely a function of elevation, however, upward of
4,000 millimetres are recorded at 300 metres above sea level in the
upper reaches of the Capilano and Seymour valleys, where air is topo-
graphically confined and is therefore convergent; rainfall at the mouth
of Capilano River is double that at Tsawwassen (40 metres above sea
level), and it is often possible to observe rainstorms in West Vancouver
(10 metres above sea level) from a dry and sunny Point Grey (90 metres
above sea level). On the southern side of the city, distance from the
mountain front is a more important determinent of precipitation than
is elevation – as real estate developers in Tsawwassen proclaim.

Because the temperature of the atmosphere normally decreases with
height, the freezing level commonly intersects the mountains during
the period between October and April. Therefore, snow rather than rain
falls at upper elevations (Figure 9). At the crest of the North Shore Mountains snow falls on about eighty days, giving annual accumulations of
about 8-10 metres in most of the higher elevations and up to 15 metres
at The Lions, where snow can lie on the ground well into August.

Compared to the much colder and drier air formed over the continental
interior in winter, and usually kept there by the mountains, maritime
air is mild. Seasonal variations in temperature are muted by the conservative thermal climate of the ocean. At the lower elevations, where
most of the present-day city is located, average daily temperatures rarely

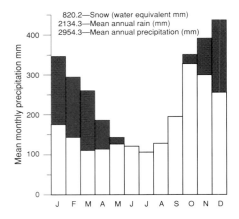

Hollyburn Ridge - Elevation 951m
(West Vancouver)

820.2—Snow (water equivalent mm)
2134.3—Mean annual rain (mm)
2954.3—Mean annual precipitation (mm)

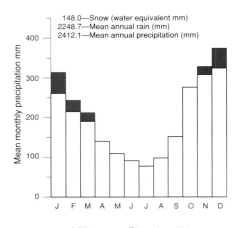

Millstream - Elevation 381m
(West Vancouver)

148.0—Snow (water equivalent mm)
2248.7—Mean annual rain (mm)
2412.1—Mean annual precipitation (mm)

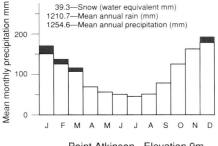

Point Atkinson - Elevation 9m
(West Vancouver)

39.3—Snow (water equivalent mm)
1210.7—Mean annual rain (mm)
1254.6—Mean annual precipitation (mm)

■ - snow
☐ - rain

Figure 9

Seasonal distribution and composition
of mean monthly precipitation

drop much below freezing (Table 1), except in midwinter when outbreaks of cold polar air from the north and/or the continental interior spill over the mountains to the coast. This cold air typically flows down the Fraser Valley, and the fjord and fjord-like valleys between Howe Sound and Harrison Lake, to occupy the valley bottoms. Vancouver then experiences climatic conditions similar to those in much of the rest of Canada. The cold, dense air does not dislodge easily. Warmer, moist air carried eastward by the regular procession of onshore storms often overrides the cooler, denser air and, with the freezing-level close to the ground, brings the lowlands their largest snowfalls. Even during the coldest weather, however, Vancouver harbour never freezes; nor is it subject to drifting sea ice.

Despite Vancouver's location in the belt of the westerlies and along a coast, strong winds are uncommon in the city (Table 1). Because Vancouver Island shelters the region from the westerlies, the strongest winds tend to come from the northwest, funnelled by the Strait of Georgia, after the passage of a cold front. Strong southeasterlies often occur before and during the passage of storms channelled through the Strait of Juan de Fuca or around the Olympic Peninsula. The mouths of valleys and inlets such as Howe Sound occasionally experience high winds due to outbursts of cold air. In the Lower Fraser Valley prevailing winds are approximately east-to-west due to the channelling influence of the valley, but local variations are considerable due to the effects of mountain topography.

If the main air flow is across the valley it may set up a rotor eddy within the valley. Then, if rain is falling and the forest canopy on the slopes is saturated, the upward rising air on the windward side produces an eerie mist and shreds of cloud that hug the slope. Large hills, such as Burnaby Mountain, Capitol Hill, and Little Mountain, cause an obstruction so that part of the flow is deflected over and part around. In either case, the speed is increased on the windward side. Shelter is usually experienced in the lee, but if there are sharp changes of slope the flow may be more gusty. The same principles apply to isolated obstacles such as the small islands in Howe Sound, around headlands such as Point Atkinson and Point Grey, clumps of trees, and even individual rocks and trees. The surface roughness and, therefore, the drag exerted by water is much less than that of land, especially when the latter is forested, and, hence, other things being equal, coastal sites are much more open to strong winds.

In spring, the upper-level westerly flow weakens and its mean position moves northward so that the frequency and intensity of weather disturbances over Vancouver declines. This transition season often includes both pleasantly mild and bright conditions and disappointingly persistent cloudy, drizzly weather as pools of cool, moist air stagnate over the area. Summer and early fall are dominated by a northward extension of the subtropical high pressure system which brings largely cloud-free

Table 1

Long-term climatological statistics (Vancouver International Airport, 1951–80, unless otherwise indicated)

	Jan	Feb	Mar	Apr	May	Jun	Jul	Aug	Sep	Oct	Nov	Dec	Year
Mean daily temp.*	2.5	4.6	5.8	8.8	12.2	15.1	17.3	17.1	14.2	10.0	5.9	3.9	9.8
Mean daily max. temp.	5.2	7.8	9.4	12.8	16.5	19.2	21.9	21.5	18.3	13.6	9.0	6.5	13.5
Mean daily min. temp.	-0.2	1.4	2.1	4.7	7.9	10.9	12.6	12.6	10.1	6.4	2.8	1.2	6.0
Extreme max. temp.	14.4	15.0	19.4	24.4	28.9	30.6	31.7	33.3	28.9	23.5	18.4	14.9	33.3
Extreme min. temp	-17.8	-16.1	-9.4	-3.3	0.6	3.9	6.7	6.1	0.0	-3.3	-12.2	-17.8	-17.8
Mean no. days with frost	15	10	9	1	0	0	0	0	0	1	8	11	55
Heating degree days (18°C)	480	379	379	277	180	92	39	42	115	248	363	438	3031
Growing degree days (5°C)	16	25	45	113	223	302	382	375	276	156	54	28	1994
Mean monthly rainfall (mm)	130.7	107.1	95.1	59.3	51.6	45.2	32.0	41.1	67.1	114.0	147.0	165.2	1055.4
Mean monthly snowfall	25.7	7.5	6.6	0.3	0	0	0	0	0	Trace	2.8	17.5	60.4
Mean monthly precipitation (mm)	153.8	114.7	101.0	59.6	51.6	45.2	32.0	41.1	67.1	114.0	150.1	182.4	1112.6
Greatest 24-hr rainfall (mm)	68.3	43.9	49.3	44.5	28.8	40.4	45.2	31.8	49.5	60.7	61.0	89.4	89.4
Greatest 24-hr snowfall (cm)	29.7	18.3	25.9	3.6	Trace	0	0	0	0	0.3	22.1	31.2	31.2
No. days rainfall over 0.3 mm	17	15	15	13	10	10	6	8	10	15	18	19	156
No. days snowfall over 0.3 cm	6	2	2	<1	0	0	0	0	0	<1	1	4	15
No. days precipitation over 0.3 mm	20	16	16	13	10	10	6	8	10	15	18	21	163
Mean monthly bright sunshine (hr)	54	87	129	181	246	238	307	256	183	121	69	48	1920
Most frequent wind direction	E	E	E	E	E	E	E	E	E	E	E	E	E
Mean wind speed (m s^{-1})	3.6	3.4	3.8	3.8	3.3	3.2	3.3	3.0	3.0	3.2	3.4	3.6	3.4
Mean no. days heavy fog	6	5	2	1	0	0	1	2	7	9	6	6	45
Mean daily solar radiation (MJ m^{-2} day^{-1}) [†]	2.9	5.5	10.0	15.1	20.2	21.8	23.0	18.6	13.2	7.4	3.6	2.3	12.0
Cloud cover (monthly mean) [‡]	8.0	7.5	7.1	6.8	6.4	6.8	4.8	5.1	5.4	7.1	7.9	8.1	6.8
Mean relative humidity [‡]	87	85	82	75	74	76	74	78	80	87	88	89	

Notes:
*All temperatures in °C
[†] Data for Vancouver, UBC
[‡] Data for 1941–70

skies (except for scattered fair-weather cumulus clouds) and warming of the lowermost one kilometre of the atmosphere. As a result, thunderstorms are relatively rare. High-pressure systems also produce rather weak pressure gradients, so winds are light. Local heating and cooling differences become accentuated and lead to local breezes. Land-to-sea and mountain-to-valley breezes operate in unison across the Vancouver region, especially on warm summer days. By midday, upslope breezes sweep up the North Shore slopes, often producing a belt of cumulus clouds aligned along the crests. More general breezes start at the coast and blow inland up the Fraser Valley at a low level. A return flow aloft occurs in the opposite direction, sometimes transporting small cumulus clouds seaward. By late afternoon, the circulation system extends inland at least as far as Abbotsford. In the evening, slopes cool and downslope breezes seep into the valleys. Towards nightfall, a land breeze blows toward the coast over a larger area. In mid-summer, when the continental interior is strongly heated, a thermal low pressure cell develops there, causing inflow of air at the coast. This draws in maritime air and partially overpowers local circulations. Proximity to cool ocean water confers a moderating influence in the Vancouver region to match the relative warmth of winter so that daytime temperatures rarely become uncomfortably high, and night-time temperatures are generally somewhat lower (Table 1). Further up the Lower Fraser Valley, this ameliorating influence declines and daytime temperatures are several degrees higher than they are on the coast. The fine weather which often persists for a few weeks at a time is interspersed with occasional disturbances bringing showers. Nevertheless, the prolonged anticyclonic conditions usually bring drought to the area by the end of the summer, and water shortages regularly afflict the Gulf Islands.

Fog is more likely in the fall (Table 1), especially in the evening in low-lying, damp areas. This is radiation fog formed by the cooling of humid near-surface air on clear, almost windless, nights. It usually disperses soon after solar heating becomes effective the following morning. On other occasions, mainly in the fall and winter, fog forms over cool coastal waters off the west coast of Vancouver Island or within the Strait of Georgia and is wafted onshore by westerly winds. This is known as advection fog. If it survives transport over the warmer land, it can become more widespread, long-lasting, and even thicken as a result of overnight radiative cooling. It may then blanket large areas of the Lower Fraser Valley, and it often appears as if the mountain blocks are emerging majestically from a bed of cotton wool (Figure 10).

At a local level, variations in relief, aspect, and surface cover produce many small-scale (local and micro) climatic variations. Temperature inversions, in which temperature increases with altitude are common in valleys, basins (eg., Port Moody), and other low areas as cool air sinks under the influence of gravity, especially during cloudless nights. These areas are also foggy and prone to frost. By day, slope and aspect affect solar heating. Southerly slopes, such as those of the North Shore, South Vancouver, and Burnaby, are warmest.

The tall, massive, and often close-packed tree canopy intercepts most of the sunshine and precipitation and exerts a drag on the wind. The forest floor is an area of reduced light with scattered sunflecks, a small daily range of temperature (relative cool in the daytime but warmth at night compared to open spaces in summer), very light winds, and a relatively high humidity. Indeed, many early visitors to the region characterized the climate of the forest floor of the coastal rain forest as gloomy, damp, and stagnant. In the early stages of a storm, the canopy provides an almost complete shield for the floor, but later, when the storage capacity of the canopy is exceeded, the floor receives a continual rain of drips and water running down the trunks. The moist floor sustains luxuriant vegetation, but, because of the lack of light, it is rich in shade-tolerant species.

Figure 10

Fog in the Lower Fraser Valley

Overall, the climate of the Vancouver region is classed as west-coast maritime temperate, with cool, wet winters and warm, relatively dry summers. If it were slightly warmer and drier it would be called Mediterranean. Vancouver has a climate similar to that of northwest Portugal and Spain, but the processes involved are different there, and the winters lack the cold snaps of Vancouver. Closer analogues may occur in the sheltered coastal areas of southern Chile.

The hydrology of the Vancouver region is a product of the interaction of climate, land, and vegetation, and rivers are its most obvious expression. The rivers of the Vancouver region are of four main types (Figure 11). The Fraser River is a low gradient, large sand bed river. Medium gradient, small sand bed rivers located within the Fraser Lowland such as the Salmon River form a second category. Third, are the steep gradient, medium-sized gravel bed rivers of the North Shore Mountains such as the Coquitlam. Finally, there are the very steep, very small debris torrent tracks of the North Shore Mountains such as Cypress Creek. In the mid-nineteenth century, fifty or so small rivers and streams, with an aggregate channel length of over 120 kilometres, ran across the site of present-day Vancouver (Figure 12). Most of these streams, which have since disappeared, were salmon-producing until the last decade of the nineteenth century. Among those that remain, Still Creek and Brunette River in Burnaby and Salmon River in Langley have long since ceased to yield salmon. The mean flow of all of these rivers is relatively small, and even peak flow events, which have been increased during the period of historic record by the influence of urbanization, are relatively benign (Table 2).

By contrast, the Fraser River drains an area of 230,000 square kilometres, which is more than a quarter of British Columbia. It is one of the

Figure 11

The four types of rivers of the Vancouver region: *(top left)* Fraser River; *(top right)* Salmon River; *(bottom left)* Lynn Creek; and *(bottom right)* Ted Creek

great rivers of North America. Its delta meets the sea along a perimeter of 50 kilometres, and its active western delta front extends south for 37 kilometres from the North Arm. Much of the snowmelt from the Coast and Rocky mountains escapes via the Fraser River. Prior to the construction of dykes, the flood hazard in the delta region was extremely high during May and June.

The medium-sized rivers of the Coast Mountains, typified by Lynn Creek, and the Capilano, Seymour, and Stawamus rivers (Figures 11 and 13),

Figure 12

The coastline and the surface drainage pattern in the late nineteenth century

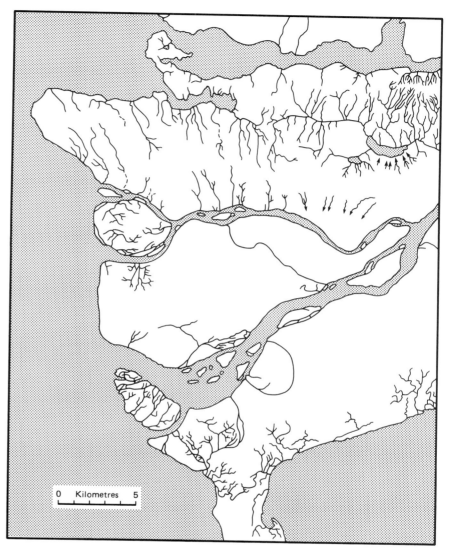

form the third category. They are steep mountain gravel bed rivers which flood in two distinct seasons each year: April and May (due to snowmelt) and October to December (because of winter rainstorms). The rainstorm floods in late fall tend to be the most severe, because they incorporate snowmelt from newly formed snowpack at higher elevations. Rivers such as the Pitt and the North Alouette, as well as Jacobs and Kanaka creeks, are transitional and have characteristics both of the Fraser Lowland and Coast Mountain rivers. When early settlers first encountered them, they were low gradient sand- and silt-bedded rivers, scarcely differentiable from other Fraser Lowland rivers.

The fourth category comprises small, almost unimaginably steep, mountain creeks (such as Ted, Alberta, Charles, and Mosquito creeks), which are subject to debris torrents (Figure 11) during the fall and winter. Most of these rivers cease to flow completely during July and August

Table 2

Flow characteristics of Vancouver region rivers

Rivers	Gauging station	Area (km²)	Years of record	Mean flow (m³s⁻¹)	Peak flow (m³s⁻¹)	Date of peak flow	Intensity of peak flow (m³s⁻¹km⁻²)	Intensity of mean flow (m³s⁻¹km⁻²)
Fraser Lowland								
Brunette R.	8MH026	68.6	34	2.71	83.0	19/01/68	1.21	0.04
Salmon R.	8MH090	49.0	20	1.42	64.6	17/12/79	1.32	0.03
Still Cr.	8MH061	29.2	13	0.48	53.8	14/12/59	1.79	0.02
Yorkson Cr.	8MH097	6.0	19	0.15	5.1	30/01/65	0.85	0.02
North Shore Mountains								
Capilano R.	8GA010	172	31	20.0	544	15/11/83	3.16	0.12
Coquitlam R.	8MH141	54.7	7	6.41	147	05/11/88	2.62	0.12
Seymour R.	8GA030	176	50	16.3	650	31/10/81	3.69	0.09
Stawamus R.	8GA064	40.4	17	3.83	113	28/12/80	2.82	0.09
Transition Zone								
Jacobs Cr.	8MH108	12.2	14	0.96	24.6	17/11/68	2.02	0.08
Kanaka Cr.	8MH076	47.7	27	2.81	146	14/12/79	3.06	0.06
N. Alouette R.	8MH006	37.3	18	2.88	162	24/02/86	4.33	0.08
Pitt R.	8MH017	515	12	54.0	597	03/11/55	1.16	0.11
Fraser Basin								
Fraser R.	8MF005	217,000	76	2,720	15,200	31/05/48	0.07	0.01

and are rarely observed at high flow because of the short duration of the flood and the fatal danger of getting too close to the event. Unfortunately, West and North Vancouver, the village of Lions Bay, and Coquitlam are in the direct path of a number of such creeks.

Vegetation

The climate and topography of the Vancouver area and the effect of the latter on precipitation totals and temperature, as well as the variability introduced by different soils and slope aspects, favoured the growth of a number of different coniferous species in the region. Because different species tend to dominate different elevations, the landscape can be divided into biogeoclimatic zones, marked by similar physiographic, climatic, hydrologic, vegetative, and animal characteristics. The principal zones of the Vancouver region are: the Coastal Douglas-Fir, Coastal Western Hemlock, Mountain Hemlock, and Alpine Tundra (Figure 14).

The lowest elevations of the Coast Mountains and the adjacent islands, such as Bowen Island, are the driest forested sites and the least likely to experience frost. Here massive Douglas firs once towered 50-60 metres above the forest floor (Figure 15). The exceptional height of these trees attests to the ideal growing conditions and lack of disturbance, particularly the lack of extreme winds, that characterizes this region. There was

Figure 13

Mean annual runoff and annual maximum daily discharge of the Capilano River

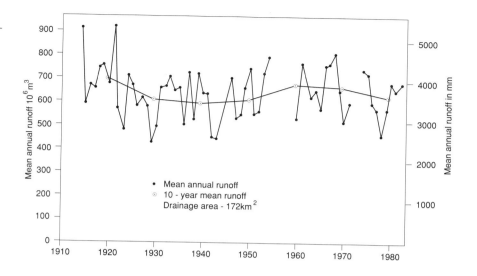

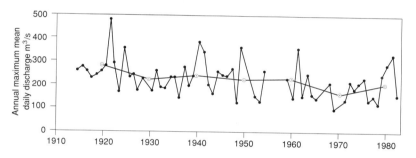

only a discontinuous shrub layer, consisting mainly of salal and Oregon grape, and a scattered herbaceous layer of shade-tolerant annuals, perennials, and sword fern with patches of mosses. Through this deeply shaded ground cover, the next generation of forest trees emerged. On the wetter sites there were no Douglas fir, but western red cedar or western hemlock flourished. Because the Douglas fir cannot regenerate in its own shade on moist sites, its domination of the coastal lowland forests indicates the past disturbance of such sites by fire. These fires killed the standing forest and revealed bare soils too dry in summer for colonization by cedar and hemlock, which need a more reliable moisture supply. Red cedar and hemlock cannot compete with Douglas fir on the driest ridge top sites, even under a closed forest. Here, fir succeeds fir, sometimes accompanied by arbutus, an evergreen broadleaf tree at the northern limit of its range in the Vancouver area.

At higher elevations precipitation increases, summers are cool, and the length and regularity of the summer drought decreases. The microclimatic factors which favours the success of Douglas fir on open sites are far less likely to occur, and hemlock and cedar dominate both the canopy and understory. Yet higher on the mountain slopes, the increasing length and severity of the winter reduces the growing season for

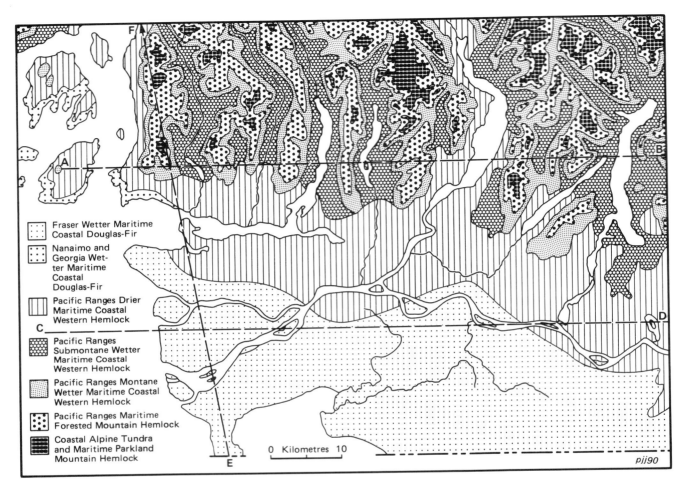

Figure 14

Biogeoclimatic subzones of Vancouver

cedar and western hemlock and eliminates their height advantage in competition. Here, the hardier, though smaller, mountain hemlock dominates.

Disturbance of these forests by lightning or fire, wind-throw, slope failure, flooding due to beaver dams, or by disease and death usually results in the appearance of deciduous trees in the otherwise conifer-dominated forest. The alder was, and still is, the first tree to colonize vacant sites. Once established, alders grow faster than conifers, and their ability to fix nitrogen gives them an obvious advantage over all competing species. However, the alder's dominance of a site is short-lived, as it has neither the ability to grow to great height nor the longevity of its coniferous competitors. Evergreen conifers take longer to establish their seed and they grow more slowly, but after forty years they begin to overtop the alders, and once the canopy is closed the light-loving deciduous species die off.

In the Vancouver area there are few extensive areas of alpine tundra or parkland above the forested zone, because glaciation left little land of

Figure 15

The lush vegetation and giant trees of the coastal rain forest

suitable slope at these high elevations. However, patchy subalpine meadows are found on the North Shore Mountains (Figure 16). Within the broad framework of biogeoclimatic zones, the complicated topography and local climatic patterns create a number of distinctive subzones. Each subzone has differing amounts of soil moisture, and although this type of variation does not affect the dominant tree species, it is usually evident in the species composition of the shrub, herb, and moss strata beneath the canopy. The valley floors of the Capilano, Lynn, Seymour, Indian, Coquitlam, and Widgeon rivers are in the Submontane Wetter Maritime subzone of the Coastal Western Hemlock zone; the steep mid-slopes of these valleys are in the Montane Wetter Maritime subzone; and the subalpine upper slopes are in the Maritime Forested Mountain Hemlock subzone. The ridges are in the Coastal Alpine Tundra and Maritime Parkland-Mountain Hemlock subzone. The strictly coastal slopes of the Coast Mountains are largely in the Drier Maritime Coastal Western Hemlock subzone, and the lowest elevations are occupied by the Coastal Douglas-Fir zone (Figure 17).

Figure 16

Subalpine meadows on Howe Sound Crest

In the Fraser Lowland, subzones of the Douglas fir and western hemlock forests are a function of detailed variations in the physiography and do not correspond to broad altitudinal variations as they do in the North Shore Mountains. The asymmetric hills or cuestas with a bedrock core (such as Burnaby Mountain and Capitol Hill) stand in contrast to narrower lowland valleys (such as Still Creek and Salmon River), to wide, flat-bottomed valleys (like the Nicomekl and Serpentine valleys), and to the more extensive flat area of the Fraser Delta and adjacent Lower Fraser Valley. All but the last of these areas were heavily forested until the latter part of nineteenth century or until as recently as the 1980s (see Chapter 9).

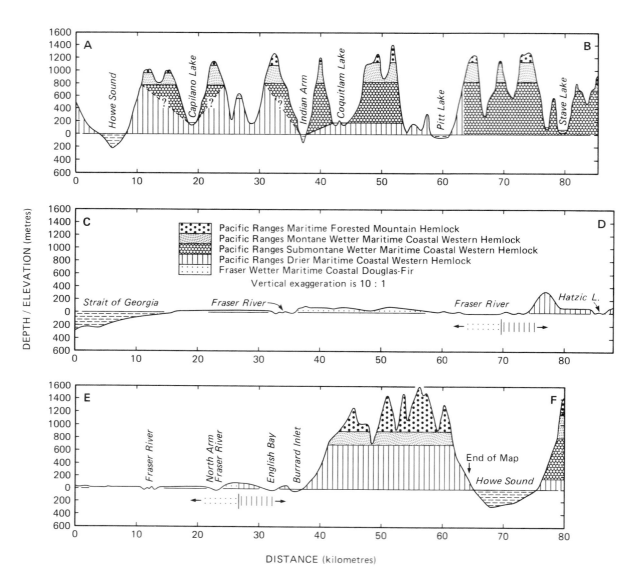

Figure 17

Altitudinal zonation of biogeoclimatic regions

The Lower Fraser Valley and its delta, and the Nicomekl and Serpentine valleys, were all subject to regular flooding from high seas in winter and from the annual freshet of the Fraser River before twentieth-century dyke construction. These lowlands were dominated by grass-like plants, such as reeds and rushes (as are found on the tidal marshes today), by grasslands, grass, and shrubs, or by bog shrub communities. The existence of non-forest communities in the lowlands was recorded by the early land surveyors, who, between 1850 and 1899, set out the legal land subdivisions prior to the first major influx of non-Native settlers.

The vegetation varied in response to the type and frequency of disturbance. Flooding from the sea or the rivers, or by the seasonal rise of the water-table, was of variable duration and more or less saline. Salt marsh occupied the most saline areas exposed to two tidal incursions daily, and tidal marsh occurred where the fresh water of the Fraser diluted the salinity of the tidal waters. Grassland extended from the high-tide line inland as far as annual sea and river flooding occurred. As the frequency and duration of flooding decreased with increasing distance from the sea and rivers, shrubs colonized the grasslands. Where the river, in flooding, had developed elevated banks, often of coarser sediments, or the sea had thrown up a high storm beach, these linear ridges, standing above flood level, provided drier sites that were colonized by deciduous trees, and, eventually, by conifers. Where depressions in the delta or floodplain sediments trapped water, or were periodically filled with groundwater, the slow anaerobic decay of plants created high acidity levels which excluded all but the most acid-tolerant bog plants, such as sphagnum moss, blueberry and cranberry bushes, and bog laurel. Peaty areas such as Burns and Camosun bogs are the result of these accumulations.

The second type of disturbance was caused by fire. In the floodplains, the fires may have spread from the adjacent dry forests that were prone to fires induced by lightning, but mainly they were set by aboriginal peoples in their management of the shrub communities on the bogs. Palaeobotanical investigation of Burns Bog shows evidence of periodic fires throughout the 4,000 years of its existence. The apparent failure of this large, raised delta bog to support a bog forest can only be explained by this constant deliberate intervention that prevented natural succession from occurring.

Conclusion

The large land surface features of Vancouver – the juxtaposition of land and sea and the absolute heights and depths of the North Shore Mountains and the Strait of Georgia – were shaped by evolutionary processes over a period of 100 million years. The last 2 million years have seen the fretting of mountain tops, the formation of fjords, and the over-steepening of valley sides by repeated glacial action. Yet much of the story of the evolution of Vancouver's setting belongs to the last 25,000 years (the Fraser Glaciation and the postglacial period), and a surprisingly large amount of that environmental modification occurred between 12,000 and 7,000 years ago. After the establishment of continuous forest cover over most of the region by 7,000 years ago, however, there was little significant alteration of the environment until the clearance of the forest in the late nineteenth century. Through almost seven millennia, climatic perturbations produced significant changes in landscapes at the margins of glaciers and at timberline but did little to alter vistas from the present site of Vancouver.

3

The Lower Mainland, 1820-81

Cole Harris

Europeans first reached the Lower Mainland only at the end of the eighteenth century and did not consolidate their hold over the area for almost another seventy years. Until 1858, the year of the gold rush to the Fraser River and the creation of the Crown Colony of British Columbia, the Lower Mainland was, overwhelmingly, a Native place in which Fort Langley, the trading post established in 1827 near the mouth of the Fraser, was an island of British control. After l858, effective Native control of the Lower Mainland collapsed very quickly. In l881, when the peoples of the region were first enumerated in a Canadian census, power lay squarely with governments and their representatives, the courts, the owners of sawmills and fish canneries, and, to an extent, local white leaders. Natives had been allocated small reserves where, their numbers much reduced and their voices unheard, they lived at the margin of a non-Native world. The state, industrial capital, and the cultural values of an immigrant population of predominantly British origin dominated the Lower Mainland.

In a few years the area had passed through a remarkable transformation: from the local worlds of fishing, hunting, and gathering peoples to a modern corner of the world economy within the British Empire and an emerging federal nation-state, the Dominion of Canada. A transition that in Europe took millennia, in the Lower Mainland took decades, which is why the Lower Mainland presents such interesting analytical opportunities.

Although Spanish and English navigators in the early 1790s, and Simon Fraser in 1808, wrote briefly about the area, the documentary record begins to expand only in the 1820s. This record plus ethnographic accounts permit a consideration of the human geography of the Lower Mainland circa 1820, when it was still Native. A second section of the chapter describes the intrusion of Fort Langley into this Native territory. A third deals with the momentous geographical changes in the area between 1858 and the first federal census in 1881. The final section comments briefly on some of the causes and implications of these changes.

The Lower Mainland: A Native place

In the winters of the early 1820s there were probably three population clusters in the Lower Mainland: Musqueam on the North Arm of the Fraser River, the Kwantlen villages just upriver from modern New Westminster, and Tsawwassen on the western shore of Point Roberts (Figure 1A). At Musqueam and the Kwantlen villages, the largest of

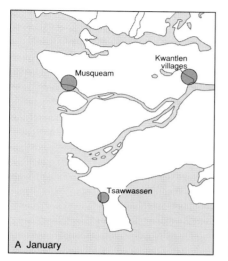
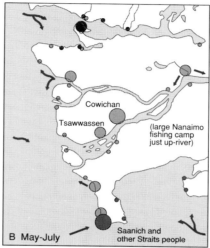
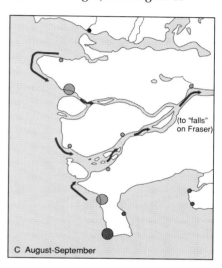

Figure 1

Generalized population distribution, about 1820

POPULATION LANGUAGE

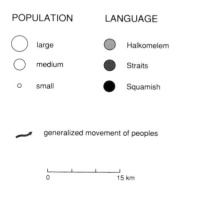

these settlements, there were several hundred people. In the 1930s, the anthropologist Homer Barnett was told that Musqueam had once comprised three adjacent 'villages.' The largest, Male, was a continuous shed structure, 500 yards long, made up (as shown in Figure 2) of many house segments; a second had seventy-six house segments arranged in a circle around a lacrosse field; and another comprised thirteen house segments. Barnett recorded the circular village in his field notes but, doubting his informant on this point, described it as semicircular in his book. Apparently there were also three adjacent Kwantlen villages. In all three population clusters, buildings were constructed of cedar posts and planks as shown in Figure 2.

Within these three population clusters, the largest unit of regular authority was the house group, most of whose members were close relatives. Those who were not were usually slaves, or people who, for one reason or another, were unable to form a house group of their own. One house

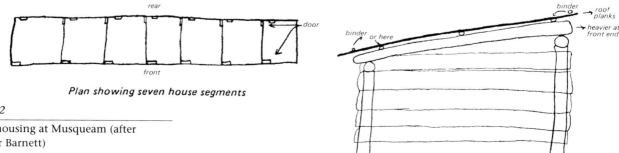

Plan showing seven house segments

End view

Figure 2

Shed housing at Musqueam (after Homer Barnett)

group might occupy several house segments within a long shed building or all of a shorter structure. Each owned house and mortuary sites and some principal resource procurement sites as well as names, songs, and rituals. Usually the head of a house group was a descendant of a founding family, sometimes a respected elder. But in the matrix of kin within which most people lived, authority depended on esteem and had little coercive power. Except as their actions affected the collective property of the house group, in which case decisions were taken by the head in conjunction with the elders, individual conjugal families did as they wished. Many of them owned important resource procurement sites. The heads of house groups met frequently, and usually one of them – because of age, eloquence, or the size of his or her house group – was particularly respected. But this person had no institutionalized authority; there was no institutionalized hierarchy of power beyond the house group. Within every village different house groups were linked by blood ties and by obligations incurred by giving and receiving gifts. Similar ties linked the three winter population clusters of the Lower Mainland with each other and with people elsewhere.

Although winter curtailed movement between settlements, the three population clusters in the Lower Mainland participated in a much larger region of social interaction (Figure 3). They spoke Halkomelem, the

Figure 3

The Lower Mainland in linguistic context

Linguistic Family
- Coast Salish
- Wakashan
- Interior Salish
- Eyak-Athapaskan
- Chimakuan

Language
HALKOMELEM

- Lower Mainland

0   30km

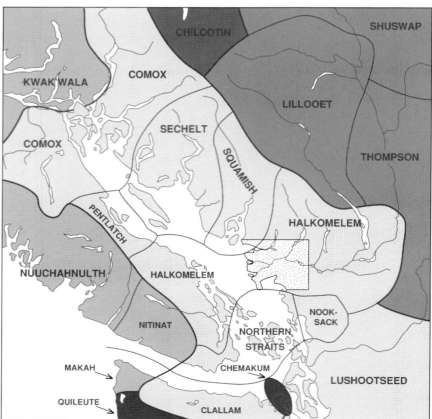

language of the peoples of the Lower Fraser from the Fraser canyon to southeastern Vancouver Island. Along the coast to the north were speakers of other Coast Salish languages, the nearest being the Squamish (who wintered at the head of Howe Sound) and the Sechelt. The most southerly speakers of a Wakashan language, the Lekwiltok, were a good 240 kilometres north of the mouth of the Fraser, around Johnstone Strait. Immediately south of the Halkomelem, along the coast, were speakers of other Coast Salish languages: Nooksack, Straits, and a little farther away, Clallum and Lushootseed. Inland, to the north and east, were speakers of Lillooet and Thompson, each an Interior Salish language. The people at Musqueam, the Kwantlen villages, and Tsawwassen, knew and, in one way or another, dealt with all these peoples. Many of them were bi- or trilingual; someone would understand almost any visitor from afar.

In March, when food supplies were much diminished, the people of the Lower Mainland began to move in conjugal families or house groups to resource procurement sites, where they often met other peoples. This, with great variation in detail, was the northwest coast pattern. Near the Fraser, the seasonal round of resource procurement was dominated by the river itself. The ethnographer James Teit concluded that, compared to the Columbia, the Skeena, or the largely overland routes from Squamish to Lillooet or from Bella Coola to the Upper Fraser, the Fraser was not a major route of Coast-Interior trade (because of the canyon). But, as one of the largest reservoirs of food on the continent, it drew people from afar and supported many more people than lived along it in winter.

Figure 1B shows the distribution of population in the Lower Mainland in early summer. Winter population clusters had dispersed. Musqueam groups were scattered around what are now English Bay and Burrard Inlet and at many fishing or gathering sites on the Gulf Islands and around the Fraser Delta; the Tsawwassens had a fishing camp on the south bank of Lulu Island; the Kwantlen were dispersed along the river and its tributaries in locations most of which cannot now be precisely known. Other peoples had come to the Lower Mainland. Squamish groups sought shellfish, fish, and birds in English Bay and Burrard Inlet. Cowichan people, Halkomelem-speakers from Vancouver Island, occupied plank-house fishing camps along the south bank of Lulu Island, where, in spring, they caught oolachan and sturgeon (in 1824 the fur trader F-N. Annance said there was one, mile-long village). Some 500 Nanaimo people, also Halkomelem-speakers from the Island, had a plank fishing camp on the Fraser near the mouth of the Pitt River, where they also took oolachan and sturgeon. By early July, Saanich and other Straits-speakers, together with some Clallums and others from Puget Sound, congregated at the southwestern tip of Point Roberts where, with rows of reef nets in shallow waters, they caught large numbers of salmon as the fish rounded toward the Fraser. Annance considered this settlement, which was uninhabited when he saw it in December, almost as large as the 'mile-long' fishing camp on the south bank of Lulu Island.

The intense early summer use of the shorelines and riverbanks of the Lower Mainland diminished abruptly in early August, when most of these people (though only the Saanich, apparently, among the Straits-speakers) moved some 180 kilometres upriver to take salmon in the rapids at the foot of the Fraser canyon (Figure 1C). People remained at the reef netting sites and at a few other locations, but Burrard Inlet and much of the Lower Fraser may have been virtually deserted. Several thousand people congregated at the foot of the lower canyon (10 kilometres above Yale). Most of them returned in late September, often transporting their dried salmon on plank platforms laid across two dugout canoes. In the last two weeks of September 1828, the chief trader at Fort Langley recorded 550 canoes of Cowichan and 200 of Squamish passing down river. Then, in early October, many peoples collected Indian potato (Wapato, *Sagittaria latifolia*) from marshes along the lower Fraser and its tributaries before there was a large social and ceremonial gathering of well-fed, well-provisioned peoples at the mouth of the Pitt River (or perhaps at the eastern end of Lulu Island). Not everyone returned to the population cluster in which they had spent the previous winter. Cowichans sometimes over-wintered at the foot of the Fraser canyon; others stayed along the lower Fraser. Apparently, a good many people passed different winters in different villages with different kin.

This flux of peoples, coupled with marriages that were almost always between people from different house groups (and often from different population clusters) and with property rights held by house groups and conjugal families, created intricate webs of socioeconomic relationships that spread well beyond the Lower Mainland. Social space had few clear boundaries. Individuals could belong by right of birth and marriage to several different house groups. People from different winter population clusters would have rights, therefore, to many of the same resources. Ties of kin crossed dialect and language boundaries that, in any case, were not sharp. The obligations incurred by gift-giving at the feasts added another layer to the web. When, for example, Cowichans came to Lulu Island, or Squamish to Burrard Inlet, or some Musqueams went to the Gulf Islands, they were not entering the territory of another people so much as moving within webs of social and economic relations that connected different individuals and groups to each other and to each other's places.

Such arrangements did not prevent conflicts. Wrongs were avenged, whether with payments or blood. The cycle of small-scale raids and reprisals was difficult to break and could occur locally or over considerable distances – anywhere within the range of peoples' contacts.

Sometimes sizeable war parties formed, reflecting a consensus among many or all of the heads of house groups in, or even beyond, a population cluster. Led by temporary 'war chiefs,' their attacks were swift and fleeting, usually just before dawn. If possible, men were killed and women and children captured, often to be ransomed by surviving relatives at the cost of most of their possessions. But there was no institutional or logistical support for prolonged assaults. Nor, given the lack of political authority beyond the house group, were alliances formed easily between different population clusters.

In the annual round of resource procurement, conjugal families and house groups fished, hunted, and gathered a large part of what they ate: fish and shellfish primarily, but also birds and land and sea mammals and a great variety of edible plants. Natives fashioned most of their tools, clothing, and buildings from locally available materials. Yet material goods circulated steadily between families and house groups in winter population clusters through informal gift-giving and through more formal giving to guests at feasts, which were held to maintain status and consolidate social claims (to a name, a marriage, a guardian spirit, a song, or a place). These were frequent winter occurrences. There was little trade as such within population clusters; the circulation of property by giving was, rather, an integral part of their social organization. Feasts involving people from different population clusters – 'potlatches' as they were known later in Chinook trading jargon – were held in late spring when the weather was fit for travelling. These feasts were large, and gifts were distributed on the basis of rank and generated reciprocal obligations (at subsequent feasts guests were expected to give more than they had received); they may have become common, as Homer Barnett surmised, only after Europeans introduced new wealth to the area. Exchanges of property over much longer distances, such as those which involved dentalium shells (*haiqua*) from the southwest coast of Vancouver Island or canoes from Johnstone Strait, apparently involved barter and bargaining combined, as far as possible, with gift-giving and kin relations. Raiding may have redistributed more property than did such trade.

Economy and society were not separate categories, nor was the environment. People lived in familiar places, named them intimately, knew the seasonal ways of flora and fauna and, within the limits of their technology, satisfied most of their material needs from local resources. Subsequent inhabitants of the Lower Mainland would not begin to match this detailed localized knowledge of a natural world. Nor would they know its spirits. The Native peoples of the Lower Mainland drew no distinction between mind and matter, animate and inanimate, human and natural, subject and object — the categories of a modern European mind. The founding figure in a creation story of the Pitt River people had taken a sandhill crane, then a sockeye salmon, to wife. The sockeye, therefore, were relatives, returning to give themselves to their kin.

A soft rain on a west wind was a mother weeping. Everything depended on spirit power. At puberty, the young sought a guardian spirit, and their subsequent lives were an unfolding relationship, often mediated by dreams, with that spirit and others – spirits interwoven with society, interwoven with the environment.

People lived within a coherent totality, a lifeworld. Life was highly contextualized within experience, and the context was largely shared, with individual stories wrapped around each other and around the creation stories. Everyday life was part and parcel of an experienced, known whole, from which it was impossible for an individual to stand aside.

It is not clear to what extent, in 1820, this Native world had been influenced by white contact. In 1808 Natives at the foot of the canyon (near Yale) told Simon Fraser that white men had already been there. He did not believe them, but David Thompson evidently did, noting on his remarkable map of l813-14, 'To this place the white man come from the sea.' When Fraser met the Kwantlen, they 'evinced no kind of surprise or curiosity at seeing us, nor were they afraid of our arms, so that they must have been in the habit of seeing white people'; the hostile reception he received from the Musqueam also implies that Europeans were known there. They cannot have come frequently, for when John Work, exploring for the Hudson's Bay Company, was on the Lower Fraser in 1824, he found very few European goods and noted that the meadows and small streams south of the river were replete with beaver, indications that the trade in land furs was not under way. The indirect effects of the white presence along the coast had been far more telling. There had been two smallpox epidemics (in the 1770s and again about 1800), while the Lekwiltok, affected by the establishment of Nawittee as a major centre of the maritime fur trade at the northern end of Vancouver Island, were raiding along the Lower Fraser for slaves.

These raids and the smallpox epidemic of 1800 both appear to figure in an account – told by elderly natives to a local white woman, Ellen Webber, who published it in *The American Antiquarian* in 1898 – of a former Kwantlen village on the north bank of the Fraser River just above the mouth of the Pitt. The villagers had fended off one attack from the river but then were surprised from behind:

> Now all was confusion. Many were killed, and many women were taken slaves. A few escaped to the woods, where they remained in hiding two or three days. Then, with the children, they came out, and with sad hearts they laid away their dead ... But misfortune followed the little band of survivors. In the swamp, near the village, lived a fearful dragon with saucer-like eyes of fire and breath of steam. The village was apparently regaining its former strength when this dragon awoke and breathed upon the children. Where his breath touched them sores broke out and they burned with the heat, and they died to feed this monster. And so the village was deserted, and never again would the Indians live on that spot.

Fort Langley in a Native world, 1827–57

An American ship was trading well into the Strait of Juan de Fuca in the spring of 1827, a few months before the Hudson's Bay Company established Fort Langley. From this time on, Natives of the Lower Mainland had direct, regular access to Europeans.

The establishment of Fort Langley brought an island of outside control and a well-developed system of nineteenth-century British trade into the middle of Native territory. The twenty men who built Fort Langley in the late summer and fall of 1827 were old hands of the Hudson Bay and Montreal fur trades who knew exactly what they were doing. A first bastion went up in two weeks. It appears, wrote the chief trader, 'to command respect in the eyes of the Indians, who begin, shrewdly, to conjecture for what purpose the Ports and loopholes are intended.' Within six weeks there were two bastions, a palisade, and gates. The traders had enclosed a small, defensible ground. Then living quarters went up, built of timber frame construction, like the bastions, with the spaces between the vertical posts filled by squared horizontal timbers. This was *pièce sur pièce en coulisse* (Red River frame) construction which was brought to the Fraser from the French settlements on the St. Lawrence by the transcontinental fur trade. Before winter, a simple fort was finished. On 26 November 'a Flag Staff was cut and prepared, and in the afternoon erected in the South East Corner of the Fort. The usual forms were gone through ... in baptizing the Establishment.' For those who made it, this small, defended settlement, like hundreds built before it in the northern continental interior, was not very remarkable.

Established to extend the Hudson's Bay Company's presence along the coast and to thwart American trading vessels, Fort Langley was an outlier of the Company's extensive, co-ordinated trading system. From it, furs, and, by 1829, salted and barrelled salmon were shipped to Fort Vancouver on the Columbia River. Supplies came from London via the Columbia River. Sometimes a Native was given letters to carry upriver to Fort Kamloops, to catch, with luck, the brigade for the Columbia. Local self-sufficiency, encouraged by the Company, was quickly achieved at Fort Langley. Sturgeon, salmon, and some game were traded from the Natives or fished and hunted by Company employees. In 1828 potatoes were found to yield well; soon there were other vegetables, grains, and livestock. By 1830 Fort Langley fed itself and had a food surplus for export.

Its small society was polyglot and multiracial. The first officers were Scots. Most of the men were French Canadians or métis (French and, soon, Chinook were the fort's most spoken languages), but others were British or Iroquois, and one was Hawaiian. All the women were Natives; some came from the Columbia River, more were local. James Murray Yale, who arrived in October, 1828, was married a month later to the daughter of a local chief, a marriage that the Natives probably considered would cement an alliance, and that the chief trader thought would be good for trade. In principle, the officer in charge of the fort, like the captain of a ship, commanded this society. In fact, his decisions dominated defensive arrangements, the deployment of contract labour, and the commercial economy but not the fort's domestic world. Some of the Native women in the fort kept their own slaves, and social networks crossed the palisade to the society beyond the fort. For thirty years after 1827, there was no mechanism by which a few whites in a fort or on ships could control Native life in an adjacent village or anywhere else in the Lower Mainland.

Yet Hudson's Bay Company introductions, not only trade goods such as blankets and firearms, became much more common. In 1829, Indian women at Fort Langley received 'a mess of potatoes' for a day's work, and soon potatoes were grown at most Native settlements along the Lower Fraser and around the Strait. Chickens were also introduced at this time. Dogs killed the first two chickens to reach a Native settlement; their owner returned to Fort Langley with the two dead birds, asking for live ones in exchange. Venereal diseases also diffused from Fort Langley (although they may also have been in the area before 1827). All such introductions affected Native life, as did the process of trade itself. The attentions white traders gave to chiefs probably raised their status in Native society. Native groups that participated directly in the trade had an advantage over those that did not. The Kwantlen largely abandoned their villages at what later became New Westminster and resettled near the fort; much of its trade soon passed through their hands, establishing a relationship that, in general, had been repeated across the transcontinental span of the fur trade. New wealth undoubtedly increased the scale of gift-giving at the feasts and, as Homer Barnett thought, may have made the larger feasts into inter-village occasions (the historic potlatch).

However, in the last analysis, Fort Langley was a bounded society. Company traders had difficulty extending their control much beyond its walls. If goods were stolen from the fort, every effort was made to retrieve them and to punish the thief. The traders had to be seen as resolute in the protection of property. They would threaten to discontinue trade, and, when they did, both the missing goods and someone who may have been the thief were usually turned over to them. When a fort wife left, claiming that she had not been well treated and 'secured with enough property,' the traders suspected that she departed to be with her Native 'paramour' and sent a party of five to bring her back. She had become property not so much of her husband as of the Hudson's Bay Company. Traders had dealt with Natives for years and knew the tactics of bluster, threat, and imperturbability. They were more than prepared to play off local Native factions. They could also adjust their prices, particularly to draw Native trade from American ships. And they could sell firearms with deliberate, geopolitical intent, specifically to counteract

Figure 4

View from Fort Langley by J.M. Alden, 1859. *Foreground:* a corner of the fort and sheds for salting and barrelling salmon. *Across the river:* a Kwantlen village, relocated to be near the fort, but with most of its houses still traditional. The sharp gradient of Hudson Bay Company power away from the fort is apparent.

the power of the Lekwiltok, of which the Natives along the Lower Fraser lived in terror. But when a Cowichan chief was said to have traded his arms and ammunition to the very Lekwiltok he was supposed to fight, the traders were helpless. They could not establish the truth of the rumour; to threaten would be to lose much of the Cowichan trade. The Chief Trader recorded the simple reality in the fort journal: 'We Cannot undertake to Control them So effectually, and must take them as they Come.' Traders did not attempt to change Native values, they did not actively interfere with the Native subsistence economy, and they did not begin to control Native territory.

In 1839 Fort Langley was moved 4 kilometres upriver to be closer to its farm. A larger fort was built. In the early 1850s the fort was less a fur-trade post than a hub of a diversified export economy and a supply depot for the interior trade (with connections upriver to forts Yale and Hope and thence by brigade trail to Kamloops and beyond). More than twenty Company servants worked at Fort Langley; its polyglot, multiracial population must have approached 200. Furs, fish, and cranberries were traded from Natives and, together with grains, butter, and meat from the Company farm (which went mainly to the Russians in Alaska), were the fort's principal exports. Native women worked in salmon salteries at Fort Langley, at the mouths of the Chilliwack and Harrison rivers, and at Fort Hope. Other Natives worked in the fields or on the bateaux that they helped track, pole, and row upriver. Such work was seasonal and was paid in kind. Trade and, to a lesser extent, seasonal work transferred large quantities of Hudson's Bay Company goods to Native hands without, so it appears, disrupting the seasonal rounds of most Natives or undermining their control of their subsistent economies. Hudson's Bay Company traders at Fort Langley in the mid-1850s were no more able to control surrounding Native societies than their predecessors had been in the 1830s (Figure 4).

Yet the context of Native life in the Lower Mainland was certainly changing. Hudson's Bay Company price competition and coastal shipping (particularly the Beaver, a side-wheel steamer introduced in 1836), eliminated American traders from the Strait of Georgia. The Company established Fort Victoria on Vancouver Island in 1843; with the border settlement of 1846 it became the Company's main depot on the Pacific. After 1846 the reef net fisheries and seasonal villages at Point Roberts were in American territory. In 1849 Vancouver Island became a proprietary colony. The Cowichan soon began to learn what this meant. When two of them were accused of murdering a white shepherd, a force of 130 sailors and marines from the British frigate *Thetis* accompanied Governor Douglas to Cowichan Bay. There, in January 1853, he told the Cowichan 'that the whole country was a possession of the British crown ... Give up the murderer and let there be peace between the peoples, or I will burn your lodges and trample out your tribes.' The Cowichan turned over one of the accused, and British sailors captured the other near Nanaimo. The two were tried by jury on the quarter-deck of the *Beaver* and publicly hanged. Three years later a force of almost 450 Royal Marines and seamen returned to Cowichan Bay to apprehend another suspected murderer, try him the next day, and hang him that evening in front of the Cowichan. By this date a colliery, and with it up-to-date industrial technology and the beginnings of an industrial town (the first in British Columbia), had been established at Nanaimo. New systems of white power were entering the territory of the Halkomelem-speaking peoples.

The effects of these developments on the peoples of the Lower Mainland are far from clear. Native trapping must have altered some seasonal rounds but probably not very much as beaver became scarce; everywhere, said a fur trader in 1847, the steel trap had done its work. Some people from the Lower Mainland began to visit Fort Victoria, where there was a larger choice of European goods than at Fort Langley as well as a Native market for slaves. After the border settlement, they also dealt with American traders in Puget Sound. With the establishment of Fort Victoria and Nanaimo, the peoples of southeastern Vancouver Island may have come in smaller numbers and/or for shorter periods of time to the Fraser River. As Coast Salish acquired firearms, Lekwiltok raiding diminished and stopped; in the 1850s there was considerable Coast Salish trade with the Lekwiltok and other more northerly tribes, principally at Fort Victoria. British military force had not been used against the Musqueam, Tsawwassen, or Kwantlen, but these people undoubtedly heard about it, in much-repeated detail, from their kin across the Strait. When Simon Fraser had descended the river some Natives in the canyon took him for Coyote, the wily transformer figure in their accounts of creation, but by the 1850s most people of the Lower Mainland must have recognized (without, of course, putting it this way), that quite alien sources of power, entirely outside their experience, were affecting life around the Strait of Georgia. Shared experience no longer fully explained their changing lifeworld.

Town, countryside, and camps, 1858–81

News of gold along the Fraser, which for a few years the Hudson's Bay Company had been trading quietly from Natives at Kamloops, became public in San Francisco in the early spring of 1858. Coupled with the increased capitalization of the California mines, the decline of small-scale, individual opportunity there, and other gold fevers around the Pacific basin in the mid-nineteenth century, it produced a gold rush to the Fraser River. At least 20,000 miners entered the mouth of the Fraser River in the spring and summer of 1858, most coming via Victoria (which suddenly became a city of 5,000) and virtually all headed for gold fields that were thought to begin below Fort Hope (150 kilometres inland) and extend upriver who-knew-where.

At the beginning of August 1858, the Cordilleran mainland north of the 49th parallel became a Crown Colony; Queen Victoria named it British Columbia. Colonel Moody, commander of a regiment of Royal Engineers sent to the new colony, selected 'the first high ground on the north side after entering the [Fraser] river' as the site for a capital, and Queen Victoria named it New Westminster. The site, Moody thought, could 'be defended against any foreign aggression' and make ready 'connections with Vancouver's Island ... as with the back country.' A defensive site near the head of marine navigation on a river route to the interior: two hundred and fifty years before Samuel de Champlain had chosen a similar site for similar reasons on the St. Lawrence.

With these developments, the exclusive trading regime of the Hudson's Bay Company, the relationships between Natives and Europeans on which it rested, and Natives' control in their own territory all began to come to an end. British civil and criminal law suddenly began to be enforced in the colony. A judge and stipendiary magistrates were appointed to interpret it, jails were built for offenders, and, if necessary, gunboats or troops could be used – a new system of administration backed by the most powerful and expansionistic of nineteenth-century nation-states.

The gold rushes also shifted the economic balance toward resource industries that were increasingly dominated by industrial capital. The industrial age was entering British Columbia in specialized work camps that combined an abundant resource, elements of industrial tech-nology, and transportation connections to distant markets. New people came, some from across the Pacific, but more, directly or indirectly, from Europe – particularly from Britain. Most of these immigrants believed in progress, private property, and the superiority of British civi-lization. Thousands of miles from home, they defended such assump-tions with the vigour of those who are dislocated and have the power of government at their disposal. Natives and Chinese, therefore, were excluded from citizenship – the latter were to be kept out or apart; the former, if they did not die off, were to be civilized. Such plans had spatial imperatives as, indeed, had the entire changed momentum of British Columbia after 1858. A new human geography was required, and more than anywhere else in the abruptly created colony of British Columbia, this new geographical project began in the Lower Mainland.

There, essentially, it established the ancient dichotomy between town and countryside and, equally basic in Canada, camps associated with staple trades. More specifically, it created a river-mouth town; farms superimposed on, but not entirely displacing, an older Native world; and canning, sawmilling, or logging operations. Within and among these settlements were particular configurations of power. Beyond them were the world economy, the British Empire, cultural values of the late nineteenth-century English-speaking world, and, less insistently overall, but extremely important individually, the particular settings from which other non-English-speaking peoples had come.

A town

New Westminster, a capital and also a river-mouth steamboat port, was the creation of both government and gold rush. The Royal Engineers established their camp (Sapperton) on the site of the former villages, just east of the new townsite, which they surveyed in a grid embellished by crescents, terraces, and public squares (Figure 5). They designed and supervised construction of most of the capital's first government buildings: Customs House, Treasury, Assay Office, Courthouse, Gaol, and Land Registry Office. They designed Holy Trinity Anglican church, a dignified Gothic revival church built of wood above the

Figure 5

Plan of New Westminster, 1862 (simplified from New Westminster Capital Plan, 1862, Royal Engineers, New Westminster, BC; drawn by J. Launders, R.E.; lithographed under the direction of Captain Parsons by order of Colonel Moody)

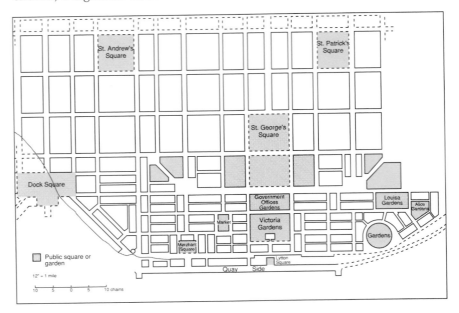

Figure 6

Two views near New Westminster.
Top: Engraving (by J.B.S., n.d.), looking up the Fraser River (probably from Sapperton), suggests how readily the English gothic imagination incorporated the wilderness.
Bottom: Photograph (1861) shows Holy Trinity Church (right centre) and the government buildings (centre) built by the Royal Engineers, as well as the commercial activities of the port (lower left, lower right). An institutional upper town and a commercial lower town, both still dominated by forest, were appearing.

fledgling town at the edge of the forest on land set aside for a public garden. But the capital being created by the Royal Engineers was also a port – a trans-shipment point between ships arriving from Victoria and river steamers. At the height of the Cariboo gold rush (1862) there were some 400 beds in boardinghouses and hotels, many stores selling outfit for the mines, and steamboat captains vying for the river traffic.

The officers of the Royal Engineers were well-educated men who held a vision of nature shaped by the romantic poets, the writings of John Ruskin, and the broad literary and artistic reaction in England to industrial society. At Sapperton they laid out a cricket ground and walks into the ravines. At the site of New Westminster, Colonel Moody had found a forest 'magnificent beyond description, but most vexatious to a surveyor and the first dweller of a town. I declare without the least sentimentality, I grieve and mourn the ruthless destruction of these most glorious trees. What a grand old park this whole hill would make.' For the eastern Canadians in the town, most of whom were Methodists or Presbyterians only a generation or two removed from pioneer conditions, forest walks and natural parks were 'useless follies.' These people considered themselves practical and were the strongest boosters of New Westminster's trade and industry. When the Royal Engineers were recalled in the fall of 1863, their vision of nature went with them and would hardly reappear in mainland British Columbia for another forty years (Figure 6).

After the Engineers left, Moody's house in Sapperton was enlarged for the new governor. When the colonies of Vancouver Island and British Columbia were united, in 1866, New Westminster briefly became the capital of an enlarged British Columbia. But with the post-gold-rush slump and, in 1868, the reversion of the capital to Victoria, New Westminster's population fell to under 500. It remained a port of entry to the mainland and a local service centre, and it benefitted from industrial developments along the Lower Fraser, growing slowly in the 1870s in proportion to the expansion of these activities. By 1881 almost 2,000 people lived there. Four canneries and three small sawmills operated in or near the town. It was the urban connection for farm families throughout the Lower Fraser Valley and, to an extent, for loggers and sawmills on Burrard Inlet, and it was the gateway for the railway construction that was beginning in the lower canyon above Yale, the site of the old Native fishery. It still housed a few government employees (government agents, an Indian agent, a constable, a customs officer, a warden and a prison guard, land agents, and a postmaster), and the Oblates, a French Roman Catholic order, administered mainland British Columbia and the Queen Charlotte Islands from New Westminster, but the town's momentum now lay with commerce and industry.

New Westminster's occupational structure was more diverse than was any place else on the mainland. It was a small, late nineteenth-century city of somewhat British cast, although in 1881 some 10 per cent of its inhabitants had been born in the United States, another 10 per cent in China, almost 20 per cent in eastern Canada, and, even excluding Natives, a good third in British Columbia. A large part of the labour force was made up of artisans and workers associated with the port or resource industries, and many of them spoke little or no English. The élite, who controlled the town, reflected bourgeois Victorian values. Their wooden houses frequently copied styles in San Francisco, their gardens were more inclined to be English, and their social conventions, such as ladies 'at home' days, were taken-for-granted parts of a Victorian image of respectable urban life. There were other, more basic assumptions: the law should serve the interests of property, men should head households, and, with only slightly less unanimity, the Queen should reign and British civilization should be reproduced in British Columbia.

From this perspective, Chinese and Natives were alien and 'other.' In the early 1860s a few Chinese lived in shacks at the edge of town; by 1881 there were over 200 Chinese in New Westminster, almost all in a 'Chinatown' situated at the western end of Front Street, near a cannery, two small sawmills, and the Natives. Racism largely created Chinatown, and the existence of Chinatown itself, and its real or imagined pathologies, reinforced the social construction of race. In the early 1860s Natives came for the spring oolachan and sturgeon fisheries and often stayed a while. 'Decent people,' fulminated a newspaper editor, were subjected to 'the intolerable nuisance of having filthy, degraded, debauched Indians as next door neighbours.' Although 'citizens' tried to keep Natives out (a small river-front reserve, established in 1862, was intended to segregate them), they moved easily through New Westminster streets and empty lots until, in the 1870s, they began to settle in shacks beside the fish canneries. For most whites, the Natives were a different, incomprehensible, racially identifiable people whose economic usefulness was acknowledged, but whose social status was not measured by any scale that included whites. Less threatening than the Chinese, they were equally alien and other.

Most of the whites in New Westminster kept in regular touch with home and the larger world from which they came. Until 1869, letters, magazines, and newspapers from London came by ship and rail across the isthmus of Panama in less than two months. With the completion of the Union Pacific Railroad to San Francisco in 1869, they took about a month. In 1865 telegraph lines connected New Westminster to the American telegraph system; a year later the first trans-Atlantic cable extended the link to Britain. Goods came via the Union Pacific and San Francisco or around the Horn and entered through Victoria, in these years British Columbia's Pacific port. By such means the outside world was accessible, much more so, really, than was British Columbia itself. In the Lower Fraser Valley the river was virtually the only artery of inland transportation: river steamboats ran to Yale at the foot of the canyon, and a wagon road and trails extended beyond. Away from the routes to the mines, most of the interior was inaccessible, and the range of New Westminster's influence there was extremely limited. Victoria continued to exert its distant authority through gold commissioners, magistrates, and government agents; and the Department of Indian Affairs in Ottawa had begun to create Indian agencies and to appoint Indian agents to administer them. The Oblates had devised a system, based on itinerant priests and Native officials, for the regular surveillance of dispersed, exceedingly isolated populations. But control was hampered by the problems of movement. Even within the Lower Mainland, movement was very restricted. It was difficult to walk in the forest. The government cut trails or seasonal roads (initially partly for defence) to Burrard Inlet and along the Fraser, but the river itself remained the main avenue of transportation. People were perched in New Westminster within a stretched, internationalized social space that had only begun to come to terms with British Columbia.

A countryside

As well as links with a town, a settled countryside required a land policy, which meant, in British Columbia, that terms had to be worked out to turn land judged to be wilderness into private property. The Colonial Office favoured compact settlement on surveyed lands sold at a good price. Cheap land and lax terms of payment would encourage 'the premature conversion into petty and impoverished landowners of those who ought to be laborers.' Therefore, shortly after they arrived, the Royal Engineers surveyed a number of ranges (at the mouth of the Pitt, opposite New Westminster, and in the Fraser Delta) and a few lots, intended for military settlers, along the American border. Each range was divided, as topography permitted, into thirty-six square, numbered, 160-acre sections. These sections were priced at ten shillings an acre or eighty pounds, half payable in advance. Few were sold, in good part because land was available across the border for half the price. The next year local officials, with the reluctant support of the Colonial Office, adopted a more American policy of pre-emption (thereby increasing the availability of land by allowing settlement to precede surveys and reducing its initial cost by allowing payment to follow surveys). Then, in 1861, in the face of Governor Douglas's arguments that 'the sturdy yeomen expected this year from Canada, Australia, and other British Colonies might be driven in hundreds across the frontier to seek for homes in the United States territories,' the Colonial Office accepted a lower price for land: four shillings, two pence (one dollar) an acre.

Figure 7

Land surveys to 1876

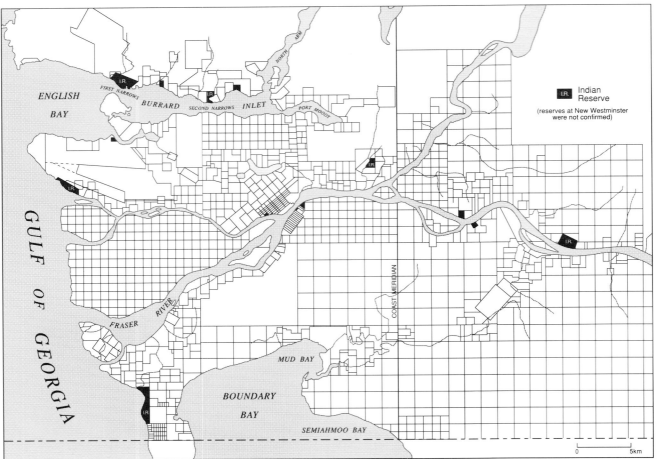

Pre-emptors of 160 acres could purchase an additional 480 acres, paying half the price at the time of sale, half when the land was surveyed. They were to be, or become, British subjects. With minor changes, such policies remained in effect until 1875, when land began to be surveyed into townships of thirty-six square-mile sections and offered to British subjects in free grants of 160 surveyed or unsurveyed acres. Additional land could be purchased for one dollar an acre. Implicit in such legislation was the denial of Native rights to land. From 1866, Natives were not allowed to pre-empt land, because, the Governor argued, the Crown could not delegate its constitutional trust over Indian lands. Rather, Natives received a few tiny reserves, held for them by the federal government. Figure 7, which shows all these land surveys, is, essentially, a map of the new balance of power in an area subdivided into properties and opened for development.

Yet the pace of land acquisition was slow. Figure 8 shows the sales and pre-emptions of Crown land in the Lower Mainland before Confederation. Almost no land was sold in 1859, when the price was ten shillings an acre. Lower land prices, pre-emptions, and the expectations

Figure 8

Land pre-emptions and sales, 1858-71

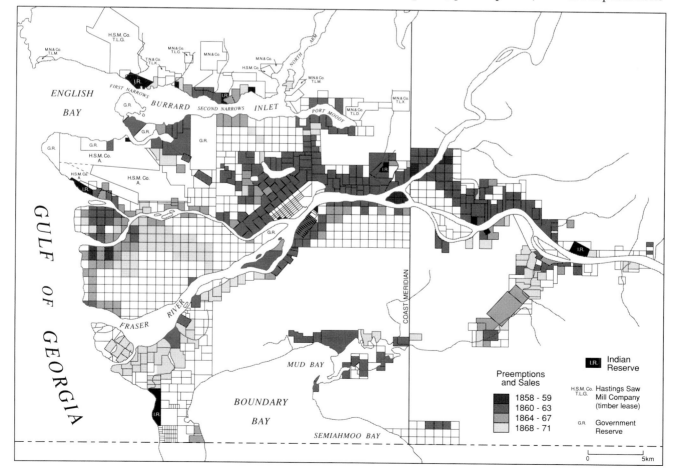

of a gold rush to the Upper Fraser improved the land market in 1860 and 1861, but the rate of alienation of Crown land declined thereafter. For the most part, lands were taken up along the river (the main artery of transportation) close to New Westminster or around Burrard Inlet. In 1868 just over 1,100 acres were under crops in the entire Lower Fraser Valley (below Hope). Most of the alienated land shown in Figure 7 was undeveloped; much of it was held by speculators, but prices for uncleared land were static. Confederation itself did not increase the attraction of land. In 1872 there were 228 pre-emptions in all of British Columbia; in the first full year of free land grants (1875) only 179 such grants were made.

The essential problem, almost as old as the European settlement of Canada, was that isolated settlements created by a staple trade were not easily converted to other pursuits. The development of alternative economies was hampered by small local markets and inaccessible distant ones. As the gold rushes waned in British Columbia, this conundrum loomed everywhere. Agriculture, which seemed to offer both economic security and social stability, was one apparent alternative, supported by a spate of agrarian rhetoric. The Lower Mainland was judged to be one of the most promising agricultural areas in British Columbia. But there were no external markets for its agricultural produce; even the Cariboo mines and the cities of Victoria and Nanaimo were better served by farmers closer at hand. Markets, such as they were, were in New Westminster and in the lumber camps and sawmills on Burrard Inlet. Moreover, there were major environmental obstacles to farming: prairies were subject to seasonal flooding, and massive forests were exceedingly difficult to clear.

In this situation, agriculture required some capital and developed slowly. Farming appeared first on the prairies where, for all the labour of dyking and draining, start-up costs were less than they were in the forest. Without hired labour, a pre-emption would not become a farm for years, but white labour was expensive, and Chinese or Native labour was commonly used. As farms developed they became mixed operations, partly because of the diverse requirements of family subsistence, partly because the local market was unspecialized. Therefore, wheat and other cereals were grown on most farms; orchards were common; and roots were raised for domestic use, animal feed, and sale. All farms had land in meadow and pasture and raised some cattle; and some of them, the most prosperous in the region, sold hay, oxen, and various foodstuffs to the lumber camps. For many, however, farming was essentially a subsistence operation, an interminable labour that men might interrupt by wage work in New Westminster, on a road-building crew, or in a logging camp or sawmill.

Farm settlers in the Lower Mainland were of many backgrounds. At Langley, where there had been a Hudson's Bay Company farm for thirty years, most heads of households in the 1860s were former Company employees: mostly French Canadians, Orcadians, or other Scots. Their wives were frequently Natives, and their children tended to intermarry. By the 1870s the Langley population was changing as new settlers arrived, most of them from eastern Canada (particularly Ontario), some from the British Isles (Irish and English as well as Scots), and only a very few from the United States. There was a good deal of chain migration – people coming to a relative already established – and the links connected east rather than south. Although only a few miles from the border, the early agricultural settlement of the Lower Mainland was not American. Seeds, livestock, and farm implements were more likely to come from the United States. Several of the Lower Mainland's farmers went to Oregon to purchase cattle and then drove them to Puget Sound for shipment to the Fraser. Some settlers imported bulls from Ontario or, occasionally, from England.

In good part, the Ontarian pioneer landscape and experience were being recreated in the Lower Mainland. Family farms were hewed from the forest with simple tools; as farms took hold, log cabins gave way to houses (in Ontario usually brick, in the Lower Mainland frame, but often of similar style and trim). The mix of peoples and religions – Presbyterians (both Reformed and Church of Scotland), Wesleyans, Church of England, and Roman Catholics – was much the same as in Ontario – the British connection was real. A settler society was emerging, based on family farms and supported by the local rural neighbourhood, churches, schools (with curricula adapted from Ontario), roads, post offices, and commercial and institutional connections with New Westminster. A few pigs, firkins of butter, vegetables, or a cow or two would be taken to New Westminster for sale. There, settlers saw about title deeds for land. Going to New Westminster, small as it was, was 'going to town.' Always, the idea of progress was in the air. If pioneering was a struggle, accidents common, and the outcome by no means foregone, individual and family progress toward economic security and social respectability tended to be measured by establishing a farm. To do so required land and hard work; the commitment to both, therefore, was enormous.

What, then, of the Natives? Like many other settlers, Fitzgerald McCleery, who came from County Down, Ireland, and settled on the North Arm of the Fraser River near the village of Musqueam, hired Native and Chinese labour. But he knew that his farm was his property, and that he, and his more prosperous uncle across the river, were citizens of a British colony. These were certainties; the winter dances at Musqueam, which McCleery saw occasionally, were curiosities from another world.

Figure 9 shows the distribution of Native population around the Lower Mainland in the winter of 1877. Government officials and missionaries had vaccinated Natives in the Lower Mainland just before the smallpox

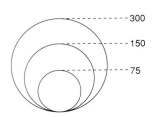

Figure 9

Native population, winter 1877

Language

- largely Squamish
- mixed
- largely Halkomelem

- 300
- 150
- 75

Circles are proportional to population.

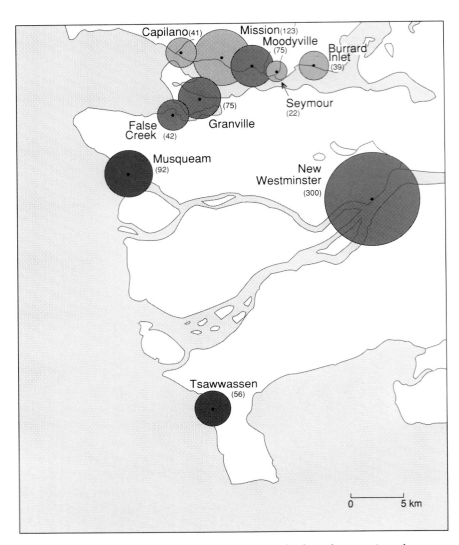

Capilano (41)

Mission (123)

Moodyville (75)

Burrard Inlet (39)

Seymour (22)

Granville (75)

False Creek (42)

Musqueam (92)

New Westminster (300)

Tsawwassen (56)

0   5 km

epidemic of 1862. Its effects, therefore, were far less devastating than they were in most of British Columbia. As people from outside the Lower Mainland settled there, its Native winter population may have been higher than it had been fifty years before. Although the old Kwantlen villages at New Westminster no longer existed (the last Natives living there were relocated by the Royal Engineers in 1859), some 300 Natives, many undoubtedly Kwantlen, lived in frame shacks near the two canneries in New Westminster. Musqueam and Tsawwassen remained, the former with just under one hundred and the latter with just over fifty people. Around Burrard Inlet and False Creek there were five small Native villages, each occupied in the main by Squamish-speakers who had come to work in the logging camps and sawmills of Burrard Inlet in the early 1860s. Some 150 Natives lived in Moodyville or Granville, the two sawmill towns on the Inlet. As in the 1820s, this winter population was augmented in spring and summer. Cowichan still came to their plank-house village on the south shore of

Lulu Island in 1867, perhaps even later, and in the 1870s many Natives still depended on the Fraser's oolachan, sturgeon, and salmon fisheries. Increasingly, however, they were becoming trespassers on their own former resource procurement sites.

Except at New Westminster, Moodyville, and Granville, all the people shown in Figure 9 were now on reserves, patches of land set aside for Native use and administered, from the mid-1870s, by the Dominion government through its local Indian agent. The first small reserves had been surveyed by the Royal Engineers. In 1864, however, Governor Douglas instructed surveyors to make reserves as large as Natives wanted, and some of them wanted a hundred or more acres per family. Posts were driven, but surveys were not completed, understandings not recorded. After Douglas retired later that year, his successors disavowed the surveyors' authority and abandoned a policy they judged to be prodigal. Neither understanding white law nor allowed access to white courts, the Natives were helpless. But they remembered. Years later, in 1913, a Musqueam chief told a Royal Commission investigating the Native land question: 'Since these posts were put down by Sir James Douglas for the Indians, the land has been lessened twice. The Indians were not notified or consulted ... and after that three persons came here to Musqueam and told some of the Indians that the posts ... meant nothing at all.' Indeed, by the mid-1870s, government sales and pre-emptions to white settlers largely prevented any expansion of Native reserves (Figure 7). In April 1878, Gilbert Malcolm Sproat, Indian Reserve Commissioner, reported a meeting with six chiefs from the Lower Fraser Valley: 'The Indians say that their lands are not sufficient in area and that, for several years, white settlers have been coming into the District and, in some cases, have been permitted to take up land which the Indians were hoping to get. They ask that this be no longer permitted ... They consider that they have been quiet and obedient to the law, and have not been well treated as regards their land.' Sproat asked the provincial government to alienate no more land near reserves until he was able to investigate the situation. The request was ignored. In June 1879, when Sproat tried to adjust Native lands in the Lower Valley, he found that the areas they wanted had been taken up. Although Sproat was sympathetic to their wishes, he did not question the rights of private property. Because holders of land titles could not be dislodged, there was virtually nothing he could do.

Government officials and missionaries intended that Natives become farmers, a civilized pursuit that would help 'turn them into Europeans' and their reserves into countryside. At Musqueam in 1877, some twenty-two acres were cultivated, a third in potatoes, vegetables, oats, and fruit trees, the rest in meadow and pasture. There were nine horses, thirty-one head of cattle, and some poultry. Such agriculture produced a good deal of food to supplement the 'large quantities of dried salmon' the census enumerator found in the village, though neither the Musqueam nor any other Natives in the Lower Mainland became primarily farmers. The reasons are obvious: a traditional subsistent economy that still produced a great deal of food, access to wage-labour in the white commercial economy, and insufficient land. Many Natives had learned farming as farm laborers, but, in their circumstances, farming could hardly be more than a part-time activity.

Natives still fished, gathered, and hunted away from the reserves. Of these activities, fishing was the least obstructed by conflicting demands for land and resources. Gathering and hunting were increasingly squeezed. Yet Native women still picked cranberries in the bogs of Sea and Lulu islands, hunters still netted birds in marshes along the river, and families still dug clams on the beaches at Point Grey. Moving as seasonally as they could through land they no longer controlled, Natives were everywhere and nowhere. But, as time went on, their off-the-reserve activities tended to focus on fewer work sites. In the 1870s Tsawwassen men worked on the river as commercial fishermen; their wives worked in the canneries. Cowichan still came to the spring oolachan and sturgeon fisheries but more to the summer sockeye fishery organized around canneries near New Westminster. The lumber camps and sawmills around Burrard Inlet depended on Native labour. Traditional patterns of spatial mobility adjusted to these new employments.

Industrial work camps

Around the Lower Mainland after 1858, as from the beginning of European enterprise in Canada, work camps appeared at sites where local resources could be exploited and shipped to distant markets. In British Columbia in these years, many such work camps became industrial workplaces.

Some of the first of them were along the river. As early as 1864, independent entrepreneurs adopted methods employed by the Hudson's Bay Company to salt-cure and barrel salmon and a few sturgeon. Little capital was required, nor was more required in the late 1860s when the first experiments began with canning. In the summer of 1871, a cannery operated on the left bank of the river just below New Westminster. Over the next decade, a dozen more were built, their locations shifting toward the rivermouth as competition for fish increased (Figure 10). For a time, salting and canning co-existed (usually in the same establishment), but canned salmon, a far less perishable product and one that reached the British market, accounted for 95 per cent of the value of fish exports from the Fraser by 1879. Cannery owners depended on commission agents in Victoria who advanced cash for supplies and labour; insured, transported, and marketed the salmon (secured by a chattel mortgage of plant, boats, and equipment); and deducted all costs and fees before the cannery owner received any return for a year's catch. Because marketing took longer than the annual cycle of the fishery, prolonged indebtedness to commission agents was a feature of

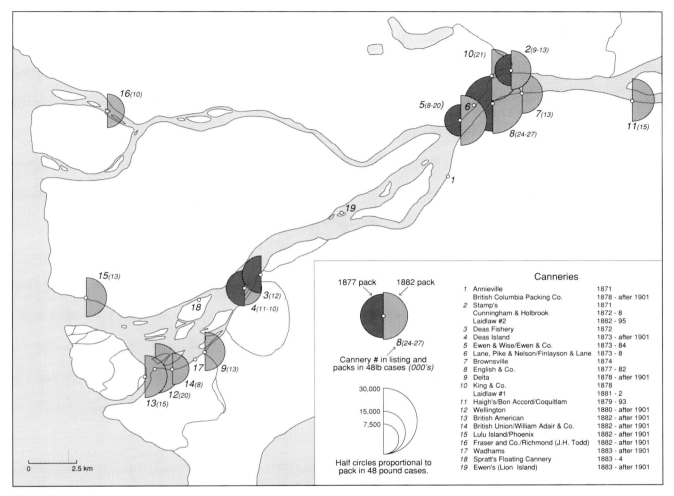

Figure 10

Canneries on the Fraser River, 1871-85

early canning that entrepreneurs of modest means operating far from markets could hardly avoid.

The cost of building and equipping a one-line cannery in the 1870s – perhaps $5,000 for the building and considerably less for equipment – was less than half the annual cost of operating it. Canneries were labour-intensive work places that employed some 130 people per canning line. Before the fishing season, Chinese crews made cans from imported tinplate, and Native women made nets from flax twine. When fish were delivered to the wharf, they were butchered and washed and cut into can-length pieces by hand. Cans were filled and soldered by hand, then lowered into cooking kettles, removed, tested for leaks, and cooked again. Eventually, they were lacquered, labelled, stored, and packed in 48-can cases, all by hand – manual methods but factory organization. In 1878 the commissioner of fisheries noted that:

> Many ingenious devices, with labor-saving apparatus of divers kinds, are eagerly adopted as necessity suggests. It is of course only by an organized system of action and the minute subdivision of labour that the operations of the industry, from the cutting of the

tin plates, the shaping, the soldering up to the final labelling of the cans, after the insertion and cooking of the contents, can be profitably or successfully carried on. It is pleasing to witness the order and regularity with which these various processes are accomplished.

The most important of these 'ingenious devices,' a steam retort that pressure cooked the salmon, was introduced in 1877. Gang knives that cut the fish to exact lengths at the push of a lever followed shortly thereafter.

Only one species of salmon, the sockeye, was canned in the 1870s. Its large runs close to the surface, and its fairly uniform size (5-7 pounds), made it relatively easy to catch and process, while its red, oily flesh (poor for drying) was favoured by the British market. In the Fraser's opaque waters, sockeye were taken in gill nets drifted behind small, flat-bottomed, two-man skiffs. In a good year, one such boat would catch 5,000 or more fish a season (about thirty days beginning in early July). Because spoilage was rapid, fishing took place close to canneries until, in the late 1870s, steam tenders began to extend the fishermen's range. With tenders, they could operate from various points along the shore taking their fish to a scow that the tender collected daily. With or without tenders, some twenty-five boats were required to serve each cannery line during the height of a run, and more were required as it tapered off. Usually canneries owned the boats, which cost some forty dollars each.

In the 1870s the canning industry on the Lower Fraser depended on Native labour, not only Halkomelem-speakers but others, such as the Thompson from the Fraser Canyon (who had never previously frequented the Lower Fraser), or Lekwiltok (who, fifty years before, had come to raid). Native families, rather than individuals, canoed to the canneries each season (as previously families had gone to resource procurement sites), and at New Westminster many stayed year-round. Under white foremen and alongside Chinese butchers, Native women did most cannery work, earning ten cents an hour. Older children must have worked as well, while younger children played and babies were carried by their mothers. Men fished, earning a wage of $1.25-$1.50 for a twelve-hour day. From the canners' point of view, such a cheap, reliable, and seasonally available labour force was indispensable, especially its women. Some canners hired more fishermen than they needed, so there would be Native women in their canneries. For the Natives, the sockeye fishery opened up a familiar resource where traditional native technology could hardly exploit it and provided a seasonal income approaching $100 per family; also the timing of the fishery lay conveniently between the spring oolachan and sturgeon fisheries and the late summer and fall runs of what the canners derisively called dog salmon – species that dry-cured easily and were the basis of the Native fall fishery.

The commercial fishery was a point of intersection of the world economy and the Native subsistent economy, of industrial work and time discipline and Native spirit worlds, and of peoples of drastically different cultures who, unless they spoke Chinook, usually could not talk to one another. For Native men these disjunctions were less severe: they worked, unsupervised, in small boats, with the river and its fish. For Native women it was quite another matter. They worked on a cannery line and were subject to industrial discipline under the eye of a white foreman. Yet Halkomelem was the principal language in a cannery, plus other Native tongues, Chinook, Cantonese, and English. Native women prepared their families' food and took sunburned and slightly spoiled fish to dry beside the tiny frame houses (apparently Native built) in which they lived. For women, men, and children, canning was a social occasion, as was the case whenever large groups of Natives briefly assembled. Then the sockeye run ended and, except at New Westminster, the Natives dispersed. Cannery buildings were left to a watchman.

The large sawmills built on Burrard Inlet in the 1860s provided, on the other hand, virtually year-round, mechanized, factory-organized work. British capital seeking to build a 'first class sawmill' arrived in British Columbia in 1859. 'The only question,' its manager Edward Stamp informed Governor Douglas, 'is where the establishment is to be fixed.' His first choice was at the head of the Alberni Canal on the west coast of Vancouver Island, but within five years the mill there closed for want of accessible timber. Stamp, having raised another £100,000 in Britain, decided that a new mill should be located in Burrard Inlet. Initially, he favoured a site on the Government Reserve at First Narrows (now Stanley Park), but in the summer of 1865 he fixed on a site 3 kilometres to the east on the south shore of Burrard Inlet, where a large, steam-powered mill was built (known, after 1870, as Hastings Mill). In 1865 a sizeable, water-powered sawmill (S.P Moody and Company) was already in operation on the north side of the Inlet near the mouth of Lynn Creek.

These mills cut almost entirely for export, selling lumber around the Pacific Rim and spars in Europe. They were up to date mechanically; when Moody converted to steam in 1868 he equipped the new mill with 'the latest machinery available.' Visiting this mill in 1875, the geologist George M. Dawson found 'a pair of large Circular saws, & a large gang saw, besides a small circular saw with long traversing table for cutting up the large planks, & others for cutting boards into lengths &c. Two planing machines, Mill driven by steam, but water power formerly used & still available, often employed to drive planers when other machinery Standing.'

Attracting capital to Burrard Inlet were magnificent Douglas fir-western red cedar forests located on gently sloping land close to tidewater and a sheltered, deep-sea port. With the creation of a Crown Colony and improvements in oceanic shipping, these forests became accessible to the international economy. Yet capitalists wanted, and virtually got, their timber free. As Stamp explained to the colonial secretary: 'After a careful examination I am now satisfied that sufficient timber exists on

Frasers River, Burrards Inlet, Hows Sound [*sic*], and the adjoining coast to justify me in a costly sawmill in that locality, I only now wait the government to grant me such concessions as are absolutely necessary to enable the company I represent to proceed with their undertaking.' These concessions were: a Crown grant of 243 acres purchased for one dollar (four shillings, two pence) an acre, the right to purchase another 1,200 acres at the same price, and 15,000 acres in leaseholds (most of Burrard Peninsula) for an annual rent of one cent (half a pence) an acre. By conservative estimate, Stamp's leaseholds contained a billion board feet of prime timber. Moody received similar concessions on the north shore of Burrard Inlet. The Native reaction to this can only be inferred. At Alberni, Stamp, who was also the Justice of the Peace, explained his sentence of a Native to four months of hard labour in Victoria as follows: 'Thefts have become so common here with the Indians; this step was absolutely necessary; I was told by the chief that I had stolen their land and they had a right to steal from us. Perhaps a good flogging would have done more good; but unfortunately I have not the necessary instrument of punishment by me; this defect I will remedy without loss of time.' In Burrard Inlet the surveyor of Stamp's proposed mill in the First Narrows Reserve found 'the resident Indians ... very distrustful of my purpose and suspicious of encroachment on their premises.' However, as the constable reported a few days later, 'they can at any time be removed. The ground does not belong to their tribe (the Squamish).' He was right, they could be removed. The full apparatus of colonial power supported the timber leases. Over the next few years fires destroyed a large part of the valuable timber on the North Shore; Moody thought that Natives had set many of them.

According to one of Homer Barnett's informants, Moody induced the Squamish to settle in Burrard Inlet. Certainly, in the early 1860s, some of their summer villages in the inlet became year-round residences as Squamish men took regular work on the docks or as general mill hands. Other natives came from much farther away and, having left their families, lived in company bunkhouses. There were also Chinese in good number, eastern Canadians, Americans, Europeans of various nationalities, and a few Hawaiian Islanders. Here, as in the canneries (and as earlier at Fort Langley), a multiracial, polyglot labour force was assembled. If there were a lingua franca, it was Chinook. Yet labour was scarce overall, and the mills required Native workers. G.M. Sproat, Indian Reserve Commissioner, estimated in 1876 that $70,000-$100,000 passed annually to Natives in Burrard Inlet, some in payment for fish, other foodstuffs, and hay brought by Natives to mills and logging camps.

A shift in the sawmills was twelve hours, the working week was six days, and pay for general labour was seventy-five cents a day with room and board or $1.25 without. Such work was superimposed on the patterns and prejudices of vastly different lives. For many, as one white mill-hand said, 'all there was to do was work, eat, and sleep.' Many of the Natives in Moody's mill walked to work from their village a mile away. The Chinese lived in shacks east of Hastings Mill, segregated from the whites who lived to the west in the mill town of Granville. Apparently there was a Chinese bunkhouse in Moodyville and, as the settlement grew, a 'Frenchtown' and a 'Kanaka Road' (for Hawaiians). Racism or sheer cultural distance, usually both, kept people apart. Space was also gendered. The sawmills, bunkhouses, and docks were male places. The Chinese shacks were male, for the families of Chinese workers were in China. The houses became increasingly female; the men, for the most part, were away or asleep. On the other hand, the Saturday night dances at the Hastings Hotel were attended by white men and Native or mixed-blood women. White women stayed away. Over these little-studied patterns of life around the sawmills on Burrard Inlet in the 1860s and 1870s loomed the mills themselves, their bosses, and their long, machine-and-clock-dominated hours of work. People who had grown up on the banks of the Squamish River within a fishing, hunting, and gathering economy, or, for that matter, in an agricultural village near Canton or on a farm in eastern Canada, would appear to have entered another world. However, we know next to nothing about the ways in which industrial discipline and more traditional forms of social organization interacted in such settings.

Companies also largely controlled logging. Loggers hired by a company or by a logging contractor felled the trees and bucked them, and as many as twenty yoke of oxen hauled logs over corduroy skid roads (made of small logs laid across the road). In the 1870s, a few operations experimented with steam traction engines run on tracks made of logs chopped flat on one side. By such means massive logs could be hauled two or three miles to tidewater, where they were made into booms and towed by a steam tug to a sawmill. The gently sloping flanks of Burrard Inlet and English Bay were ideal for such operations. There were, for example, five logging camps at different times in what is now Stanley Park, each with a bunkhouse (and perhaps a cookhouse), stables for oxen, and a network of skid roads. Such operations were expensive; a principal skid road itself might cost almost $1,000.

There were some independent hand loggers, who usually worked in pairs, logged virtually at tidewater, and lived with Native women ('bought' for fifty dollars) in isolated cabins around the inlet. The Natives quickly learned from such loggers and began hand logging themselves. In 1875 the Sechelt cut and boomed one and a quarter million board feet of timber, which the mills purchased for three dollars a thousand and towed to Burrard Inlet. Sproat urged the provincial government either to allow Natives in non-agricultural coastal areas to take out timber leases or to grant them good-sized reserves for the purpose of logging. The provincial government did neither, which meant that, except as loggers working for wages, they were excluded from commercial access to the forests.

Toward an interpretation

In 1881 the census enumerator in the Lower Mainland travelled through a human geography that had not even begun to exist there twenty-five years before (Figure 11). Much of it, of course, had existed elsewhere. Neither town nor countryside nor resource camps nor innumerable details within them were invented in the Lower Mainland. All these components of the outside world were somewhat rearranged there, but most of the elements of the composition had been introduced.

Figure 11

The peoples of the Lower Mainland, 1881

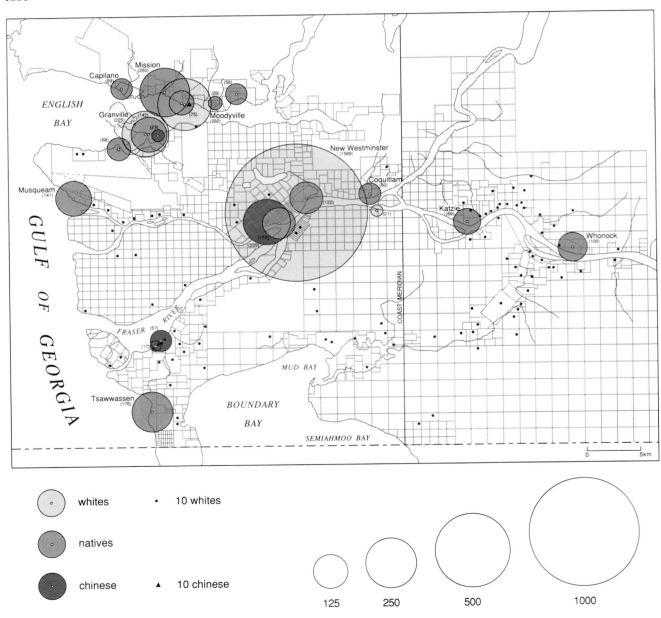

whites   • 10 whites

natives

chinese   ▲ 10 chinese

125   250   500   1000

Circles are proportional to population.

The Lower Mainland was suddenly part of a much larger world, part of the ways in which it organized and controlled space. The traders at Fort Langley had not sought or been able to exert such control, and when Natives dominated the Lower Mainland they organized it in ways that, in most respects, were virtually the opposite of those that held sway in 1886.

The complex of power that had suddenly assumed control in the Lower Mainland ultimately depended on brute geopolitical force. It was what Thomas Hobbes would have called sovereign power, vested in a monarch and expressed, in this case, through a Colonial Office backed by the Royal Navy and, briefly, the Royal Engineers. Edward Bulwer Lytton, the colonial secretary, told Colonel Moody that while civil societies should manage themselves, their internal stability required 'the unflinching aid of military discipline' in the background. Moreover, 'wherever England extends her sceptre, there, as against the Foreign enemy, she pledges the defense of her sword.' This was the language of imperialism and of sovereign power, and, as the Cowichan heard it from Governor Douglas, the language of conquest. Battles were unnecessary; shows of force and a few summary executions did much to establish the new realities. In a newly acquired territory where other forms of control were unavailable, the quick, brutal, episodic application of sovereign power established its authority, and fear bred compliance.

Once the realities of sovereign power were established, other less brutal and more disciplinary forms of power could begin to be put in place. These included codes of law, judges, police, and jails; and also public schools (and residential schools for Native children), industrial work discipline, institutionalized religious practices, and a land system — almost all of which white immigrants took for granted. In a newly settled place where, as far as immigrants knew or cared, there was no prior system of landed property, the land system itself became the most powerful single agent of disciplinary power. It defined where people could and could not go as well as their rights to land use, and it backed these rights, as need be, with sovereign power. The two Cowichan who killed a shepherd they thought was on their land experienced the power with which the newcomers defended their own conception of property. The point so made, the land system itself became powerfully regulative. Survey lines and fences were pervasive forms of disciplinary power backed by a property owner, backed by the law, and requiring little official supervision. Moreover, Natives could not acquire land of their own and many of them lived on reserves, an imposed spatial discipline which, given the mobility of former Native ways, had a profound capacity to modify Native life.

The de facto establishment of sovereignty and a regime of property with their inherent opportunities and controls provided protection for development. Farmers could acquire land knowing that title was secure and the colony safe for settlement. Townspeople could buy lots and live within the familiar guarantees of civil society. An immigrant society, especially its élite, could begin to put its world back together. But not, of course, entirely, if only because the setting had changed, and because societies and their settings are not separate categories. Immigrants on pioneer farmsteads or in lumber camps lived in unfamiliar relationships with the land; all immigrants lived without a local past and amid a strange mix of peoples. Such mixing brought ideas of ethnicity and race to the fore, weakened somewhat the idea of class, and tended to turn what in other more homogeneous settings were the unremarked details of everyday life into explicit and increasingly symbolic elements of difference. Of these, whiteness became the most generalized and powerful symbol, and, as it did, racism was built into the landscape of settlement. The Lower Mainland was certainly not a replica of any other place, yet its emerging human geography did convey a complex of power that did come, broadly, out of the English-speaking North Atlantic world of the mid-late nineteenth century.

Native power over the Lower Mainland and, to a considerable extent, earlier Native lifeworlds had collapsed. Many Native settlement and spatial routines were radically different from those which existed a short generation before. Most Natives were ostensibly Roman Catholics, and at the church-run residential school at Mission, not far up river from Langley, Native children were being taught in English and meticulously disciplined so that 'savagery' would yield to civilization. How much the Native cognitive world had changed is another question. Most Natives still spoke little or no English. Spirits still haunted Burrard Inlet near Deep Cove; Natives would not go there. For some of them, Christianity offered a new trail to a familiar land of shades and dancing ghosts. The most that can now be said, perhaps, is that changes in the Lower Mainland were such that Native cognition would not long survive unaltered.

In 1881 the future of the Lower Mainland was at hand. The land had been restructured and, as it had, way had been made for a railway, ever-more-modern sawmills and canneries, a largely British and eastern Canadian middle class, a city, and a metropolis. Native peoples, pushed to the margins, would turn increasingly to the heart of their problem, the land issue, and, in so doing, run squarely into the geographical reality of a place remade by others.

4

The rise of Vancouver

Graeme Wynn

*F*ew cities have grown as spectacularly as Vancouver. In 1870, the newly surveyed settlement of Granville on Burrard Inlet had fifty residents. Early in the next decade, approximately 300 people – most of them unmarried men who worked in the Hastings sawmill – lived there. In the spring of 1886, in anticipation of the arrival of the transcontinental Canadian Pacific Railroad on Burrard Inlet, a thousand or so residents won incorporation of their frontier village as the City of Vancouver. A year later, local boosters claimed that no other city in the world had enjoyed such prosperity or held such bright promise for the future as did theirs. In some minds, this burgeoning place with sixteen real estate firms, twelve grocers, and almost 3,000 inhabitants was well on the way to becoming 'the metropolis of the west, [and] the London of the Pacific.' Within five years, Vancouver's population topped 15,000, and in 1901, it was 27,000. By 1911, the city and its neighbouring munici-palities (Point Grey and South Vancouver) included 120,000 people. Ten years later, the same area had 163,000 inhabitants. In 1931, two years after the amalgamation of Point Grey and South Vancouver with the central city, there were approximately 250,000 Vancouverites. By that most simple and revealing of all measures of city size – population – Vancouver had grown almost a thousand-fold in fifty years.

Such was the momentum of expansion that the city of 1931 spilled well beyond its legal bounds. Burnaby, to the east, counted some 25,000 residents, there were 18,000 on the north shore of Burrard Inlet and, across the Fraser River to the south, Richmond and Surrey each had approximately 8,000 inhabitants. Yet further afield, the populations of Coquitlam and Delta approached 5,000 and 4,000, respectively. New Westminster, the largest city in mainland British Columbia in 1886, had been dwarfed by its neighbour, although its own population had risen almost six-fold, to 17,500.

During the next twenty years, relative growth rates in these outlying districts far exceeded those in Vancouver proper: Burnaby, Richmond, and the North Shore more than doubled their populations between 1931 and 1951; Coquitlam tripled in size; and Surrey quadrupled.

Figure 1

Vancouver in 1948. New Westminster in the right foreground; continuous development along Kingsway to the downtown core, but a good deal of green space to the north, south, and west

Vancouver, by contrast recorded a mere 71 per cent increase. But both the city and its neighbouring municipalities added approximately 100,000 people through these two decades. By mid-century, Greater Vancouver had more than 550,000 residents – almost half the population of British Columbia (Figure 1).

This is an impressive story. But simple numbers hardly do it justice. However concisely they illustrate the transition from 'Milltown to Metropolis,' they fail to illuminate the causes of this change. However neatly they outline the spatial extension and demographic expansion of Vancouver, they reveal little about the intricacies of those processes or of their impacts upon land and society. However effectively they suggest the dynamics of urban growth, they fail to capture the tensions, challenges, and satisfactions of life in the city. To illuminate these things, more than a simple narrative of growth is required. The intricate web of reality must be broken apart and its pieces examined systematically in order to understand how and why Vancouver developed as it did. Only pointed inquiry will reveal the physical and human consequences of urban expansion and uncover the mix of ideas, ambitions, possibilities, and constraints that gave form to the city as well as meaning to the lives of its inhabitants.

Recognizing as much, this chapter has four major sections. The first, and longest, deals with city space. It offers several perspectives on the physical fabric of Vancouver at a variety of scales. Among its concerns are the impact of city growth on the surrounding countryside, the development of transport networks, the sorting out of urban land-use patterns, and the processes of city-building. In sum, it explains the evolving spatial and architectural form of the developing urban region. The second section examines the urban economy and explores something of the material basis of life in Vancouver, including the structure of employment, the seasonality or (ir)regularity of work, standards of living in the city, the role of the port, and the relationship between the city and its hinterland. Next, attention is turned, briefly, to the planning and servicing of the city. This section explores something of the intellectual backdrop against which Vancouver evolved, and illustrates that, for all its newness, the city inherited a considerable past. Indeed, many essential elements of Vancouver's form derive from urban experience elsewhere. To appreciate as much is to gain a new perspective on the rise of Vancouver. The fourth section illuminates the social geography of the city. Questions of ethnicity, class, and community are brought to the fore in accounts of the very different neighbourhoods occupied by the city's rich and powerful families, and by its predominantly male Chinese population. Together these vignettes capture the diversity of life in this complex, dynamic place and remind us, finally, that imagined geographies are as important as are the material realities of population growth and areal expansion to understanding the story of Vancouver and its region.

'More room, more homes ... more people'

These words, observed a traveller who ventured *Through the Heart of Canada* in 1912, were the common cry of Vancouver. Wherever one turned in the city and its surroundings, there were signs of 'the battle ... being waged against forests and stumps by the makers of homes.' The outcome was never in doubt. A relentless rhythm seemed to govern the march of urban expansion. 'Chop. Chop. Chop. The blessed forests came down,' recalled novelist Ethel Wilson, 'and interested passers-by watched ... then speculated on their past and their future. The forest vanished and up went the city.'

The theme was a familiar one in the development of New World towns. From the plains of Missouri to the streets of Melbourne, contemporaries marvelled at the replacement of wilderness by civilization. But in Vancouver, as in few places before, the process ran across space with astonishing speed. By 1914, the city and its suburbs spread over 30 square miles of land. Forty years later, the tentacles of the metropolis extended over much of the Lower Fraser Valley; Haney, Cloverdale, and White Rock, each some 40 to 50 kilometres from the centre of the city, housed appreciable numbers who worked and shopped (at least occasionally) in the city, and these communities were clearly part of the urban fringe. The consequences are perhaps best revealed by comparison. In 1860, Toronto had a population of 50,000 people in little more than four square miles, and densities were described as 'not particularly high'; fifty years later, twice as many Vancouverites occupied approximately seven times as much territory. In 1921, population densities in Montreal and Toronto were three times those in Vancouver, South Vancouver, and Point Grey, and in 1929, when these administrative units amalgamated, most of the Burrard peninsula, some 88 square miles of land, had been turned to urban uses.

Streetcars, interurban railways, buses, and automobiles were the technological keys to this sprawl. Initiated a mere three years after the incorporation of the city, Vancouver's streetcar network comprised 20 kilometres of track by 1891. Sixty-four kilometres were added in the next eighteen years and some eighty more between 1909 and 1914. Although there was little expansion thereafter, cars serviced a dense network that extended from western Point Grey to Burnaby and south to the Fraser River (Figure 2). At the same time, more remote areas were

Figure 2

The technological keys to urban expansion: streetcar, interurban, and ferry routes, 1889-1928. A few years after the incorporation of the city, streetcars circled False Creek. Early in the twentieth century they ran east to Burnaby and west through Kitsilano. By 1920 they extended south to the Fraser River. Ferries linked the North Shore settlements to Vancouver and interurban railways ran to Steveston and Chilliwack.

Streetcar tracks construction period		Single track kms
1889 - 1899	————	25.9
1900 - 1909	– – – –	59.9
1910 - 1919	– - – - –	80.8
1920 - 1928	··········	16.6
total		183.2

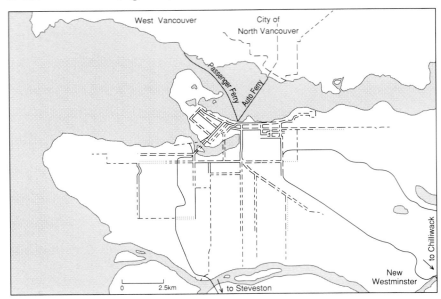

integrated with the city by interurban lines. Vancouver and New Westminster were linked in 1891, and a spate of construction in the early years of the twentieth century carried interurban service from Vancouver to the fishing settlement of Steveston; along the North Arm of the Fraser River from Eburne to New Westminster; from Vancouver to New Westminster via Burnaby Lake; and beyond New Westminster to Abbotsford and Chilliwack. Serving the needs of salmon canneries, saw and shingle mills, and farmers, these lines carried a good deal of freight as well as passengers. Far less successful was the 20-kilometre line that ran along the north shore of Burrard Inlet from Horseshoe Bay to Lonsdale Avenue. Built in 1914 as the first stage of the Pacific Great Eastern Railway's planned route to Prince George, it was unable to draw enough passengers away from the ferries that crossed Burrard Inlet from several points in West Vancouver to turn a profit. Lacking funds, and awaiting a promised subsidy to bridge the inlet at Second Narrows, its parent company fell so deeply into debt that it suspended operations in 1928.

By this date, passenger traffic was falling on interurban lines south of the Fraser. In 1930 it was little more than half of what it had been at the beginning of the decade. Vancouver and Ladner had been linked by bus and ferry since 1914. In 1919, provincial government crews had begun to rebuild and pave arterial roads in the valley, and motor bus services were quickly extended to outlying communities (Figure 3). In the 1920s subsidiaries of the British Columbia Electric Railway company (BCER), which operated Vancouver's streetcars as well as most of the region's interurban railroads, developed bus routes along both sides of the river: through New Westminster and Port Moody to Haney on the north, and via New Westminster to Chilliwack (and then to Hope) on the south. Buses also served the immediate periphery of the city, and by 1930 new and improved bridges across the Fraser River and Burrard Inlet at Second Narrows allowed connections to Richmond, North Vancouver, and West Vancouver. These links were further enhanced by completion of the Pattullo and Lions Gate bridges in 1937. Although few secondary roads beyond the city were paved, trucks and private automobiles accounted for a steadily increasing proportion of freight and passenger movements within the region between 1920 and 1950. By 1926, there were 44,000 cars (almost two-thirds of all cars in the province) in the Vancouver-New Westminster area. In British Columbia as a whole the ratio of people to cars fell from 116 in 1911 to slightly more than 6 in 1931 and 4 in 1951. Recognizing the reality of changing circumstances, the BCER terminated its passenger and express services in September 1950. Much of the compensation it paid municipalities affected by this decision went to improving local roads.

By 1950, rapid population growth, a massive elaboration of the transportation network, and striking improvements in the ease, cost, and flexibility of movement had transformed much of the region. Over 600,000 people lived where barely 6,000 had been in 1881 (Figure 4).

1886

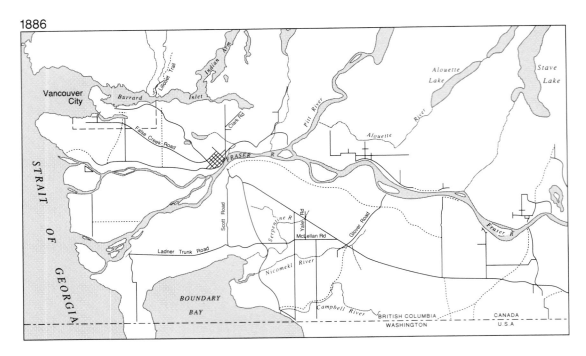

1905

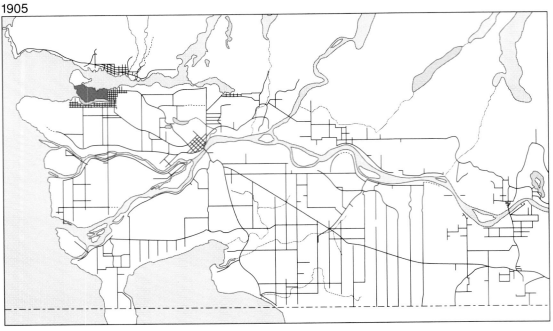

Figure 3

The technological keys to urban expansion: roads for wagons, buses, and automobiles, 1886-1966. By 1905 the open mesh of early roads and trails had been extended into a relatively broad grid on the Burrard peninsula and southeast of New Westminster. Twenty-five years later the peninsula was densely covered by roads, and new cross streets linked the arterial routes beyond. By the mid-1960s a tight intricate grid extended across most of the region.

1931

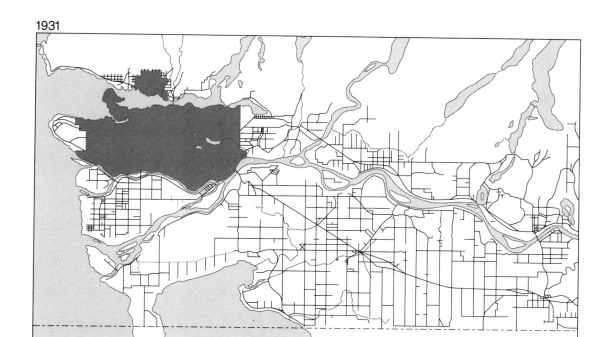

1966

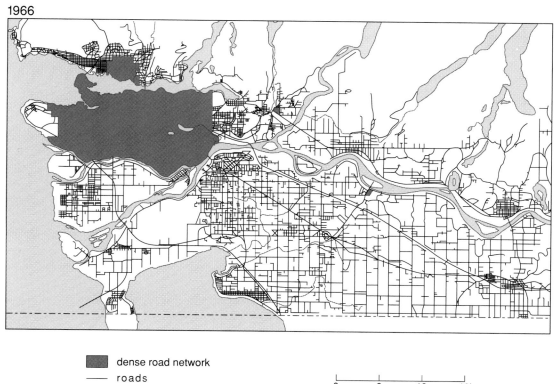

dense road network

——— roads

·········· trails

0   6   12   18km

1901

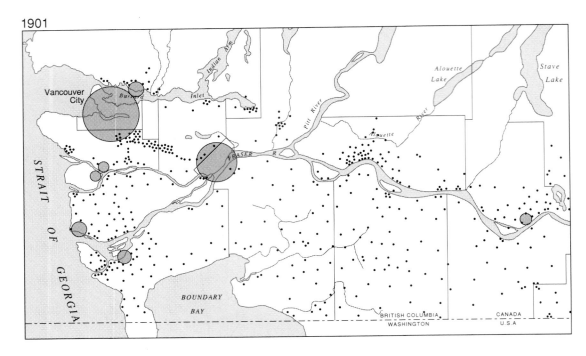

1921

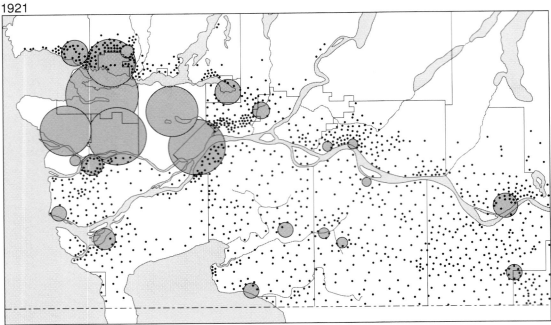

Figure 4

Sixty years of population growth. These maps reveal the urbanization of the region in striking fashion. In 1901 Vancouver and New Westminster were the only administrative districts with sizeable population concentrations; by 1941 population densities were relatively high everywhere northwest of the Fraser and Pitt rivers; twenty years later the entire western half of the region was densely peopled and numbers to the essentially rural east were considerable.

1941

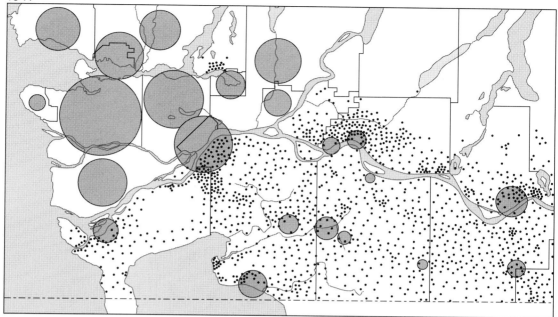

1961

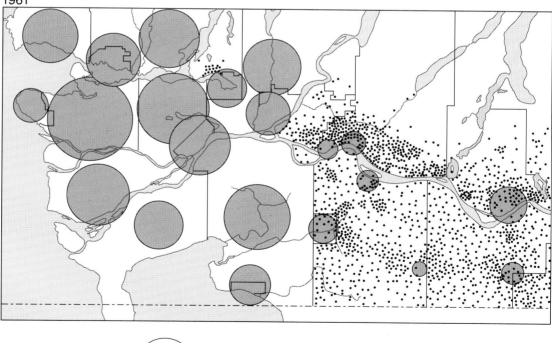

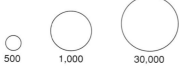
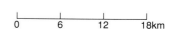

Circles are proportional
to population

One dot represents
25 people

500   1,000   30,000

0   6   12   18km

East and west across the narrow funnel of relatively flat land flanking the Fraser River, the changes of seventy years had created a strikingly new landscape. Much of the forest that dominated the area in the late nineteenth century had fallen to loggers, fires, and settlers. Most of the region's original inhabitants – the Natives – lived on small, scattered reserves, their communities islands in a sea of newcomers whose values, beliefs, and aspirations were generally very different from their own. Lakes and swamps had been drained, and dykes protected much of the valley from all but exceptional floods. Approximately 300,000 acres of farmland yielded a wide variety of produce – including small fruits, vegetables, and dairy products – for markets in Vancouver. On average, rural land values were relatively high; in very broad terms they had increased almost 5 per cent a year since 1900. Yet the area in farms had fallen in the 1940s and continued to decline by about 2,000 acres a year through the 1950s.

As the city spread, speculators acquired and held properties on its distant fringe in expectation of lucrative gains when they were converted to urban uses. The practice was by no means new. For seventy years people had sought profit from Vancouver real estate by buying cheap on the periphery, sitting tight, and selling at a premium. And the essential tactics had been much rehearsed in cities the length and breadth of North America in the nineteenth century. But as urban limits expanded and the internal combustion engine provided new freedom of movement, an inherently haphazard process became even more hit or miss. Far more land than would be needed for suburban development in the mid-term future – much of it of prime agricultural quality – was tied up by speculators. By one estimate, in the 1960s, 20 per cent of the good soil of Surrey was used neither for agriculture nor urban purposes. Unoccupied, it lay idle in anticipation of a speculative harvest. By and large, of course, these changes had the greatest impact on the western part of the region, where residential, commercial, and industrial uses had encroached decisively on forest and farmland, and where speculators had the best prospects of quick profit. But these changes also affected much of the extensive plain to the east.

Peripheral expansion: Richmond and Maple Ridge

The impact of Vancouver's rise to metropolitan status on the surrounding countryside is best revealed by brief case studies of the Richmond and Maple Ridge districts. Each illuminates much of what occurred, to a greater or lesser degree, over a far wider area. Sea and Lulu islands, which were incorporated as the Township of Richmond in 1879, were first settled by immigrant homesteaders in the 1860s (Figure 5). As the population of this area increased in ensuing decades, large sums were spent on dyking and draining the low-lying delta land. Still, flooding was frequent until the construction of larger, fortified dykes began in 1906.

Figure 5

'One of the most lovely spots in British Columbia.' The McRoberts home on Sea Island, 1862, drawn for Jennie McRoberts by an admirer

Figure 6

Steveston: A City Someday. 'Six hotels, an Opera House, a theatre': The main street of 'Salmonopolis' in 1908

As bridges and roads were built, Richmond expanded. On the southern edge of Lulu Island, a considerable settlement – sometimes light-heartedly described by contemporaries as Salmonopolis – had grown up around a dozen or so canneries at the mouth of the Fraser River. By 1910, this town, Steveston, had six hotels, an opera house, a theatre, and a main street lined by false-fronted buildings (Figure 6). But its urban character was exceptional. Even in 1930, half of Richmond's 8,000 people lived on farms; two-thirds of the municipality's land was agricultural, and less than 5 per cent was given to transportation and urban uses. Two extensive peat bogs, amounting to approximately 8,600 acres, accounted for the remainder (Figure 7). Over most of Sea and Lulu islands, large, well-kept dairy barns rose from extensive fields growing grass and grain. Smaller, multi-crop holdings and intensively cultivated market gardens were scattered here and there. Residential development was concentrated in Steveston; in the vicinity of the municipal hall, interurban station, and horse-racing track at Brighouse; and near the small cannery and sawmill at Bridgeport on the North Arm of the Fraser River. Each day, Richmond farmers sent fresh milk and cream, in bottles and in bulk, to Vancouver and New Westminster.

Figure 7

The transformation of Richmond. Here, early surveyors charted forest, shrub, and grassland. In the 1920s, dairying, grain growing, and market gardening were the most common forms of land use. By 1963 Richmond was markedly suburban.

1859-90 vegetation

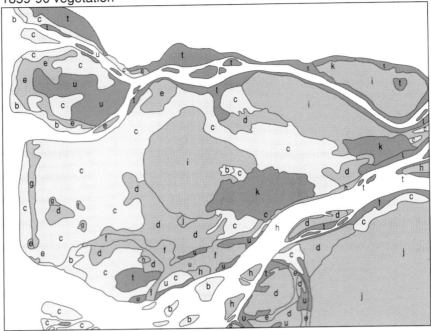

1930 land-use

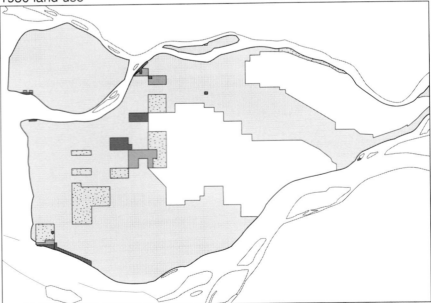

In season, the same markets took a large share of the vegetables grown on the islands. Grains (mainly oats) went to local mills and to feed the dray horses that could still be seen on city streets. Closely tied to the city though it was, Richmond remained essentially rural.

Twenty-five years later, Richmond was markedly suburban. In the 1930s, a few big farms were broken into small holdings where, reported

Herbaceous vegetation
| b | marsh |
| c | grassland |

Shrubs
d	grass with shrubs
e	mixed shrubs
f	willow scrub
g	crabapple scrub
h	alder scrub
i	cranberry swamp
j	bog

Forest
k	moss with trees
t	mixed coniferous
u	spruce forest

0 ⊢———⊣ 2.5km

agricultural
small-holding
residential
industrial, commercial institutional
peat extraction
vacant

1949

1958

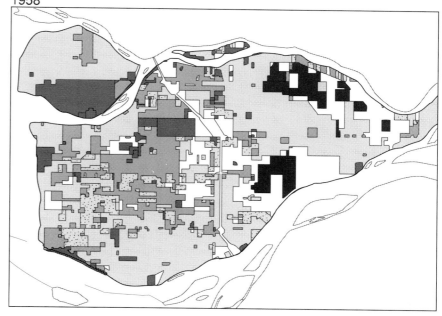

the *Marpole-Richmond Review* of 27 May 1936, 'the man with a job in the city makes himself a comfortable home with chickens, a cow, small fruits and a garden to occupy his spare time.' Bulb-raising, seed production, bee-keeping, and rabbit-raising were other sidelines that small-holders 'found profitable and pleasurable.' Scattered across the municipality, such enterprises encompassed almost 3,000 acres of

Figure 8

Richmond in 1948

Richmond land in the early 1950s (Figure 8). As their number increased so, too, did the area in residential use. Some of this growth was initiated under the Veterans' Land Act, which provided homes for ex-servicemen and their families, but more was due to the efforts of individual developers. By 1958, housing subdivisions ran, almost without interruption, from Bridgeport to Brighouse and formed discrete island 'neighbourhoods' across much of the municipality. Quite typical of suburban developments on the fringes of most North American cities in

Figure 9

Suburban development in Richmond: the A.B. Dixon School and its surroundings

Figure 10

Canneries, wharfs, refrigeration units, storage facilities: the Steveston waterfront in 1953

the prosperous postwar years, these were essentially bedroom communities, offering relatively accessible housing (in one of a small handful of styles) to young couples whose major source of income was generally a job in the city (Figure 9).

On the southern edge of the municipality, Steveston formed a distinct community. A substantial proportion of its residents were Japanese-Canadians, who had returned, often with little material wealth, after internment during the Second World War. Their settlement was heavily dependent on the fishery and closely tied to the string of wharfs, dry docks, canneries, reduction plants, refrigeration units, and storage facilities that lined the river (Figure 10). Poorly built shacks; rows of small, wooden company houses; patches of overgrown, debris-strewn, vacant land; and a commercial core showing many signs of dilapidation as a result of competition from newer commercial establishments elsewhere in Richmond led some to describe it as a 'dismal' and 'confused' place (Figure 11).

Figure 11

Poorly built shacks and company houses. Residential landscapes of Steveston

As the population of Richmond grew (it reached 26,000 in 1956), so commercial, institutional, and industrial activities expanded. Shopping centres, commercial nurseries, golf courses, theatres, and corner stores were built to serve residents (as well as recreation-seekers from Vancouver), a reflection of the growing integration of the municipality with that city. New schools, parks, and playgrounds also appeared. The airport, established on Sea Island in 1929, expanded its area and its operations; and an RCAF station, Department of Transport buildings, businesses providing aircraft supplies, and flying schools clustered nearby. New lumber and shingle mills as well as a plywood plant were built along the North Arm of the Fraser, which quickly became an industrial zone. To the south, two food processing plants (a rice mill supplied from the southern United States and a flour mill to process Prairie wheat) took advantage of cheap land with access to shipping lanes. Elsewhere, the municipality had a cement works, a cardboard box and paper converting plant, and several establishments (such as those producing pet food) that needed large sites remote from residential areas, from which liquid waste could be emptied into the river. Even the extensive, long-vacant peat bogs on the eastern half of Lulu Island were being made over. Heavily utilized in the production of magnesium during the Second World War, spaghnum moss was extracted, dried, and pulverized as a soil conditioner after 1945; and depleted areas of the island were turned over, in part, to cranberry and blueberry farms.

By 1958, barely half of Richmond land was agricultural; in twenty-eight years, 4,400 acres had been converted to other (mainly urban) purposes. Together, residential, commercial, institutional, and industrial uses occupied 6,250 acres of Richmond land (compared with 560 in 1930) and accounted for better than 20 per cent of the municipal area. An expanded road and rail network occupied three times as much space as it did in the 1930s (fully 7.5 per cent of the municipality). There was also a good deal less vacant land.

Through all of this, good quality agricultural land was lost to development more quickly than was poorer quality land. Unencumbered by zoning restrictions, individuals and real estate companies were free to subdivide and develop land almost anywhere in the municipality. In the western section of Lulu Island, few of the extensive grain, hay, and pasture fields of the 1920s remained. Urban and agricultural uses were thoroughly intermingled; fields were generally small and on many of them vegetables were the primary harvest; here and there idle farms went to seed, awaiting subdivision and construction crews. To the south, where urban growth was less vigorous, small fruits and vegetables were the most important crops on a substantial number of farms owned and operated by people of Chinese descent. Dairying, once dominant in the western part of Lulu Island, was concentrated in the east. At mid-century only 451 of Richmond's 19,000 residents were farm operators; five years later less than 10 per cent of the municipality's population lived on farms. Substantial demographic growth, and the changes it brought through three decades, had produced a disjointed, fragmented landscape, the essential characteristics of which are best described as urban sprawl.

Maple Ridge lies on the north bank of the Fraser River, some 50 kilometres east of Vancouver. In 1891, the district had fewer than 250 settlers. They occupied scattered farms established in the previous three decades and clustered around the Canadian Pacific Railroad (CPR) stations at Port Hammond and Port Haney. Much of the district was heavily wooded, but there was a sizeable tract of scrub (grass and wood) land in the east, and extensive areas along the river had been at least partially burnt. For the next forty years, the population of the district increased relatively slowly. Mixed farming (with some emphasis on dairying), poultry-raising, and fruit-growing supported rural families, and a small handful of sawmills and brickyards employed several hundred local residents. Similar numbers worked in Maple Ridge logging camps. Still, the census of 1931 counted fewer than 5,000 people here.

Economic depression limited growth through much of the 1930s, although completion of the Lougheed Highway to Vancouver broke the area's dependence on the river and the railroad. On average, the population of the district increased by about 150 people each year, and most of them occupied newly established smallholdings. Many of these newcomers were Japanese-Canadians who combined small fruit and berry production with work in the logging and salmon fishing industries. Rather fewer were retired Vancouverites, attracted by cheap land and the natural beauty of the area. In the 1940s and 1950s, however, a more prosperous provincial economy, better access to Vancouver, and the wartime displacement of the district's Japanese population, considerably quickened the pace of change in Maple Ridge (Figure 12). Between 1941 and 1951, the population increased by 50 per cent, to almost 10,000; by 1961 it was 16,750. Commercial establishments once located on the river moved to the highway; Haney, reported the magazine section of the *Vancouver Sun* on 1 September 1951, had 'Walked "Up Hill"' as the Fraser front was abandoned. Many of the smallholdings left vacant on the internment of their Japanese-Canadian owners were subdivided for suburban residential development. Others were taken over by former city dwellers attracted by rural surroundings, low taxes, and the possibility of raising fruit and vegetables for the Vancouver market. Many of them also worked for wages beyond the municipality. At the same time, new establishments – schools, retail businesses, and a correctional institute – provided new jobs within Maple Ridge and encouraged new housing development. In the ten years after 1948, the value of residential building permits granted by municipal authorities more than quadrupled. Late in the 1950s, Haney and Hammond had

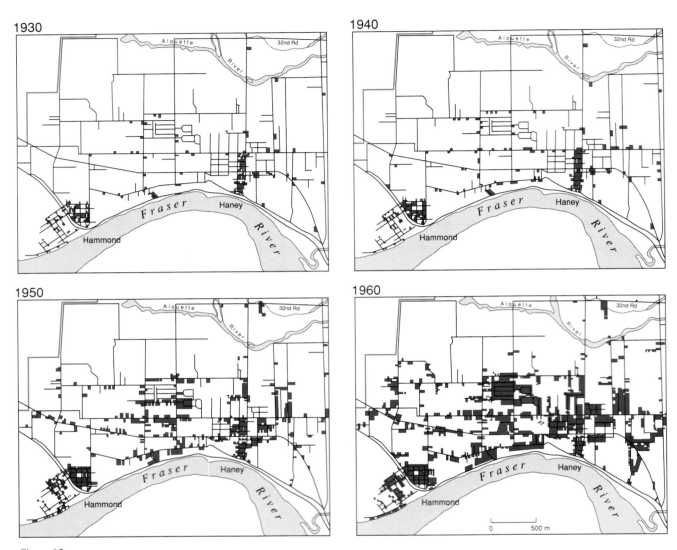

Figure 12

The suburbanization of Maple Ridge

begun to merge into a single, sprawling settlement, and residential properties were scattered, seemingly indiscriminately, along the main highways and connecting streets of the district.

There was more pattern to this sprawl than first impressions suggested. Here (in the Alouette valley), retirees lived in comfortable 'cottages,' each on an acre or more of land. There, hobby farmers kept a horse or two, and larger 'ranches' (as well as an 'equitation centre') raised thoroughbreds, stabled show-jumpers, and offered riding lessons. Elsewhere, 'country estates,' with large houses on at least an acre of carefully landscaped ground tended by a full-time gardener, clustered on view sites away from the main highways. At the other end of the scale, a significant number of Maple Ridge residents made do with barely adequate accommodation in company housing erected by the Hammond Lumber Company in the late nineteenth century or in former summer cottages converted to permanent dwellings. By the late 1950s, the municipality was at a crossroads. Recognizing that 'the speculative sub-division of land for suburban development on a large scale' was well under way, the Lower Mainland Regional Planning Board considered it 'almost inevitable' that 'with the rapid growth of Greater Vancouver ... Maple Ridge will ... become increasingly dominated if not absorbed, by the metropolitan area,' and warned that with this type of development, 'the scale of problems ... is correspondingly greater.'

A city growing like 'Topsy'

Growth within the city was equally haphazard. After reviewing the development of Vancouver and its suburbs through the late 1920s, one observer explained that he had made 'no mention' of the exercise of 'any human foresight ... in the control or direction' of expansion in 'this British Columbia metropolis,' because its history was 'that of Topsy, it just growed.' Certainly it might have seemed that way. Although the general pattern of urban expansion was shaped by the technology of transportation (in 1929 fourteen of every fifteen residents of Vancouver, Point Grey, South Vancouver, New Westminster, and Burnaby lived within 400 metres of streetcar or interurban tracks), the processes that created the urban landscape were enormously complex. Many who experienced them firsthand considered them both remarkable and chaotic. On the peninsula, as later in Richmond and Maple Ridge, urban growth was the product of numerous stimuli. The city took shape piece by piece, each fragment forged from some combination of speculative adventure, corporate strategy, private interest, government regulation, and personal whim. Year by year, expansion was orchestrated by the press of demand – for space, shelter, and locational advantage – to a tempo set by fluctuations in local, provincial, national, and international economies.

In bold outline, the consequences of this shifting play of forces are clear enough. The CPR terminus, strategically located west of the Granville townsite, at the north end of an enormous block of land (amounting to almost 6,500 acres) granted the railroad company, became the hub of the city. Hotels, banks, offices, department stores, and theatres clustered near the station at the foot of Granville Street. Around this core, a newly surveyed grid of streets provided the basic template for later expansion. CPR spur-lines and sidings served a growing concentration of wharves along Burrard Inlet. A roundhouse and railroad repair shops on the north side of False Creek gave that area an industrial cast. To the west, where the CPR owned approximately one-quarter of the peninsula that extended to the military reserve which became Stanley Park in 1888, the substantial dwellings built by company officials and other prominent citizens formed the nuclei of Vancouver's first prestigious residential district. To the south of False Creek, CPR surveyors extended Granville Street through the centre of the company's land, established the grid of numbered east-west avenues and named cross-streets that would eventually stretch to the Fraser River, and opened lots for settle-

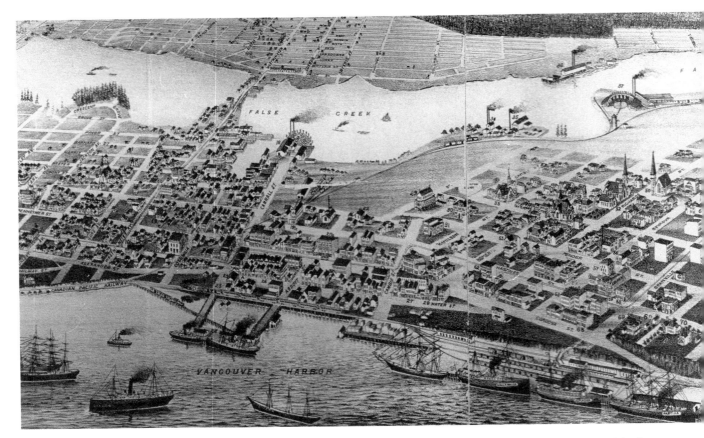

Figure 13

Vancouver in 1890. A bird's eye view produced for the *Vancouver Daily and Weekly World* in 1890

ment in carefully controlled succession. Land beyond the CPR block was also brought to market by private developers, real estate firms, and land companies, first in the vicinity of the original townsite, and then, with the growth of population and the spread of street-car service in the early twentieth century, further afield to east and west of the railroad's holdings (Figure 13).

As the city grew, the differentiation of space increased. Large retail, financial, and service activities concentrated downtown, where they were most accessible to the population at large by virtue of their location at the hub of an essentially radial transportation network. Competition for space, rising rents, the availability of the elevator, steel-frame construction techniques, changes in the organization of office work, and the symbolism and prestige associated with prominent buildings quickly raised the downtown silhouette. In the boom years immediately before the First World War, the process gained fabulous momentum. In 1909, the city's Dominion Trust Building, built near the original courthouse on the edge of the Granville townsite, was proclaimed the tallest building in the British Empire; a year or so later, it lost that distinction to the nearby 17-storey World Tower. But the financial and business heart of the city had already begun to move west. By 1914, a 'temple' bank, and a 'Parisian-influenced' post office framed views of the CPR's second, 'chateau-style' station. An elegant, neoclassi-cal courthouse, designed by Francis Mawson Rattenbury, the province's leading architect, stood on Georgia Street just west of Granville; the second Hotel Vancouver – a massive, ornately decorated stone structure with 500 guest rooms – loomed to the east; and large and splendid office buildings clustered nearby. Reflecting the need to impose some limits to growth, new city bylaws limited buildings to a height of 120 feet, with a tower covering no more than a third of the ground area up to eight storeys higher.

Meanwhile, new suburbs were established further and further from the centre of the city. 'It is a trait noticeable in the people of Vancouver,' observed the magazine section of the *Province* newspaper in June 1910, 'that the homemaker would rather live in a large comfortable residence a few miles out of town than exist in a shack in the heart of the city.' If few of the new suburbs were socially or economically homogeneous, they were nevertheless differentiated by price, prestige, and popular perception. In broad terms, the middle and upper classes settled in the west, and working men and women congregated in the east. But there were many nuances to this picture. In several sections of the city – and especially in those areas immediately beyond False Creek known as Fairview, Kitsilano, and Mount Pleasant – socioeconomic strata over-lapped. When these areas were developed between 1890 and 1912, no one knew the niche that they would occupy in the booming city. In this fluid, uncertain environment, developers divided their holdings into lots of different sizes. Builders seized on one or another asset of a site (perhaps the view or the proximity to the workplaces of False Creek) in determining the type of structure to erect upon it. Soon, sub-stantial homes stood alongside modest cottages; bungalows beside 'batching' cabins. Artisans and entrepreneurs, labourers and profession-als shared the same streets. But beyond these 'inner suburbs,' time and experience (to say nothing of regulations) produced rather more com-partmentalized landscapes.

Straddling Granville Street atop the hill south of 16th Avenue, the CPR's Shaughnessy subdivision was planned and protected by restrictive covenants as a prestigious enclave for the truly wealthy; on the distant fringes of development in Point Grey, less affluent but nonetheless well-to-do home-seekers were lured by the prospect of a home 'among the cedars' of Crown Heights or by the advantages of a Kerrisdale location 'just far enough away from the noise and bustle of the city for peace and contentment.' 'There will be no unsightly shacks in Bryn Mawr,' promised the developers of this west-side neighbourhood, 'because there is a building restriction of $2,500 on each lot. This restriction is your protection and is ample assurance that your neighbours will be desirable.'

East-side neighbourhoods were even more clearly distinguished one from another in plan, appearance, and image. Strathcona was the city's first residential district. Close to the wharves, mills, and refineries on Burrard Inlet and immediately east of the developing downtown core, it was an area of boarding houses, bachelor cabins, and modest houses on relatively narrow (25-foot) lots. Turn-of-the-century Grandview was recalled by one of its inhabitants as 'a wilderness ... stumps, stones, humps and hollows were everywhere; only a few streets were opened; none were graded and macadamized. Sidewalks, where there were any, were of the 3 plank variety.' South Vancouver was a place of 'neat cottages and gardens and chicken yards,' a place where real estate advertisements continued to offer dwellings such as a 'four-room bungalow on three lots, 12 fruit trees, small fruit and garden' well into the 1920s. Here, according to local boosters, 'the rays from the sun in Heaven... [could] enter without restraint' and workers and their families could enjoy fresh air and prosperity. In several parts of this district, there were no more than one or two houses per acre.

Still the pace and magnitude of change in this and most other parts of the rapidly expanding city were arresting. By one account at least, the years between 1913 and 1918 were 'harrowing times indeed' in South Vancouver. Chief among the reasons for this were 'the enormous and sudden flowing in of population and the haphazard, non-directed manner in which these people were allowed to settle, according to individual whim, throughout the length and breadth of the District' in the early years of the century, as well as the disruptions caused by the Great War. Heavy obligations and falling tax revenues left the municipality of 25,000 (in 1916) heavily in debt. Ironically, the most obvious solution to the problems of growth was more growth, which would increase the tax base without a concomitant increase in the cost of providing services. By 1929 the 9,200 acres of South Vancouver housed over 45,000 people; and approximately 1,600 residential and commercial buildings had been erected there since 1924. So, too, in Grandview the number of houses climbed approximately sevenfold in eight years to exceed 1,700 by 1912; in less than a decade, an area 'all trees and logs lying on top of one another and buried' gained a reputation as a comfortable location for skilled workers. Within the city of Vancouver proper, 4,633 buildings, 85 per cent of them residences, were erected in little more than four and a half years after January 1923.

Developing the urban fabric

Again, the ways and means by which the fabric of the city took shape are best (if in the end only partially) revealed by example. Consider first the small (2-square-kilometre) tract southeast of False Creek, officially referred to as District Lot 301 (DL 301) until its annexation to the city in 1911. It reveals the way in which a good deal of suburban land was developed; the close connection between residential development and the extension of transportation lines; and the role of banks, investors, and speculators in the city's dynamic land market. At the turn of the century, DL 301 (also known as northern Hillcrest) was a semi-rural locale with a few scattered clusters of working people's homes. Owned by wealthy New Westminster merchant H.V. Edmonds since 1888, it was first subdivided in 1890, when Edmonds and his partners incorporated the Westminster and Vancouver Tramway Company. Over the next several years Edmonds disposed of much of the area that he had retained in his own name in 1890. By 1897, a handful of settlers and small investors owned approximately a quarter of the district. Their individual holdings were rarely more than a city block in extent. Most of these were clustered in the northwest corner of DL 301, near Vancouver. Much of the remaining land had been acquired by the English-based Bank of British Columbia through foreclosures on defaulted loans against which Hillcrest properties had been lodged as security.

Seven years later, the Yorkshire Guarantee and Securities Corporation, a prominent loan and mortgage company, was the major landowner in northern Hillcrest. Early in the century it acquired, in sequence, the remainder of Edmonds' holdings, a substantial part of the Bank of British Columbia's land in the area, and a major financial interest in the city's interurban railroad. In all, it held more than a quarter of the district's 127 blocks. Rather more than half the remainder was in the hands of the BC Land and Investment Agency and two private investors: Vancouver financier J.W. Horne and Vancouver wholesaler James Adams. Approximately thirty blocks were owned by small investors and settlers. These were divided in almost equal proportions among six or fewer property owners, seven to twelve owners, and more than a dozen owners. By 1910, however, there was only one large landowner in the district: James Adams held eight blocks. A dozen other investors owned single blocks or large parts thereof, but most land in the district was held in single 25- or 33-foot (8- or 10-metre) parcels by individuals.

After twenty years of subdivision and sale, the undeveloped block of land that Henry Edmonds acquired on the forested fringe of early Vancouver was home to almost a thousand households. A rough tract of stumps, bushes, salal, and small trees had become a suburb. Rather more than 1,800 lots flanked the area's streets. But it remained, in the memory of one resident, 'a wild place very sparsely settled up.' Only two of northern Hillcrest's roads were graded or paved. Little more than half of the district's lots were occupied. Although title to the area had passed from Edmonds to 1,265 people, over 40 per cent of the new owners were absentees. Even in the northwest corner of the district, residents constituted a clear minority of property owners throughout this period; three of thirty-five in 1897, they accounted for about a third of the total in 1904 and slightly more than a fifth in 1910 (Figure 14). Although a relatively consistent grid of streets imposed some basic order upon the landscape of northern Hillcrest, the district had a

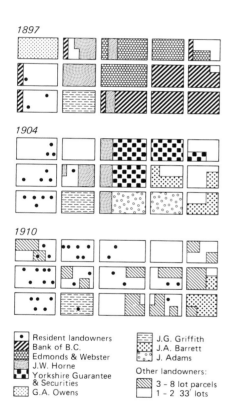

1897

1904

1910

•	Resident landowners
▨	Bank of B.C.
▦	Edmonds & Webster
▨	J.W. Horne
▪	Yorkshire Guarantee & Securities
▨	G.A. Owens

▦	J.G. Griffith
⬚	J.A. Barrett
⬚	J. Adams

Other landowners:

▨	3 – 8 lot parcels
☐	1 – 2 33' lots

Figure 14

Land ownership in the northwest corner of DL 301, 1897-1910. Across the city, corporations and individuals who held large blocks of land sold them off in much the same manner as revealed here. Even when most land was held in one or two lot parcels, however, much of it remained vacant or was rented out.

Figure 15

The early suburban landscape

markedly snaggle-toothed appearance (that would persist for several years after the First World War). Parcels of land of various sizes had come on the market in dribs and drabs; some had been built on, some held for later sale. Most housing had been erected by the settlers them-selves, or by developers who built in small groups of three, four, or six. Typically, therefore, dwellings stood alone, or in small clusters, sur-rounded by vacant land (Figure 15). Here and there, smallholdings and farmlets established in the late nineteenth century persisted amid the developing suburban landscape. Yet modest cottages and more substan-tial two-storey houses became the norms of this predominantly blue-collar suburb, although the latter dominated the 25-foot lots on which they were often built. In many parts of northern Hillcrest the 'yards for people to play in, and [the] gardens' that were so central to contempo-rary images of suburbia were humble indeed.

In the eyes of many early Vancouverites, real estate promotion and development were alluring wheels of fortune. Land was, undoubtedly, the most important commodity in the city. Visitors and locals alike marvelled at Vancouver's frenetic real estate market. Newspapers afforded enthusiastic publicity to examples of spectacular gains from property speculation. The large, assured claims of real estate promoters were constant reminders of the profitable land market. Consider, remarked the *News Advertiser* of 22 January 1905, that a small downtown lot bought for $480 in 1887 sold for $600 three months later; early in 1888 it brought $1,500, in 1904 it sold for $4,000, and now it was offered for $6,000. Two-hundred-acre lots in Mount Pleasant, bought for $16.20 at a tax sale in 1895, were on sale for $275 in 1899. Two years later a block of land in Grandview was offered for $1,500, with the promise that 'a quick profit of $2,000' awaited the buyer. Before the end of the decade, city newspapers carried advertisements for 'Grandview money makers! 4 50' lots $6,000,' in which the incentive for investment was spelled out in the phrase: 'We will show you how to make a nice turnover on this property.' Such were the prospects of handsome returns that a few years after Vancouver was incorporated, the city had thirty real estate firms; between 1910 and 1912, approximately 170 companies were directly involved in the land market, and 650 Vancouverites listed themselves as estate agents. Hundreds, even thousands more were lured into the game in hope of financial reward from quick sales and rising land prices.

Together, these players formed a richly varied group. Although it is impossible to draw clear lines between residents and investors, bona fide settlers and real estate speculators, those who speculated in the Vancouver real estate market included wealthy entrepreneurs, professionals, and working men and women with barely enough capital to purchase a lot or two on the distant periphery of the city. Looked at in other ways, their numbers included long-time residents, new arrivals, and some who lived in other places as well as aggressive gamblers and hopeful investors. From yet another perspective, those who speculated in Vancouver real estate might be divided into those who won and those who lost, for this was a game in which some gained fabulous fortunes and others lost their shirts. But this obscures the less spectacular middle ground occupied by the majority, whose real estate ventures were part and parcel of an ongoing struggle to secure a foothold and gain a modicum of financial security in the developing city. For members of this group, success rarely meant the accumulation of a vast fortune, and failure spelled continuing struggle rather than ruin and disgrace.

Whatever their individual fates, those who speculated in Vancouver land left an enormous collective imprint on the face of the city. Together they did much to shape the geography of Vancouver, the character of its neighbourhoods, and the appearance of its streets. For example, the growth and development of Grandview, a streetcar suburb laid out on the eastern fringe of the city in 1888, owed much to the efforts of George McSpadden, sometime building contractor and city alderman, who was employed as a building inspector by the city between 1900 and 1907. Possessed of a keen business eye, McSpadden recognized the potential for development in this area at a time when it contained 'only a few crude shacks,' and 'immediately began ... [its] exploitation.' He built 'one of the first attractive houses in Grandview' and, with his partner Harry Devine, claimed to be 'well qualified to advise upon the purchase or sale of any description of Real Estate.' After the better part of a decade of active real estate promotion, McSpadden used some of the profits of his agency to move to a more prestigious address in suburban Point Grey. The Grandview that he left behind was a mixed neighbourhood, with a ribbon of fine large houses on Victoria Drive, small cottages on 25-foot lots, and more substantial frame houses on slightly larger properties. Apartment houses and small tenements were scattered through the area, and light industrial uses encroached on its edges. McSpadden, proclaimed a contemporary biographical dictionary, had been 'one of the greatest single forces' in the growth of 'this beautiful locality.'

When streetcar service was extended to Kitsilano in 1906, prominent Vancouverite J.Z. Hall, the city's first notary public, began clearing a 4-acre parcel of land south and west of the Bayswater-4th Avenue intersection. But several years elapsed before building began there, although Hall and his wife moved, in 1908, into 'Killarney,' a grand stone mansion overlooking English Bay a few blocks away. When subdivision plans for the tract were approved in 1913, the Vancouver economy was in recession, and then the war dampened activity in the real estate market. Construction began in the 1920s, and by 1922 there were twenty-three occupied houses on 5th Avenue, which ran through the middle of Hall's former property; by 1925 the block was virtually filled with thirty-one houses. All had been erected by a single builder, who generally had between four and six houses under construction at a time. By the end of the decade, title to these properties was widely dispersed. Ten were owned by their occupiers; a Mrs. L.M. Kirk held three houses; and others were owned by residents of Port Alberni, Abbotsford, and Calgary as well as by the Hudson's Bay Company and the City of Vancouver. For all that, this small area bore, as it still does, the marks of its delayed development and the fact that it was opened up as a single parcel. Within the bounds of Hall's tract, Fifth Avenue is wider than it is elsewhere; it is slightly out of alignment with the street grid to the east; and it is lined by almost identical houses, built in a style common in Vancouver only for a few years before and after the First World War. Similar unobtrusive patterns – an essential uniformity in the size and appearance of half-a-dozen houses on a block, or irregularities in an otherwise consistent street pattern – occur across the city, the lasting landscape signatures of otherwise forgotten counterparts of J.Z. Hall, whose real estate and construction ventures helped to build the fabric of the metropolis.

Hundreds of others, whose real estate dealings were fundamental to their survival in Vancouver, left more slender legacies in the cityscape. But, taken as a whole, their imprint on the development of Vancouver and its suburbs was substantial, if difficult to describe, because photographs, official documents, city directories, and other fragments of evidence provide only occasional, tantalizing glimpses of the ways in which these individuals (who might be regarded metaphorically as the carpenters rather than the architects of urban growth) shaped the city. Among them was James L. Quiney, who came to Vancouver in 1907 after serving with British troops in South Africa during the Boer War and homesteading on the Manitoba prairie. Penniless on arrival after having his pocket picked on the train west, he settled his family in a tent at the foot of what is now Balaclava Street in Point Grey. Several months later, Quiney purchased land on the northeast corner of 4th

Figure 16

Opening up the suburbs. James Quiney promotes real estate opportunities in the vicinity of the newly opened W. 4th Avenue streetcar line.

Avenue and Dunbar Street and built a house in the forest. Lumber came from the buildings of the recently demolished English Bay Cannery, water was drawn from a well, and provisions were packed in on foot. Within two years, however, the BCER's newly opened streetcar line along 4th Avenue terminated at Alma Street a couple of blocks to the west. The Quiney family moved a short distance northeast, and James entered the real estate business (Figure 16). As the surrounding area was sold off, and substantial houses were built on relatively large lots, Quiney prospered. Then, in 1912, the real estate market collapsed. Three years later Quiney went off to war in Europe. He returned an invalid. Unable to sell at a profit or maintain payments on the land he had acquired at inflated pre-war prices, Quiney lost his house and real estate. In the spring of 1919, he and his family moved to a squatter's shack on the edge of Spanish Banks beach.

A year or so after James Quiney began his real estate dealings in Point Grey, Sidney T. Wybourn arrived in Vancouver from a village in South Wales. During the next several years he followed a path trodden by countless others through their early years in the city. Initially, Wybourn rented a cabin in Mount Pleasant and worked in the BCER streetcar barns. Within two years he had savings enough to make a $200 down payment on two lots on East 43rd Avenue in South Vancouver. On one of these he began to build a small bungalow. His own labour, and that of friends from the BCER (for whom he toiled in return) as well as the work of hired men, allowed timely completion of the foundations and first floor of the new dwelling. But the second floor and an additional bedroom were finished only after Wybourn had accumulated more savings. Siding was not added to the house until 1914. And it was 1916 before the trees were cleared from this lot and its neighbour. Vegetable and flower gardens were established and a wall built between the two properties. Then, in 1920, Wybourn sold his vacant second lot, and the purchaser erected his own house upon it. Other urban homesteaders built on their extra properties before selling them. But wherever similar patterns of self-building and the investment of sweat equity in small parcels of land predominated, mixed and irregular streetscapes, with houses of different ages and styles set back varying distances from the curb, were likely to result.

So the city was built. Year by year, and district by district, scores of Vancouverites enacted variants of the parts assumed by Edmonds, McSpadden, Hall, Quiney, and Wybourn. The play was never quite the same – nor could it be with local scripts and many unrehearsed actors – but its main themes were always evident, and the essential lines of its development were repeated in one city neighbourhood after another. In this way, patterns of growth and change in most parts of Vancouver conformed, loosely, to a norm defined by the conjunction of personal ambition, speculative opportunity, and population increase in the rapidly expanding city. As a result, close analysis of almost any small quarter of Vancouver can reveal much about the creation of the urban fabric as a whole.

Shapes on the ground: Dunbar

Dunbar was one of several residential districts created during the first thirty years of the twentieth century, when vast areas were transformed from forest to suburbia in a wide arc encompassing Vancouver's earliest peripheral neighbourhoods: Strathcona, Fairview, and eastern Kitsilano. The area took its name from the American-born land speculator Charles Trott Dunbar, who acquired several large tracts in the western part of Point Grey at the turn of the century. Promoting his property by distributing free calendars advertising 'Dunbar Heights' and paying the BCER $35,000 to extend streetcar service to his land, Dunbar made a fortune from real estate and other ventures. As early as 1906, newspapers reported that his lots were 'selling like hot cakes.' Yet there was little building before the First World War; in all of DL 139, an area of some sixteen city blocks hard by the southwestern corner of the city's original boundary, there were but nine occupied houses in 1918 (Figure 17). As areas closer to the centre of the city filled up, however, the pace of development in Dunbar quickened. Improvements in streetcar service, the paving of roads, the laying of water mains and sewers, and the construction of a school enhanced the attractiveness of the area (Figure 18). By 1925 a municipal land sale in Dunbar was billed as: 'An opportunity to secure Homesite Bargains in Vancouver's Finest Suburb.' A

Figure 17

The development of Dunbar, 1915-30.
The first houses in this area were
erected in 1914. By 1930, almost 80 per
cent of all lots were occupied. Building
was especially intense between 1925
and 1929. Originally the northeast corner of this tract was laid out in the regular grid, although the land fell away
steeply between W. 18th Avenue and
Blenheim. Only one of these lots was
occupied before 1928, and the area was
replotted to take account of the local
relief on the recommendation of
Harland Bartholomew and Associates
(see Figure 32).

■ Commercial

■ New

▨ Existing

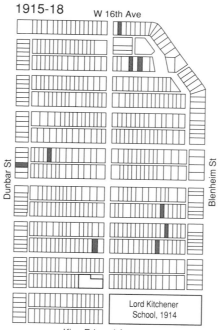

1915-18

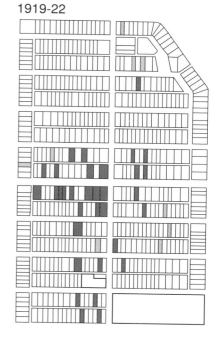

1919-22

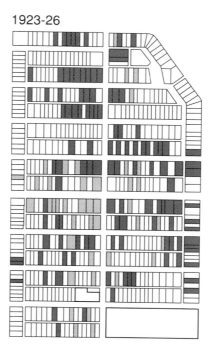

1923-26

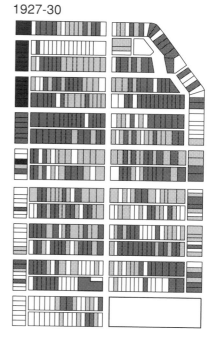

1927-30

Figure 18

A suburb in the making: Dunbar Street
looking south from 16th Avenue, 1929

year later, approximately half of the lots in DL 139 – most of which had
a relatively modest 10-metre frontage – were occupied. A large brick
building had been built alongside the original six-room frame structure
to accommodate the growing number of pupils at Lord Kitchener
School; there were two churches in the area, one of which included a
large hall/gymnasium for community events and church use; and many
properties had views of the city and the North Shore Mountains. 'Any
lot ... represents a good investment,' advised one notice of land sales in
the area. By 1930, barely 20 per cent of the 583 lots in DL 139 remained
vacant. A small cluster of retail businesses on the edge of the area –
many of them owned and operated by local people – served the everyday needs of residents. And a double-tracked streetcar line on Dunbar
Street provided frequent connection to the city core.

This was a modest middle-class suburb, settled in the main by young
married couples and sold by appeal to the fond associations of that
magic word H-O-M-E. 'Picture Your Own First Home/ The Fondest
Memory of All' coaxed an advertisement for a 'dear little' house in
Dunbar Heights. Some 80 per cent of those who lived in DL 139 before
1929 worked in white-collar and skilled blue-collar jobs: approximately
a third were employed in management or sales; fully a quarter followed
trades; and one in five was a clerk. Most had moved into the area from
locations west of Granville Street, and the majority worked downtown.
If the evidence of population movement is any guide, however, many
Dunbar residents were, indeed, more attached to the memory than to
the reality of their first home. On several sample blocks in DL 139, at
least half, often two thirds, and sometimes four of every five houses
changed hands at least once in every decade between 1921 and 1961
(Figure 19). Still, an appreciable number of residents remained at the
same address for many years; several co-workers were near neighbours;
and the recollections of a handful of people who lived in the area
before 1951 suggest that some, at least, felt a firm identification with
the neighbourhood.

Figure 19

Changes in occupancy of several 'dear little' houses in Dunbar, 1922-61

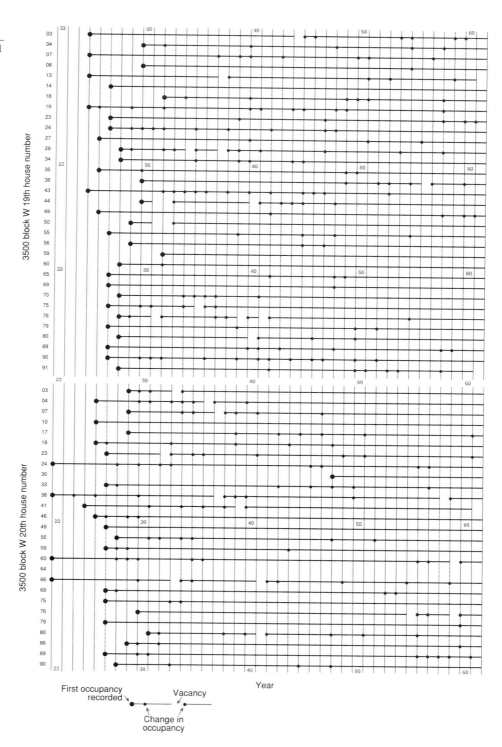

For all its rapidity, the development of DL 139 was highly uneven (Figure 17). Even in 1930, some blocks included several vacant lots. Construction proceeded piecemeal, and almost all of the houses in the area were built by or for their owners or for sale by small contractors. In this respect, the development of West 19th and West 20th avenues was entirely typical (Figure 20). Although the first house on these streets was erected in 1914, and two or three others were built in the early 1920s, almost all of the dwellings in these four blocks were erected between 1923 and 1929. In 1926, when building was perhaps at its peak, six houses were constructed on the north side of the 3500 block of West 20th, each by different owners and contractors, and fifteen were completed on both sides of the equivalent section of West 19th, all but a handful of them by individual contractors. Small speculators also influenced the area's development, in the manner of the BCER agent who bought two adjacent lots on West 20th in 1927. On one, the house (built in 1923) was let to tenants; on the other a new dwelling was constructed in 1929. Speculative developers also took part in shaping the landscape. In 1924, the Ideal Lumber Company of Granville Island built

Figure 20

Building suburbia. Patterns of development on W. 19th and W. 20th avenues, Dunbar. Micro-geographies are often revealing. Here we see the characteristic scatter of houses among vacant lots in the early stages of development, the quickening pace of growth, and the activity of speculative builders. Note also the 'anomalous' placement of some of the houses built in the second decade of the century, before early zoning regulations controlled set-backs from the street.

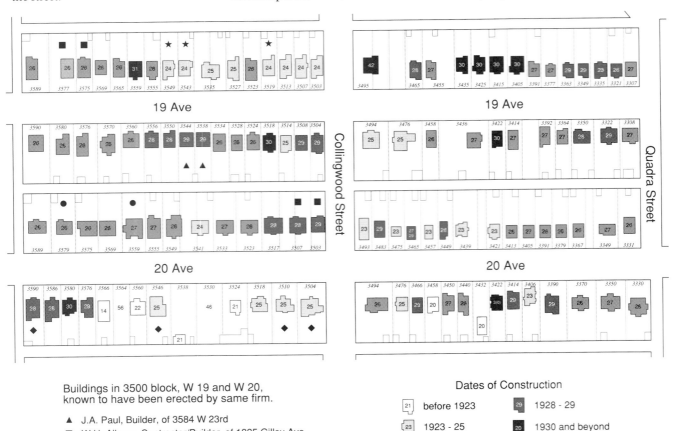

Buildings in 3500 block, W 19 and W 20, known to have been erected by same firm.

▲ J.A. Paul, Builder, of 3584 W 23rd
■ W.H. Allman, Contractor/Builder, of 1825 Gilley Ave, Burnaby (1926) and 822 W 63rd (1928-9)
● Pt Grey Realty Company, of 2541-7 Alma Road
◆ Durrant and Durrant, of 615 W Hastings Ave
★ Ideal Lumber Company, of Granville Island

Dates of Construction

21	before 1923	29	1928 - 29
23	1923 - 25	20	1930 and beyond
27	1926 - 27		

3466 — house number

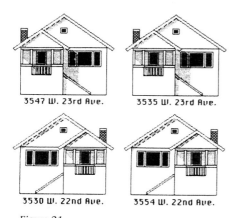

3547 W. 23rd Ave.   3535 W. 23rd Ave.

3530 W. 22nd Ave.   3554 W. 22nd Ave.

Figure 21

Matching houses built in Dunbar by E.G. Baynes

three virtually identical dwellings at 3543, 3549, and 3519 West 19th. On West 20th, L.C. Stephens, a prominent builder in the Dunbar area, who worked in conjunction with the downtown firm of Durrant and Durrant, erected at least five houses, each in a different style, between 1925 and 1928. Directly across the street from two of Stephen's houses, contractor/developer W.H. Allman brought two houses of his own construction to market in 1929. In many such cases, otherwise identical dwellings were differentiated by slight variations in exterior detailing or by simple changes in floor plans, but the ploy was often as transparent as it appeared on neighbouring streets, where developer E.G. Baynes built a pair of identical houses on West 23rd, then reversed the plan and facade for another pair on West 22nd (Figure 21).

The product of this seemingly haphazard building process was a diverse, but not disorderly, landscape. Houses of several styles and sizes lined the streets, but almost all were a storey or a storey-and-a-half in height. Utility poles and wires were largely confined to back lanes. Sidewalks, boulevard trees, and the generally uniform set-back of dwellings from property lines – orchestrated by zoning regulations introduced by the municipality of Point Grey in 1922 – gave definition to streetscapes. For all its newness, by 1930 Dunbar shared with much of the west side of the expanded city a relatively trim, uncluttered appearance. Broadly similar processes had left their imprint on much of south and east Vancouver, although the landscape in many parts of this area – where regulatory ordinances had been fewer and self-building was more common – was less coherent than was that to the west. Here, too, almost every street included a disparate mix of houses, and the landscape took its essential character from this fact.

Houses: Mirrors of status and fashion

Yet the diversity of Vancouver's neighbourhoods was not infinite. For all the minor differences that set Vancouver houses apart in appearance, and despite the countless permutations of their juxtaposition on every block, most dwellings in the city of the 1930s were variants of one or another of approximately a dozen distinctive house types. Examples of each of these basic types could be found in most parts of the city, although almost all were more common in some areas than in others, because price and changing fashion were important determinants of stylistic choice among builders and buyers. Thus, particular neighbourhoods look the way they do because they were built when they were. Moreover, all of these houses reveal something of the aspirations and preferences of Vancouverites and the wider context of attitudes and ideas within which the city grew.

Most houses built before 1910 were variants on one of four styles. Simple cabins, one storey high and one room deep with little decoration, and similar, slightly larger structures (some of which displayed gingerbread trim or machine-turned pillars on the porch) were erected

Early Vancouver house types:

Figure 22a

the cabin

Figure 22b

the prefabricated cottage

Figure 22c

turn-of-the-century homes for the middle ranks of society

across the growing city (Figure 22a). Relatively cheap and easy to build from readily available dimension lumber, they were the first – and sometimes only – homes of many of Vancouver's labouring and artisanal families. Prefabricated cottages, produced by the BC Mills and Timber Company of New Westminster (with an eye to the Prairie market) were also popular in Vancouver between 1905 and 1911. Distinguished, in perhaps their most common form, by a pyramidal roof, an indented porch, and an attic dormer, these, too, were essentially utilitarian structures, offering basic accommodation to those who could afford nothing more elaborate (Figure 22b).

More ornate two- or two-and-a-half-storey houses accommodated most of those in the loosely defined middle rank of Vancouver society (Figure 22c). Designed for relatively narrow city lots, these dwellings typically displayed steep-pitched roofs, decorated gables, scalloped or patterned shingles, and elaborate trim to the street; some included a bay window, others a more substantial projection of the façade and a small second floor balcony. All were strikingly plain to the side, and their plans followed designs published in contemporary trade magazines. According to one detailed study, houses of this type were 'carpenter versions of Gothic Revival ... architecture, inspired by ... pattern book[s], and embellished by trim from the [city's] sawmills.'

Generally, only the wealthiest residents of the city had custom-built, architect-designed houses. Clustered in, but not entirely confined to, the West End, these were characteristically exuberant in style (Figure 22d). Turrets, gables, balconies, verandahs, polychromatic shingles, and elaborate decorative trim were assembled in unique combinations to produce the complex façades favoured by late-Victorian fashion. The results were often sensational. According to novelist Ethel Wilson, one of these houses 'had a bastard architectural excrescence at its corner which resembled both a Norman turret and a pepper pot but was neither.' Yet none of these dwellings was entirely singular. Their counterparts in what architects have termed the Stick and Queen Anne styles can still be seen across North America.

By 1910, however, domestic building in Vancouver was responding to new influences. Late in the nineteenth century, influential designers in Britain and the eastern United States had begun to challenge the triumph of the machine and the standardization of production in industrial society. In England, the reactionary standard was borne by William Morris, who inspired the Arts and Crafts movement. In New York, it was carried by Gustav Stickley, a severe critic of conspicuous Victorian eclecticism in architecture, who promoted a return to simpler forms of building that became known as the 'Craftsman' movement. Different as these movements were in detail, they were oriented to broadly similar ends, and their impact upon domestic architecture was profound. 'Comfortable,' 'cosy,' 'pretty,' and 'artistic' quickly became the watchwords of the construction business. And for fifteen or twenty years,

Figure 22d

the exuberant homes of the wealthy

Figure 22e

the California bungalow

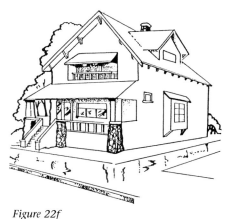

Figure 22f

the Swiss chalet

most Vancouver houses echoed something of the concern (shared by Morris and Stickley) for a return to natural materials and the replication of pre-industrial textures and forms. Following well-established leads, Vancouver builders of the second and third decades of the century erected thousands of dwellings patterned on that epitome of the craftsman house, the California bungalow, or were influenced by the Arts and Crafts-inspired turn to the roots of English architecture known as the Tudor Revival.

In its common Vancouver form, the California bungalow was essentially a one-storey building with a full-width porch and a high basement to accommodate the furnace necessitated by the British Columbia climate (Figure 22e). A larger two-storey version, with roof-dormers and elevated balconies (sometimes termed the Swiss chalet style), was also built for wealthier buyers before 1913 (Figure 22f). Almost all of these dwellings were constructed according to plans taken and adapted from countless pattern books published in California and elsewhere. Once built, they were often praised for their faithful adherence to craftsman principles. Smooth, regular brickwork (redolent of industrial regimentation) was scorned. 'If points stick out, or if the pointing lines…waver like a snake, and if broken ends show, it is all the better, from the artistic standpoint.' Pillars of fieldstone or vari-coloured and misshapen bricks, distinctive eave trims and roof brackets, patterned shingles, and stained glass windows, all suggested traditional building materials and craftsmanship. Even a chimney could be a thing of beauty, observed a writer in *BC Magazine* in 1912, 'an ornament to be displayed, and not an unsightly gaunt stack of bricks to be hidden away in the rear.' There was, of course, great irony in all of this. Minor differences in style – 'a small cobblestone stuck in here and there' was a 'much admired variation' according to one account – allowed purchasers to feel that they were acquiring a house especially suited to their particular desires, although the basic product was highly standardized.

The Tudor Revival came relatively late to Vancouver. Several 'late-medieval English manor houses' were built in Britain and the United States in the last quarter of the nineteenth century, and by 1910 muted forms of half-timbering had begun to appear on relatively modest suburban dwellings on both sides of the Atlantic. Houses in this style had broad appeal. They suggested simpler times, peace, and harmony. Sheltered by trees and surrounded by gardens, they nestled into the landscape, more like works of Nature than human intrusions. Their basic attraction was well summarized in an American guide to the style, which characterized its sixteenth-century originals as buildings that had 'no acquaintance with the paint shop or the planing mill.' They were, rather, 'offsprings of the soil, with their brick and mortar from the fields, and rough hewn timbers dragged from the forests.' No matter that modern versions evoked this organic quality with dimension lumber, stud frame construction techniques, and a veneer façade.

Figure 22g

the 'ideal house for Vancouver': the Tudor Revival mansion

Figure 22h

a modest version of the 'ideal house'

In Vancouver, a handful of architects married current refractions of old English models with local materials to produce a considerable number of large and distinctive homes for wealthy clients. They were built in widely scattered locations across the city. View sites along the northern edge of the Point Grey peninsula, rural settings to the south, and relatively remote and picturesque locations amid the forest to the east attracted those who carried the logic of retreat toward its obvious extreme by withdrawing from the city. But most were concentrated in the CPR's élite Shaughnessy Heights subdivision opened in 1909. There and elsewhere, architects characteristically combined stylistic references from several different periods of English architecture on the same building. They utilized fieldstone as well as half-timbering on exterior surfaces, lavished attention on detail, and broke roof-lines with dormers and bell-cast forms. The result was a style at once spacious and homelike (Figure 22g). Redolent of tradition and stability, houses of this type spoke volumes about their owners' success in the fluid society of the growing city. More than this, they suggested the permanence of those achievements and provided links to the past in a place where so much was new.

Little wonder that, by the 1920s, lesser variants of these 'instant old' mansions were being held up as examples of the 'Ideal house for Vancouver.' With the publication, by local architects, of a booklet of plans for *British Columbia Homes*, bungalows and small houses with stucco walls, steep roofs, and a suggestion of half-timbering gave a markedly English cast to many city suburbs (Figure 22h).

The resemblance was further enhanced by the construction of hundreds of stucco cottages with sloping buttresses and mullioned windows on the west side of the city during the 1920s. Relatively plain by comparison with common Tudor Revival stylings, these dwellings derived their pedigree from designs by English architect C.F.A. Voysey – designs that were intended to evoke the cozy, close-to-the-ground appearance of whitewashed cottages in English villages. Most often built on lots with a 15- or 20-metre frontage, Vancouver's Voyseyesque houses generally had at least one prominent cross-gable. An asymmetrical treatment, in which one side of the gable was curved down below the roof-line (to form what is colloquially known as a cat slide) was also common. To foster the desired rustic image, eaves were curved to imitate thatch, tiny windows were let into roofs, and false buttresses were added to disguise the basic rectangularity of the house, although they usually had to be breached by arches to allow access to the rear of narrow urban lots (Figure 22i).

By the late 1920s, California bungalows were no longer in vogue in Vancouver, and houses in Tudor Revival and Voyseyesque styles were generally beyond the reach of working- and lower middle-class families. To meet their housing needs, less expensive dwellings were built on 10-

metre lots across the city. Box-like in appearance, and generally one and a half storeys high in height, these structures typically had unfinished basements, two bedrooms on the main floor, and an unfinished or partly finished upper storey (Figure 22j). On these builders' specials, the influence of the Arts and Crafts movement, so evident in more affluent parts of the city, was muted indeed. On some, stucco was preferred to wooden siding. Others displayed leaded glass windows. The chamferred or jerkin gables that were popular for a few years before 1930 echoed English precedents, albeit ever so faintly. But there was no escaping the essentially utilitarian character of these buildings: they provided relatively cheap basic accommodation and the opportunity for later improvements with only the slightest concession to contemporary stylistic concerns.

Beyond this, a significant number of striking homes were built in less popular styles. Among them were American colonial houses; Dutch colonial dwellings; mansions patterned on the great plantation houses of the southern United States; Gothic, Georgian, and Italianate revivals; French baronial castles; and Spanish missions. But they were somehow unusual, even 'exotic.' In this new place, where so much was borrowed from elsewhere and there was no truly indigenous style, consensus had defined a norm. Through half a century, developers, architects, builders, and home-buyers created suburban landscapes that were remarkably consistent in appearance and form. The grid streets and frontage lots, favoured for the efficiency with which they allowed the subdivision and sale of land, imparted a regular geometry to the rapidly expanding area of the city. On this template, street after street was lined with dwellings, each much like another nearby. Then, as now, suburbs offered people space and shelter in new houses that reflected current conceptions of the appropriate setting for family life. Light, air, and comfort, English charm and Californian convenience, these were the axes around which the residential neighbourhoods of early Vancouver developed. Their appeal lingers yet (Figure 23).

Figure 22i

the Voyseyesque cottage

Figure 22j

the builders' special of the late 1920s

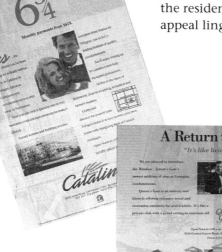

Figure 23

English charm and California convenience

Prosperity by sea and land

On the day that the first transcontinental train reached Vancouver, the editor of the *News Advertiser* could 'see nothing...wanting' in the future of his city. Steel rails connected it to the east, steamships linked it to the Orient, there were abundant forests and rich mines in its hinterland, it had 'fertile valleys on one side and the sea teeming with all kinds of fish on the other.' In this view, the little settlement on Burrard Inlet was destined to become the metropolitan centre of the British Columbian resource economy. Its motto – By Sea and Land We Prosper – simply affirmed the obvious. But many were less sanguine. Several American visitors saw obstacles to growth in the slow pace of life and 'British stolidity' of Vancouver's residents; others, who found Vancouver 'a plucky and enterprising city,' qualified their enthusiasm with such condescending phrases as 'even if it is not very big.' And the heady aspirations of Vancouverites ran counter to the interests of Victorians, who strongly resisted the erosion of their city's financial, commercial, manufacturing, and political pre-eminence in the province.

Before the CPR broke through the mountainous spine of the province, British Columbia looked seaward. People, mail, and manufactures generally arrived via the Pacific, and the province's most important exports – salmon, lumber, gold, and coal – went to market the same way. San Francisco, the terminus of a transcontinental railroad since 1869, was the province's most direct and important link to the east, and California was far and away its most important trading partner. During the 1880s almost 60 per cent of British Columbian exports (by value) went to the United States; Britain took slightly less than 30 per cent; and most of the remainder was shipped to destinations in and around the Pacific. Victoria was the hub of this trade. Its location on the Strait of Juan de Fuca, a good harbour, and its early importance as a centre of Hudson's Bay Company activity, had given it a pivotal position in the gold rush of the 1850s, and, with this head start, it benefitted from the late nineteenth-century expansion of the provincial economy. Bankers, wholesalers, brokers, and shipping agents established themselves in the city. Their capital – invested in salmon canning, sawmills, coal lands, secondary manufacturing for local markets, and even Vancouver real estate – as well as their political influence gave Victoria a considerable degree of provincial hegemony. Victoria businessmen, concluded a perceptive historian of this period, dominated the governments that encouraged 'exploitation of regional resources through the provision of public concessions to private interests' and were the chief beneficiaries of these policies. At the census of 1891, the island city had 3,000 more residents than did Vancouver and employed twice as many manufacturing workers as did Vancouver and New Westminster combined. Two years later, the value of its trade was almost three times that handled by Vancouver, and the provincial government chose to erect a fine new legislative building overlooking the harbour around which British Columbia's Pacific economy had turned for half a century.

Yet Victoria's metropolitan dominance was broken within the decade. Aggressive advertising by Vancouver interests won the city a large part of the frantic traffic generated by the Klondike gold rush after 1898. The spectacular expansion of Prairie settlement after 1896 opened new markets for British Columbia's wood products. Settlers moved into the Okanagan and other parts of southern British Columbia. Improvements in transportation tied these areas more closely to the coast. And Victoria's maritime links to San Francisco declined as California's railroads began to use oil, rather than Vancouver Island coal, for fuel.

During a decade of dramatic expansion at the end of the nineteenth century, British Columbia's salmon canning industry, another important cornerstone of the province's Pacific economy, was also reshaped. Canneries remained scattered along the coast, of course, but two of the large consolidated companies that dominated the industry in the late nineteenth century – the British Columbia Canning Company and the Anglo-British Canning Company – had their head offices in Vancouver. This emerging pattern was only reinforced in 1902, when 29 of 48 canneries on the Fraser and another dozen up the coast were brought together in the British Columbia Packers' Association, financed by eastern Canadian and American capital and based in Vancouver. The railroad that linked the salmon canneries of Steveston with the wharves and CPR terminus on Burrard Inlet was a symbol of the way in which the provincial economy was being turned from its focus on Victoria and its orientation to the Pacific. By 1900, Vancouver was the province's leading port of entry.

As British Columbia's importance as a supplier of lumber and fruit to the Prairies increased, and its economy expanded, central Canadian financial and other interests moved into the region. In the final years of the nineteenth century, a handful of eastern banks established branches in Vancouver. They were followed by insurance companies, wholesale firms, and manufacturing enterprises. Early in the twentieth century, there was also an enormous surge in American investment in the provincial lumber industry. Much of this was in timber holdings, but Vancouver soon emerged as the management centre of the industry. By 1910, the city accounted for three-quarters of British Columbia's industrial output, much of it from the large sawmills on Burrard Inlet, False Creek, and the Fraser River that made the lumber industry the city's largest employer. Most of the province's rapidly increasing imports came through Vancouver, and the city was also the banking and service centre of British Columbia. At the census of 1911, Vancouver had 28 per cent of the provincial work force but over 40 per cent of all those engaged in finance, the professions, real estate, and iron and steel manufacturing. A few years later, two-thirds of the 1,700 or so companies operating in British Columbia from headquarters outside the province had their local offices in Vancouver; and fully half of the province's 3,422 indigenous firms did also. By any of these measures the city was the hub of the province.

Still, many parts of this rugged territory lay beyond the economic tentacles of Vancouver. Although almost a thousand commercial travellers were based in the city, and Vancouver merchants made concerted efforts to increase their trade with the Prairies, the costs and difficulties of movement through the mountains limited their penetration of the interior. High freight rates on railroad traffic through British Columbia enabled Winnipeg to supply many commodities more cheaply than could Vancouver to customers as far west as Revelstoke. Trade in the important mining districts of the Kootenays and Boundary regions of southeastern British Columbia still ran to Winnipeg and Spokane, and Vancouverites ate soft fruit from Washington State, not the Okanagan. Many central and southern parts of the province remained relatively inaccessible from the coast until the second decade of the century, when new railroads, such as the Canadian Northern (from Edmonton through the Yellowhead Pass to Kamloops and Vancouver) and the Kettle Valley line (from the Boundary Region to the Fraser River), were built.

Railroad construction and feverish anticipation of the benefits of the Panama Canal fuelled the fires of development in Vancouver after 1909. In 1913, reflecting the optimistic, expansionist mood of the times, residents agreed to transfer the eastern limits of False Creek to the Canadian Northern Railway. In return, the company undertook to fill the creek east of Main Street and to construct a station, marshalling yards, deep-sea shipping facilities, and a couple of hotels on the land. At the same time, the provincial government acquired the 80-acre Kitsilano Indian Reserve at the mouth of False Creek for what critics considered a fire-sale price, with the intention of developing a grand new harbour there. And the federal government established the Vancouver Harbour Commission, with a mandate to improve the Burrard Inlet port by constructing new wharves and a grain elevator to handle Prairie wheat. Not to be outdone, Richmond also formulated grand plans for the construction of freight and passenger terminals on the seaward side of Lulu Island (Figure 24).

The economic crash and sharp recession of 1913, as well as a re-orientation of priorities during the war that followed, delayed implementation of many of these plans and derailed others completely. But the increase in shipping traffic on which they were predicated was realized in the 1920s. In little more than a decade, the number of ocean-going vessels entering the port of Vancouver all but tripled. New wharves – including the large Ballantyne Pier – were built, and new elevators at the eastern end of the harbour better than quadrupled grain storage capacity in Burrard Inlet. More than $60 million were invested in the port in the

Figure 24

Plans formulated by the Vancouver Harbour and Dock Extension Company for the development of new harbour facilities on the seaward edge of Lulu Island, 1912

dozen years after 1918. By 1931, sixty-two ocean-going vessels could be berthed in Burrard Inlet at any time. Every vessel that entered the port pumped substantial revenues into the urban economy. By one estimate, purchases of provisions, services, and labour averaged $5,000 a visit, and there were also charges for pilots, tugs, moorings, and the like. In addition, shipping agents, customs brokers, and others who managed and facilitated trade occupied new premises near the waterfront, augmenting further the impact of the port upon the city.

By 1930, Vancouver handled more export tonnage than any harbour in Canada. Two-thirds and more of this traffic was in wheat. In 1928-9, some 95 million bushels were shipped from the city; this was almost a sixth of the entire production of western Canada and nearly a quarter of Canadian wheat exports. Four years later, the seven elevators on Burrard Inlet held approximately 18 million bushels of wheat, nearly double the capacity of all the elevators on the west coast of the United States. No longer second fiddle to Victoria, Vancouver had risen, in little more than thirty years, to rank among the leading grain ports of the world. Year-in and year-out, between 60 and 85 per cent of the wheat shipped from Vancouver went to the United Kingdom. Shipments across the Pacific, to China, and, in lesser quantities, to Japan fluctuated equally widely, but between 1921 and 1933 these markets took almost a quarter of the city's wheat exports and some four-fifths of its much smaller trade in flour. Lumber shipments, which almost doubled during the 1920s, were also a significant component of the city's trade, although other provincial ports (particularly New Westminster, which had benefitted from extensive dredging, reclamation, and other works along the Fraser River since the First World War) accounted for most of the 360 per cent growth in provincial wood exports during these years. Much of this expansion was attributable to the Panama Canal, which gave west coast wood access to new markets in eastern North America and across the Atlantic. Metals, canned salmon, and fish meal and fertilizers were among the other commodities exported from Vancouver, but they ranked far below wheat and wood in importance.

After a decade of dramatic expansion, the 1930s and early 1940s were difficult years for the port of Vancouver. Grain shipments fell sharply from their 1932 peak as depression and drought racked the Prairies and world commodity markets contracted. By 1938, they amounted to approximately 25 million bushels. Four years later, wartime shipping shortages and the concentration of Allied tonnage on North Atlantic routes had reduced Vancouver grain shipments to a mere 2 million bushels. Most other commodity movements were also down appreciably. For almost twenty years, foreign freight movements through the port of Vancouver remained below the levels of 1930. Then trade quickened with the expansion of provincial, national, and global economies in the 1950s. By 1963, Vancouver led all Canadian ports in tonnage handled, and several new facilities that would increase and extend this pre-eminence into the 1990s were under development.

Although politicians and businessmen often promoted industrial growth in Vancouver, tertiary activities (transportation; wholesale distribution; and government, business, recreational, and personal services) dominated the local economy through the first half of the twentieth century. In 1931, for example, fewer than 13,000 of the Vancouver region's 380,000 people worked in manufacturing. Over half were employed in the food and beverage and wood-products sectors. Printing and publishing, and the manufacture of metals and machinery, were the only other sectors with more than 1,000 employees. In sum, the Vancouver region accounted for barely 2.5 per cent of the value of Canadian manufacturing output and a similar proportion of the manufacturing work force.

Shipbuilding, aircraft, and munitions contracts brought new industrial jobs to the city in the early 1940s. At their peak, some 40,000 Vancouver men and women worked in wartime industries, but the decline was rapid after 1945. New industries, tuned to the demands of the expanding provincial economy for such products as structural steel, reinforcing rods, wire rope, and logging equipment, absorbed some skilled labour in the late 1940s, and shipbuilders produced tugs, barges, and fishing vessels for the British Columbian market. Still, in 1951, two in every five of those who worked in manufacturing in the Vancouver area were engaged in processing raw materials. In Toronto, the ratio was less than one in ten. Population growth and technological change quickened diversification of Vancouver's manufacturing base during the 1950s, however. By the end of the decade, the city sold western consumers a wide range of relatively low-value manufactured items that could not carry the cost of shipment from eastern factories. Some specialized items, such as power saws and drilling equipment developed for local extractive industries, went to national and international markets (Figure 25).

Despite the rapidity of growth in early Vancouver, the city's labour market was highly volatile. Boom times alternated with periods of economic recession as the province's resource-based economy tracked the ups and downs of external markets and the often frenetic activity of city speculators exaggerated fluctuations in the trade cycle. Seasonal shifts in employment in the canneries, camps, and mines of the city's hinterland also periodically brought waves of job-seekers into Vancouver. Even ostensibly good times saw cycles of mild to moderately high unemployment in the city. So the years between 1900 and 1912 – which city boosters saw as prosperous and expansive, and during which the annual value of city building permits increased some thirteenfold – included many difficult months for hundreds of working people and their children. Labour was often abundant as the population of the city

Figure 25

Market distribution of goods manufactured in Vancouver

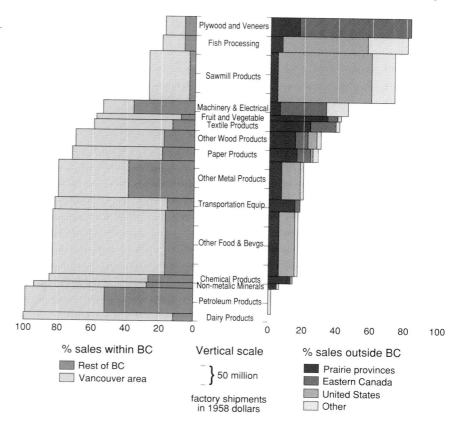

climbed and as men drifted into the city between hinterland jobs in search of urban work or to spend the winter in a relatively equable climate.

In broad terms, 1906 and 1907 were the best of the prewar years. Lumber production soared with the demand generated by the reconstruction of San Francisco after the earthquake of 1906. Wages rose more rapidly than at any other time between 1900 and 1914, and the *Labour Gazette* reported a shortage of workers in the city. But this propitious interlude followed several years of labour surplus. Strikes in the goldmining town of Rossland BC, Seattle, and San Francisco as well as among coal miners and railway construction workers brought sizeable numbers of job-seekers to Vancouver in 1901 and 1903. Slumps in mining and logging had much the same effect in 1902 and 1904. And the boom caused by the earthquake was short-lived. Tightening world money markets limited the availability of capital and slowed construction in 1907. Employment levels dropped and wages fell.

Responding much as it had in similar circumstances a decade before, Vancouver City Council provided relief work, shelter, and sustenance for the unemployed during the winter of 1907-8. Perhaps half of the 5,000 men out of work were skilled artisans, and, by some accounts, as many as three-quarters of the city's carpenters and electricians and two-thirds of its bricklayers were jobless. 'Is there an Englishman in all British Columbia who could tell another where he could get work ...?' asked a disillusioned immigrant in the pages of the *Province* in October 1907. A year or so later, the city mayor advised Prairie dwellers that 'The labour market in Vancouver is rather overdone, and no inducement can be offered to labourers before at least March or April. We give this notice in view of the rush of labourers to this city from the prairie district last fall. The city will only take care of actual residents who have resided here at least six months.' After the bubble of rapid expansion burst in 1913, many workers found themselves in even more straitened circumstances. Stand near the Labour Temple between seven and eight o'clock in the morning, members of the 1914 Commission on Labour were told, and see Powell and Carrall streets 'jammed with people scanning the boards of employment offices in search of employment.'

Hiring practices and labour disputes also contributed to the volatility of the Vancouver labour market. Large numbers of construction workers, including the skilled, were hired on a short-term basis. The wage books of one Vancouver contractor, who was probably quite typical of the city's small builders, show that only two of the twenty-eight men he employed in 1898 worked full-time; the others were hired for periods ranging from a few days to six weeks. On average then, turn-of-the-century carpenters, plumbers, or electricians might have considered themselves fortunate to have eight or ten months work a year. Strikes also took their toll on workers' incomes. According to Department of Labour statistics, there were forty-six strikes in Vancouver between 1901 and 1911. In 1903, approximately 1,800 men were off the job demanding union recognition and shorter working hours; among them, 1,000 transport workers were on strike for almost four months. Eight years later, 7,500 workers (some 15 per cent of the labour force) struck. More than 5,000 of them were off the job for two months or more.

Structural factors also brought average wage levels in Vancouver down by comparison with those in other major urban centres. For all its importance in the city, the wood products industry paid relatively low wages to the two-thirds of its work force that was unskilled. On an annual basis, they were little better off than were domestic and personal service workers, who comprised some 15 per cent of the city's 1921 work force. All in all, therefore, although hourly wage rates in some sectors of the Vancouver economy were relatively high by comparison with those in other Canadian cities, many Vancouver workers earned no more over the course of the year than did their counterparts across the country. So the $418 average earnings of Vancouver manufacturing employees in 1911 were below those of similar workers in Calgary, Winnipeg, Toronto, Hamilton, Montreal, and Ottawa. Ten years later, 20- to 45-year-old manual labourers in Vancouver reported average weekly earnings below those recorded by their counterparts in all large Canadian cities except Victoria and Saint John.

Yet early twentieth-century Vancouver was widely considered to be an expensive place to live. In 1901, woodworkers in the city claimed that they would be better off employed at twenty cents an hour in the east than at thirty cents an hour in British Columbia 'where the cost of living more than made up the difference.' Certainly, average retail prices calculated for the major cities of each province were considerably higher in British Columbia than in Ontario in 1912, and commodity prices published intermittently in the *Labour Gazette* suggest that Vancouver food costs were some 50 per cent above those in Toronto at this time. By another account, food, fuel, and lighting costs were higher in Vancouver than any of Canada's largest cities, except Montreal.

Many of Vancouver's working-class families walked a narrow financial tightrope. When economic downturns led to wage cuts or shorter work days, even workers in relatively secure employment suffered. During the depression of 1915, for example, street railwaymen had their hours cut. Conductors, said their union representative, could earn no more than $766 a year under these arrangements. Yet, he claimed, a working class family needed $1,233 to live decently. Some stretched their budgets by subletting rooms in their houses; others – perhaps hundreds – barely met the mortgages on their homes and land. In this context, strikes and other interruptions of work had serious consequences. According to one account from within the carpenters' union, its executive officers were at their wits' end within three weeks of calling their men off the job: rank and file members were so concerned about their mortgages that their leaders were haunted by the spectre of the next payment.

Wartime inflation, the example of the Russian Revolution, and the rise of unemployment and prices after 1918 generated new tensions between labour and capital in Vancouver. Opposition to conscription provided a focus for those on the Left, who protested the draft of working men while wealth ran free, and many resented the way in which the province's coal miners had been forced back to work after striking for higher wages. Longshoremen, shipyard workers, metal workers, letter carriers, bakers, and the city's outside employees called work stoppages in 1918, and in August of that year Vancouver's Trades and Labour Council declared a twenty-four hour 'holiday' to protest the fatal shooting of labour activist and draft resister 'Ginger' Goodwin on Vancouver Island. Although the walkout was far from general, there were several violent confrontations between 'holiday-makers' and veterans of the European war. The mayor threatened martial law, and others lamented the rise of 'Red Socialists with pro-German ideas.'

When workers in Winnipeg joined a general strike in May 1919, many Vancouverites prepared for the worst. The mayor worried about a fraction of the population 'decidedly inclined to Bolsheviki tendencies' and reported to the premier on the 'serious feeling of unrest' in the city. By the end of the month, workers in Vancouver (as well as Victoria and Prince Rupert) were off the job. For the better part of a month, the city was without street railways and two of its regular newspapers; although police and firemen continued to work, most other branches of labour were out at least part of the time. Yet there was little violence. Strikers organized soccer matches among themselves. Private cars picked up paying passengers, volunteers produced the *Vancouver Citizen* to compete with the *Bulletin* put out by striking typographers, and volunteers from the Citizens' League maintained telephone service when the operators struck. With the collapse of the Winnipeg strike, the Vancouver protest came to an end. For all the very public consternation of the city's establishment in 1918 and 1919, few candidates from the political Left received significant support from Vancouver-area voters in the elections of 1920.

Better economic times brought a more peaceful labour climate during the 1920s. From Victoria, Premier John Oliver sought to turn the economy of British Columbia away from its heavy dependence upon extractive industries. Funds were directed to a wide range of projects, several of which had a substantial impact upon the Vancouver region. To encourage farming, Sumas Lake and the great marshes that surrounded it in the Fraser Valley near Chilliwack were drained. To improve communications, there were investments in roads and in the completion of the Pacific Great Eastern Railway from the Cariboo-Chilcotin district of the interior to salt water at Squamish. And to ensure the future availability of a corps of technical experts who would serve the developing provincial economy, money was found to move the University of British Columbia from its obsolete and crowded accommodations on the slopes above False Creek to a magnificent new site on the tip of Point Grey. Each of these initiatives heightened the importance of Vancouver in British Columbia and stimulated growth in the city. New suburbs were built and levels of home ownership rose. With the introduction of federal and provincial industrial disputes acts at mid-decade, improvements in the Workers Compensation Act, and the implementation in 1927 of a means-tested pension plan for those over seventy, the sting of labour unrest was drawn and, overall, standards of living in the city improved. Average wages climbed, to approach thirty dollars a week in the summer of 1929. With predictable but not outlandish overstatement, a contemporary survey of the 'existing state of industry in Vancouver and its future possibilities' exulted over the city's 'experienced, reliable, home-loving, home-owning and anti-radical workers.'

The stock market crash of October 1929 quickly destroyed such complacency. Before the year was out, the Department of Labour reported that every district in British Columbia had 'a surplus of men,' and that Vancouver was 'the Mecca of the unemployed.' Churches established soup kitchens to feed the destitute. Breadlines formed and lengthened. 'Hobo jungles' – sad clusters of wretched, makeshift, and inadequate shelters – sprang up under the Georgia Viaduct and on railway and harbour board lands to 'house' those who could find no better alternative. By the late summer of 1931, the shelters were being described as seedbeds of social, moral, and physical decay, and their occupants were evicted. Across the province, one in ten of the population was out of work; by 1933 employment levels in Vancouver were barely three-quarters of those in 1929.

Marches, protests, and confrontations ensued. 'Strikes' in the relief camps established to provide work for single men in remote parts of the province spilled over into the city. When dissidents protesting conditions in the 'slave camps' damaged the Hudson's Bay Company store, and a sympathetic crowd gathered in Victory Square in April 1935, the mayor read the Riot Act. A month later, after a large May Day parade and the occupation of the Hotel Georgia, the post office, and the art gallery by relief camp protestors, the city's waterfront was shut down by a strike. To Mayor Gerry McGeer, it seemed that the people of Vancouver had to choose between 'constituted authority' and 'Communism, hoodlumism and mob rule.' The 'red menace' had shown its face, and many prominent Vancouver residents wanted it crushed. Communists, said the *Vancouver Sun,* were 'nameless vagabonds of chaos.' As the rhetoric grew more inflammatory, positions hardened. In June, 1,000 strikers marched on Ballantyne Pier. Police met them with tear gas and batons. Sixty people were injured in the ensuing riot. Some, including the mayor, saw signs of revolution in all of this. But the militancy of the 1930s was driven by the severe difficulties of the unemployed, the susceptibility of workers to exploitation by their employers, and the seeming indifference of the wealthy and secure to the plight of their less fortunate brethren. At least one columnist in the *Vancouver Sun* recognized as much. 'Some 800 of Vancouver's longshoremen are men married with families,' he wrote in early June 1935. 'One hundred per cent of the membership of the Vancouver and District Waterfront Workers' Association have worked continuously on the Vancouver docks for...five to forty years. Are these the men who, we are told, are helping fomenters of revolution? The question does not merit a reply.' Talk of an impending proletarian dictatorship was hyperbole, the 'red scare' wildly inflated. The issue was not communism but whether working men and women would find the power to defend their interests in collective action. It is an issue that lingers yet.

The city: healthy, beautiful, efficient

Although many contemporaries regarded the nineteenth century as an 'age of great cities,' it was clear, well before 1900, that urban growth had its price. For all the magnificent achievements of London, Manchester, New York, and Chicago (in architecture, culture, commerce, production, and a score of other spheres), squalor, misery, disease, and vice almost everywhere accompanied the concentration of population. Cholera, typhus, and other epidemics decimated urban populations. Powerful descriptions of urban slums, and of those who inhabited them, appeared with arresting frequency on both sides of the Atlantic after 1850. Largely unanticipated (we thought, said English novelist William Thackeray, revealingly, that a city of 100,000 was no more than ten cities of 10,000), the problems of crowded urban living quickly became an important focus of reform concerns. By the final decades of the nineteenth century, the importance of clean water and effective sanitation to urban survival were well understood; engineering journals devoted pages to the details of sewer construction and water supply schemes, and imperial officials charged with laying out new towns at the farthest reaches of the Pax Britannica implemented their wisdom as best they were able. So, too, the clutter and disarray of urban landscapes that had taken shape in a helter-skelter scramble of unregulated competitive development (and that led prominent Canadian architect Percy Nobbs to compare downtown Montreal to a Chinese harbour after a typhoon) had produced its own reaction in a movement to build the city beautiful. And the poorly built, densely peopled, underserviced tenement and slum districts, so common in British industrial cities and the large urban places of eastern North America, both precipitated the flight of many to peripheral suburbs and generated a vigorous campaign for the marriage of town and country in what British reformer Ebenezer Howard called, in the very first years of the twentieth century, 'Garden Cities.'

Health

For all its newness and physical distance from the problem cities of the day, Vancouver was shaped by these developments. By drawing upon experience elsewhere, city residents were able to implement relatively modern strategies of sewage disposal and water supply almost from the first. Moreover, although time and population growth eventually rendered inadequate some of the city's earliest initiatives in these spheres, necessary improvements have been built, in large part, upon earlier foundations.

Until 1887, Vancouverites drew their water from shallow wells and disposed of their sewage in the ground. But the dangers of doing so as the city grew were clear, and sewer construction proceeded in tandem with the layout of new streets. Soon, some 6,000 acres of land were serviced by storm-water/sewage disposal systems that drained into Burrard Inlet and False Creek, although less-populous Burnaby responded rather more slowly to the need for effective sanitation and did not build its first sewers until 1908 or so. Massive population growth early in the twentieth century quickly revealed the inadequacies of these systems, however. In 1911, the city's beaches were so badly polluted that they were closed to bathers. Two years later, the Vancouver and Districts Joint Sewerage and Drainage Board was established to implement the recommendations of a report (prepared by a prominent sanitary engineer from Montreal) on the disposal of sewage and surface water from Vancouver, Burnaby, Point Grey, and South Vancouver. Separate storm-water and sewage drainage systems were developed; the trunk lines were designed to accommodate subsequent expansion; and sewage outfalls were removed from English Bay. In sum, these initiatives cost approximately $9 million over three decades and provided the area with a system as sensitive to the environmental and ecological consequences of sewage disposal as were most in early twentieth-century North America. Yet, in the late 1940s, Vancouver's beaches and some inshore waters were again heavily polluted. Raw sewage was entering Burrard Inlet and the Fraser River from a dozen communities. Again the problem was investigated, this time by the chief engineer of Los Angeles County. His report led to the establishment, in 1956, of the Greater Vancouver Sewerage and Drainage District and recommendations for the establishment of sewage treatment plants to serve Vancouver and the North Shore. Substantial new tunnels were necessary to divert the outfalls from Point Grey and Kerrisdale, but when the new treatment facilities were opened early in the 1960s at Iona Island and Lions Gate, they essentially served the system laid down and extended after 1914.

Interest in developing a reliable water supply for Vancouver was heightened by the fire that destroyed much of the early city in 1886. Almost immediately, private investors vied for the contract to develop a water system. Initially, Council favoured the Coquitlam Water Company, which planned to serve both New Westminster and Vancouver from Coquitlam Lake. The electorate opted for the Vancouver Water Works Company's riskier, but less costly, plans to bring Capilano River water into the city through a pipeline across Burrard Inlet. Early in 1888, a stone-filled timber dam created a 14-million-gallon reservoir some 10 kilometres above the mouth of the river. Within twelve months, a 12-inch main had been laid across First Narrows, and through Stanley Park and Coal Harbour to Vancouver. Purchased by the city in 1892, this system served the growing population of Vancouver, Burnaby, South Vancouver, and Point Grey for almost two decades. Then additional supplies were brought in under the Second Narrows from the Seymour River. In 1911 a large new reservoir was built on Little Mountain to provide gravity-fed water to the city. And in the late 1920s control of the Capilano, Seymour, and Coquitlam systems was vested in the Greater Vancouver Water District. This body encompassed most of the inhabited Lower Mainland by 1931, but the sources of supply remained those established before 1910. New dams increased storage capacity, and closure of the tributary watersheds ensured a clean, high-quality water supply that required no chemical or other treatment (besides screening) until the U.S. Navy insisted on chlorination of water taken aboard their vessels in Vancouver during the Second World War.

For Vancouver, then, a combination of factors – topography and climate, experience with the problems of city growth elsewhere, and the operation of the legally distinct Water District and Joint Sewerage and Drainage Board with a common staff under a single director – enabled the efficient provision of high-quality water and sanitation services at the metropolitan scale.

Beauty

On a visit to Vancouver in 1912, the distinguished British landscape architect, Thomas Mawson, spoke to the Canadian Club on the topic of 'Civic Art and Vancouver's Opportunity.' The city, he began, was a 'most marvellous creation of twenty-five years.' Yet it remained rich in promise. Vancouver, said the visitor introduced as 'perhaps the greatest living authority on city planning,' offered 'a crowning opportunity, greater than I have seen in any part of the world and distinctly greater than any work which the great artists of the past have accomplished.' Little wonder that his opening words were greeted with applause. But Mawson had more than boosterism on his mind. He had discovered, he continued, that the people of Vancouver could be divided into two classes. There were those who loved Nature. And there were those 'interested only in commercial pursuits, with no idea or imagination that ever rises above that level; those who would do anything to destroy Nature, who do not possess the reverent soul, and who do not hear the music of the spheres.'

Vancouver seemed to mirror these divisions. The city itself was 'a purely artificial creation.' Much of it bore the ugly scars of unconsidered development. 'Your telegraph poles and electric light poles in Vancouver are not beautiful,' said Mawson to the laughter of his listeners, 'they are primitive, antiquated and apparently inefficient.' Vancouver shared

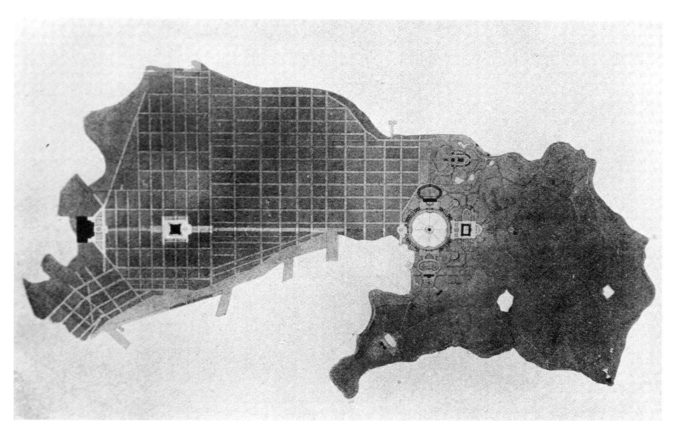

Figure 26

Georgia Street. The Champs-Elysées of Vancouver

with many other places the unfortunate 'modern idea of a great city… [as] a mere overgrown village without the connections that tie the whole together in one great conception.' Yet there was Stanley Park, a work of Nature, an area that compelled reverence, a place 'known to every schoolboy in the Old Country' as Vancouver's greatest asset, on the very doorstep of the city. The challenge – the opportunity – inherent in this pointed contrast was to integrate the dramatically different landscapes of the city and the park. Success would require bold action, but it could not be postponed if Vancouver's potential was to be realized. City residents had to develop 'the proper connection between the artificial part [of their region] and that which… [was] purely natural.' In Mawson's view, Coal Harbour was the obvious pivot of this reconciliation, 'the place where Art, Nature and Science must meet and arrange terms.' Imagine, he urged his listeners – to more laughter – Georgia Street as the Champs-Élysées of Vancouver. Widened and restricted to pedestrians, it would be the grand axis linking city and park, commerce and nature (Figure 26). In the downtown core, four blocks would be partially cleared to create a central square around which the most important civic buildings would be grouped. At the foot of the hill, midway between the square and the park, there would be a Place de la Concorde, and then, said Mawson, continuing his visionary comparison of Vancouver with Paris, 'in Coal Harbor I would have my Grande Rond

Figure 27

Art, Nature, and Science meet in Coal Harbour

Pond, and the park beyond as the Tuileries, though not humanised' (Figure 27). Surrounded by a promenade, and with a statue of Captain Cook or Captain Vancouver atop a 50-metre-high pillar at its hub, the pond would become 'the great social centre of Vancouver, where rich and poor, the rough man and the spring poet, would have to rub shoulders and pass along in comfort and ease.' It would also be a most beautiful base from which a great classical building – perhaps a 'grand museum of the natural history of British Columbia' – could arise. With the forested park and 'the magnificent setting of your seven sisters' its backdrop, it would surely remind Vancouverites 'of Browning's saying of "a thing suitably received into the bosom of Nature".'

Presented with verve and conviction, this spectacular plan seemed to have grown from the very bedrock of Vancouver's unique circumstances. But its roots lay elsewhere. Mawson's 'great composition' of axial lines, closed vistas, and 'ordered balance and symmetry' embodied the principal ideas of fashionable planning thought, commonly known as the City Beautiful movement. Drawing inspiration from the 'Great White City' designed for the Chicago World's Fair of 1892 by Daniel Burnham, and the same designer's comprehensive 'Plan for Chicago,' published in 1909, the movement captured the interest of many Canadians between 1910 and 1914. Less a coherent planning philosophy than an enthusiasm for visual coherence and a certain classical aesthetic articulated in the construction of civic centres, parks and parkways, public squares, and monumental bridges, the movement spawned more or less ambitious plans for several Canadian cities and influenced the design chosen for the new University of British Columbia campus on Point Grey in 1910. It also inspired a number of Vancouver's leading citizens to form the City Beautiful Association, with the intention of developing a civic centre for the city. Yet it left little direct mark on Vancouver. The economic depression of 1913 undercut the optimism that Mawson had found in British Columbia, and the First World War turned planning energies in other directions. Mawson's grandiose blueprint for Coal Harbour was never realized. Although a design competition produced an elaborate plan for a city hall, public library, technical college, museum, art gallery auditorium, and public spaces in the area bounded by Pender, Robson, Beatty, and Homer streets, nothing came of it (Figure 28). The Vancouver City Beautiful Association changed its name to the Vancouver Planning and Beautifying Association and spent at least part of its energies, early in 1914, on a campaign to make Burrard Street attractive to 'the hundreds of thousands of tourists and visitors' (expected with the opening of the Panama Canal) by planting shrubs and flowers. By the fall it was struggling to meet the very modest costs of its annual 'Garden Competition.' Furthermore, local architects Sharp and Thompson were forced to trim their designs for the university before construction began, with the result a much more mundane plan than the original. Although the aesthetic concerns that excited Canadian urban designers in the heady years after 1910 would never entirely disappear, the utilitarian spirit of tougher economic times was more pervasive after 1913, and wide roads, fine groups of buildings, large open spaces – the essentials of City Beautiful enthusiasm – were increasingly regarded as dispensable luxuries of urban life.

Figure 28

Plan for a Vancouver Civic Centre, 1913

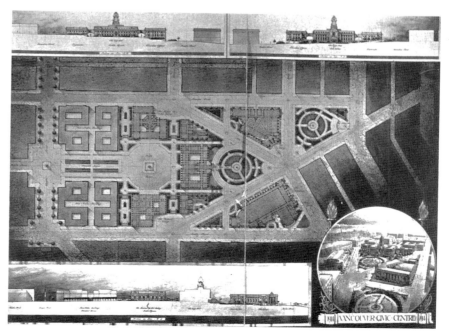

Efficiency

Thomas Adams, first president of the Town Planning Institute of Great Britain and, from 1914, Town Planning Advisor to the Canadian government's Commission of Conservation, was the herald of this new outlook. Called to Vancouver soon after his arrival in Canada to judge the civic centre competition, he made it plain that he considered the development of a comprehensive plan for Greater Vancouver a far more urgent priority than the design of ceremonial spaces. 'In the first place,' he wrote, 'a city is a manufacturing or administrative centre and the first concern of a town plan should be to provide for the proper and efficient carrying on of business.' Vancouver and its region suffered 'in a special degree from haphazard growth and speculation in real estate.' He argued that a commission should be established to prepare a topographical map of the area, to survey existing conditions, and 'to suggest a comprehensive scheme showing the best lines for main arterial roads, desirable railway and harbor improvements, suitable industrial areas and general provisions for convenience, amenity and proper sanitation.' With this done, city administrators would realize how money could be saved and would be in a position to spend wisely when spending became necessary. In Adams's view, successful planning required that all land 'be developed for efficiency, convenience, health and amenity.' It was, essentially, a bureaucratic process, best left to experts who could apply 'sound economic principles...to the administration of civic business.'

These ideas quickly took root in Vancouver. Local businessmen formed a Civic Improvement League in 1916 and established a Civic Bureau within the Board of Trade before the end of 1917. Soon, activists were lobbying for a Town Planning Act (to give legislative muscle to their reform concerns) and seeking abolition of the ward-based system of city government in favour of a small Board of Control elected at large (and intended to make civic administration more efficient and businesslike). Neither initiative won immediate success. Early in the 1920s, however, municipal councils (including Point Grey and South Vancouver but not the city, which had its own charter) were given authority to implement a form of land-use zoning by establishing housing densities, regulating noxious industries, and designating areas for residential, commercial, or industrial use. At the same time, local enthusiasts – many of them successful businessmen – established a Vancouver branch of the Town Planning Institute of Canada. Largely through their efforts, the Town Planning Act of British Columbia was passed into law in December 1925. Many advocates of planning legislation were disappointed by the act, which failed to require planning and left considerable discretionary authority with elected politicians rather than vesting power in autonomous appointed boards. But, in most respects, it echoed the Town Planning Institute's view of the planning process as 'the scientific and orderly disposition of land and buildings...with a view to obviating congestion and securing economic and social efficiency.'

Within months, Vancouver and Point Grey established advisory planning commissions. Intended to address questions of public health and safety, functional efficiency, and prosperity, but with much of their membership drawn from the commercial community, these bodies quickly made the maintenance of property values one of their central concerns. In a series of joint meetings, they agreed upon the importance of a large-scale city plan for rational and efficient development and recognized the need for expert assistance in its preparation. Calls for tenders were placed in newspapers, and several Canadian and American planning firms – almost all of them known as 'city efficient' planners – were approached. By August 1926 the prominent St. Louis, Missouri, firm of Harland Bartholomew and Associates had been hired to produce a comprehensive plan for Vancouver.

Following well-established procedures that paid little attention to the particular characteristics of the city, but which sought to imprint a set of 'desirable' features upon its fabric, the Bartholomew company moved quickly to formulate its recommendations for Vancouver. Convinced that a proper city plan addressed six topics – the major street system, transit arrangements, rail and water transportation, public recreation, zoning, and civic art – the planners prepared reports on each. Time and again, existing patterns were compared with the ideals defined by Bartholomew's planning principles, and strategies for improvement were suggested (Figure 29). According to Bartholomew, Vancouver needed new and improved radial and crosstown thoroughfares that focussed circulation on the downtown core. Because accessibility and need were best served by a hierarchy of public recreation facilities, each square mile of Vancouver residential territory should include four one-acre playlots, a three-acre elementary school playground, six acres of recreation fields, a neighbourhood park (of twenty to thirty acres in extent), a mile of 'pleasure drive,' and thirty to fifty acres of outlying park (Figure 30). To ensure an orderly, efficient use of space, zoning ordinances should orchestrate patterns of land use and give clear shape and coherence to the urban area (Figure 31). According to Bartholomew, Vancouver should be much the same as the other cities that he had planned, with a commercial core surrounded by successive rings of apartments, two-family homes, and single- family dwellings. To facilitate local shopping, commercial and multiple residential uses might be allowed along major streets, and smaller commercial areas could be scattered through single-family districts. Industrial uses would be confined to those areas with good rail and water access. Finally, the planners proposed means by which the city might be made more pleasing to the eye. Here their suggestions ranged from the construction of 'a great civic centre' through the design of streets, curves, and curbs to proposals for improving the apppearance of 'home grounds' (Figure 32).

Figure 29

'Administration of Play & Recreation Grounds.' Bartholomew and Associates' view of the changing circumstances of family life in the early twentieth-century city

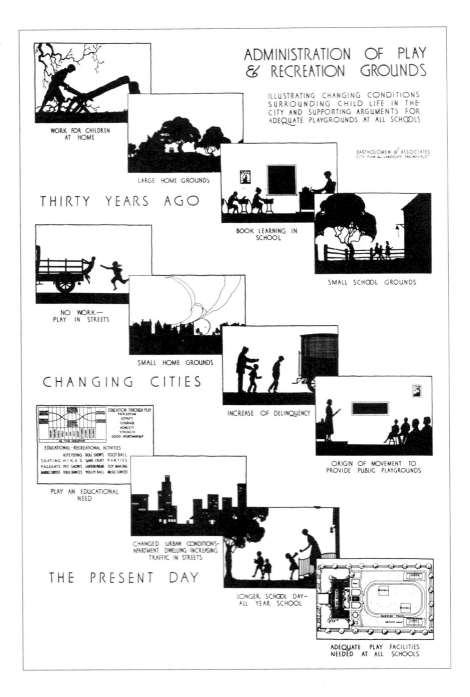

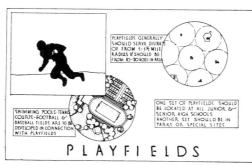

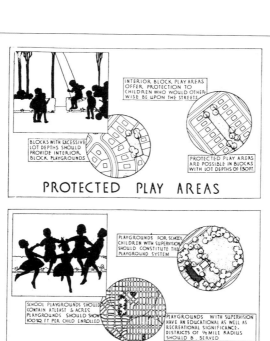

Figure 30

A hierarchy of recreation facilities for city dwellers. Bartholomew and Associates' scheme for the provision of playgrounds, parks, community centres, and pleasure drives in the city

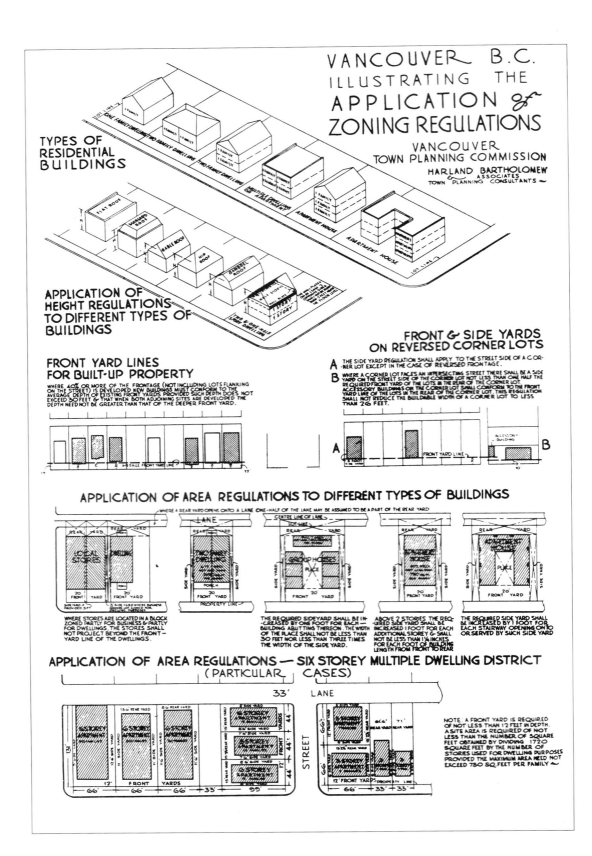

Figure 31 (left)

Zoning for order and efficiency. Bartholomew and Associates illustrate the details of preferred zoning practices

Figure 32 (right)

Making the city more efficient and pleasing to the eye. Bartholomew and Associates replot streets in response to topographic and traffic needs, and define a landscape aesthetic

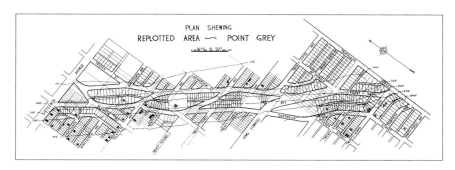

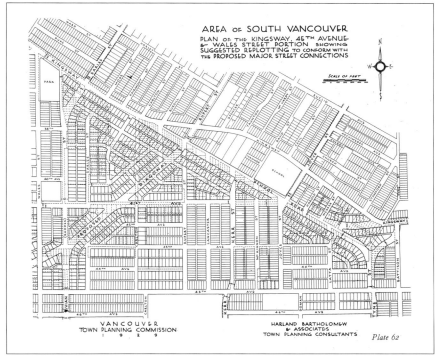

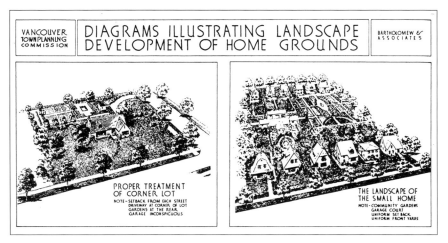

Submitted to the planning commissions of Vancouver and Point Grey in December 1928, and followed by a similar report on the former municipality of South Vancouver, commissioned after the amalgamation of that area with Vancouver and Point Grey in 1929, Bartholomew's plan was a remarkable document. In many respects rather mechanical and echoing the recommendations of earlier Bartholomew designs for other cities (in some places word-for-word), it was never entirely adopted by city council, and yet it left a firm mark upon the fabric of the city. When the councils of Vancouver and Point Grey implemented Bartholomew's zoning recommendations in the late 1920s, they established the basic lineaments of urban land use over the next several decades. Amended only slightly with the amalgamation of 1929, they shaped decisionmaking well into the 1950s, and their imprint is still evident today (Figure 33). Further, although the economic depression of the 1930s meant that other aspects of Bartholomew's plan were shelved, they were not forgotten. At the end of the decade, several of the plan's recommendations were advanced again by the Town Planning Commission, and in the years since, several elements of Bartholomew's vision have informed decisions about the development of urban space. Today, the network of major arterial and crosstown roads

Figure 33

Imposing the lineaments of land use. A commercial core surrounded by apartments, duplexes, and single-family homes, with industry tied to wharves and rail yards, and commercial ribbons for local shopping. Bartholomew and Associates' zoning plan for the city in 1930

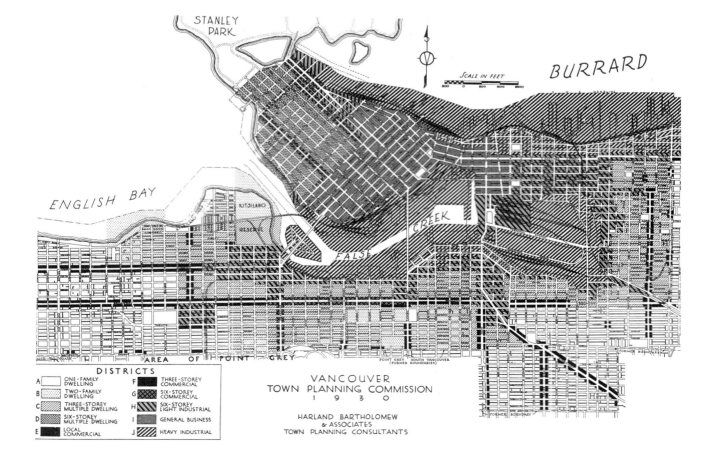

in the city closely resembles that outlined in the comprehensive plan, and the distribution of playing fields, parks, and community centres in Vancouver bears a fair resemblance to that suggested in 1929. In these ways, modern Vancouver bears the mark of early twentieth-century debates about what a city ought to be and manifests in some degree the triumph of utilitarian over aesthetic planning ideas during these years.

Yet, if few early twentieth-century Vancouverites wanted much more than prosperity and efficiency for their city, there were always other voices ready to articulate a different vision. F.E. Buck, an English immigrant, professor of Horticulture at the University of British Columbia, and founding member of the Vancouver branch of the Town Planning Instiute of Canada, held fast to the ideals of the English Garden City movement, made his cause the provision of single-family housing for all, and argued that planning had a sociological rather than a business purpose. More pointedly, during the 1930s, Alfred Buckley of Grandview toyed with the idea of speaking to the local CCF club on 'The Impossibility of Ideal Town Planning under a Capitalist Regime.' Similar sentiments have echoed down the years and can still be heard today, as Vancouverites continue to wrestle with the tensions between nature and development, betterment and beauty that Thomas Mawson identified for his listeners in 1912.

Insiders and outsiders

In early Vancouver, a rapidly expanding city of newcomers, many found it easy to assume that an essentially egalitarian society was taking shape on the edge of the Pacific. Promoters and visitors offered stories of working men who arrived in Vancouver with few savings only to build comfortable homes within a year; they wrote with great frequency of the city's 'beautiful bungalows with flower gardens and lawns,' and they encouraged the belief that hard work and perseverance were all that was necessary to ensure the prosperity and security of the individual. By these accounts, every man and woman could anticipate a fair share of the city's magnificent opportunities. In many respects their visions of Vancouver mirrored that of Ethel Wilson's character Aunt Topaz in *The Innocent Traveller,* who recalled the city's West End as a place in which 'the houses all had wooden trimmings and verandahs, and on the verandah steps when day was done the families came out and sat and talked and counted the box pleats on the backs of fashionable girls' skirts as they went by; and visitors came and sat and talked ... And then they all went in and made a cup of cocoa. It was very pleasant and there seemed to be no trouble anywhere upon the face of the earth, that you could discern.'

Others were less sanguine. Harry Archibald, a Nova Scotian who came to Vancouver early in 1910, was surprised to find that the city was 'full of Socialists,' and that they were 'nearly always foreigners – Englanders,

Germans, Americans etc.' He was also troubled by the city's 'remittance men,' English immigrants who assumed their superiority to others and believed that they were, as Archibald expressed it, 'the whole cheese.' Indeed, the people of early Vancouver were far from homogeneous; substantial numbers who were of Japanese, Chinese, and central European origin might have been added to Archibald's list of those who were not Anglo-Saxon Canadians like himself. By 1911, the city, South Vancouver, and Point Grey included more than 30,000 people – a quarter of the population – whose 'racial origin' was other than British; among them, Chinese, Scandinavians, and Germans numbered approximately 3,500 each, there were some 2,500 Italians, Japanese, and French, almost 1,000 Jewish people, and nearly 750 'Hindus.' Moreover, Vancouverites were divided along economic, ideological, and social lines in far more complex ways than Archibald suggested. Racial, ethnic, and economic divisions were reflected in patterns of spatial segregation within the city, and language, politics, and prejudice maintained meaningful, if not impermeable, boundaries between the different elements of Vancouver's social fabric. Individual opportunities to participate in the life of the city were largely configured by group affiliations within this mosaic and made Vancouver a far more pluralistic place than the boosterism of the times and an essentially suburban residential form implied.

'Wearers of purple and fine linen'

New and fluid though it was, the population of early Vancouver included a powerful and conspicuous élite. Little more than a decade after the incorporation of the city, the area between Granville Street and Stanley Park was described as 'that special portion of the city where dwell the "upper ten," the wearers of purple and fine linen, those upon whom the smiles of good fortune have fallen.' Here, enthused the *Province* newspaper in November 1900, stood 'the stately homes of the Terminal City, all and singular presenting externally the chief attributes of prosperity – elegance, artistic refinement, and luxurious comfort.' By 1908, the city had its *Elite Directory,* which rather ostentatiously certified the status of those considered worthy enough for inclusion in its pages; seventeen of every twenty lived in the West End.

For a quarter of a century after 1886, this was the city's most prestigious residential neighbourhood. Here, ever more extravagant and expensive houses proclaimed the wealth and importance of their occupants. The equation was clear, but the *Province* made it explicit. Just as a person's clothes provided 'a very fair index' to their character, so the homes of the élite provided a useful guide to the 'tastes, habits and aspirations' of their occupants (Figure 34). The mansion built by company president R.V. Winch in the economic recession of the 1890s included ten tons of steel beams, antique marble bathtubs, mahogany staircases, and solid oak billiard tables. A decade later, B.T. Rogers' Davie Street home

Figure 34

Mansions all in a row. The West End, 1906. Looking east on Burnaby Street, from Jervis

'Gabriola' was built of stone brought from the island of that name. It had eighteen fireplaces, a fine wood-panelled interior, and the first concrete basement in Vancouver. In this fast-developing city – where fortunes and reputations could be made overnight – large, architecturally striking homes in wide gardens commanding splendid views of English Bay or Burrard Inlet were highly effective signifiers of status. Together they constituted a closed and controlled social world, an enclave sheltered from the social and spatial discord of the tumultuous city beyond. Here the city's successful families found privacy for their families and forged a community of the 'well-to-do.' Following the lead of élites in late nineteenth-century London, New York, and other important cities, they adhered to formal rules of etiquette to structure social interaction, fostered group cohesion by establishing private clubs and schools for themselves and their children, and laid the foundations upon which a sense of individual and collective attachment to place could be built in the fast-changing city.

By the first decade of the century, the rituals of proper social interaction were well established. Garden parties filled the summers, allowed friends, peers, acquaintances, and newcomers to mingle and meet, and confirmed the style and status of the hostess. According to the *BC Saturday Sunset* magazine of 12 June 1909, 'the first garden party of the season' was given by Mrs. Sulley (whose husband was a successful manufacturer and financier) 'at her lovely home on Davie Street.' Capturing the ambience of comfort and refinement that marked such occasions perfectly, the magazine noted that 'the grounds were looking their very best and with plenty of garden chairs and gaily decorated tables placed about, a prettier sight would be hard to imagine. Many beautiful gowns were shown to advantage on the smooth green lawn ... The polished table was exquisitely done with lilies of the valley in silver vases on billows of chiffon ... Harpur's orchestra gave great pleasure to the guests.'

In similar vein, society hostesses also held 'At Home' days. These semiformal occasions allowed members of the city's élite to visit one another without a specific invitation, according to a schedule printed in the Elite Directory. Generally, guests were treated to tea, cakes, and conversation in the drawing room. Careful orchestration of the timing of these events ensured that they served to knit together the larger community of the city's élite. So residents of the original core of the West End (on the bluff north of Georgia Street) were 'At Home' on Friday, those west of Denman on Monday, and those south of Davie (overlooking English Bay) on Tuesday. Thus, residents could maintain contact with friends and former neighbours who lived at a distance and further their acquaintance with others from different parts of the élite district. West End homes served, in short, as centres of social interaction, information flow, and status reinforcement.

Private clubs served the city's leading businessmen in similar ways. Side by side on West Hastings Street, the Vancouver Club (which drew almost three-quarters of its 1908 membership from the West End) and the Metropolitan – later the Terminal City – Club (with a similar proportion from the West End and downtown) were important venues for daytime meetings of the city's leading entrepreneurs and financiers. In addition, the Vancouver élite created relatively exclusive social clubs for recreational pursuits. So the original, rather informal, Vancouver Lawn Tennis Club became a private organization in 1897 and developed a club house, croquet lawn, and grass and cinder courts on Denman Street. Six years later, the Royal Vancouver Yacht Club appeared. In the next two decades its membership produced two civic mayors and a score of presidents and vice-presidents for the Board of Trade and the Vancouver, Terminal City and Canadian clubs. By 1910, Stanley Park was home to cricket and rowing clubs, and the Vancouver Riding Club had a building near its entrance.

Private schools and voluntary organizations were other important elements in the picture. Intent on imparting social skills as well as book learning to the young, private schools paid special attention 'to the manners and general bearing of pupils,' sought to maintain a high standard of 'proficiency and discipline,' and, at Miss Gordon's establishment (later Crofton House), attempted to preserve a 'healthy tone' among the girls. To attend one of these institutions was to gain a heightened sense of one's place in society and to have the opportunity of establishing friendships with children from other wealthy families. Similar reinforcement of the webs of familiarity and support among the Vancouver élite was provided by the participation of its members in voluntary and charitable organizations. Linked in their corporate affairs by interlocking directorships, the leading businessmen of the West End also served, with their neighbours and fellow members of their clubs, as governors of the Vancouver General Hospital and as officers of the Board of Trade and Canadian Club. At the same time, their wives joined Ethel Wilson's Aunt Topaz in the activities of the Ladies' Aid, the Women's Auxiliary, the Women's Council of the YMCA, the Victorian Order of Nurses, the Anti-Tuberculosis Society, and the Ladies' Minerva Club.

However self-contained this world of 'the progressive euchre party, the stag dinner, the ice cream and souffle after the ball' was, its close-knit geographical expression soon proved vulnerable to the very forces that had created it. Built from the profits of urban expansion and commercial acumen, and located on the edge of the rapidly expanding central business district of the city, the prestigious West End succumbed to the intensifying pressure for economically efficient land use. By 1910 apartments and commercial premises were being built in the low-density neighbourhood. Three years later there were 50 apartment blocks west of Burrard Street; most were three-storey 'walk-ups' with fewer than 20 suites, but two had over 70 suites each (Figure 35). Between 1908 and 1914, approximately a third of the West End élite moved out of the area. Many realized substantial returns from the sale or redevelopment of their West End properties and were able to re-establish themselves in new, and sometimes even more commodious, mansions in suburban neighbourhoods distant from downtown. Shaughnessy Heights attracted fully 60 per cent of those who left the West End between 1908 and 1914; others went to Kitsilano and Point Grey (Figure 36). Quite typical were realtor and manufacturer W.L. Tait and hardware merchant W.C. Stearman. Tait demolished three single-family homes to build a 54-suite apartment block across the lane from his Thurlow Street home and then left the West End for the ornate mansion 'Glen Brae' in Shaughnessy; Stearman, a resident of Barclay Street, led the movement to western Kitsilano when he occupied a grand new house on the crest of the hill at First Avenue and Larch Street. When the *Vancouver Social and Club Register* appeared in 1927, barely one in five of those listed in its pages lived in the West End. Never again would the Vancouver élite display

Figure 35

Apartments, infill housing, and commercial properties crowd the West End. The view from the Hotel Vancouver in 1939

the high degree of spatial concentration that emerged between Burrard Street and Stanley Park before 1905. But dispersion did not destroy the group's socioeconomic cohesion. The Vancouver and Terminal City clubs continued to serve the expanding downtown business community. Many organizations founded in the West End – the tennis club, Crofton House school – moved to the suburbs with their patrons, and new, but essentially similar, institutions appeared to accommodate the most 'successful' of the growing number clamouring for inclusion in the inner circles of power and status as the city expanded.

Across the Great Divide

In 1921, the psuedonymous Hilda Glynn-Ward exposed another side of the city in *The Writing on the Wall*. Reflecting the xenophobic excesses of the postwar years, and directed toward the same ends as the recently revived Asiatic Exclusion League, this lurid novel decried the growing Chinese presence in the city and identified 'a great divide' between the Orient and the Occident. In Glynn-Ward's view, this division was racial and cultural, 'intangible and insurmountable' because 'Orientals' could not be 'brought to live by the customs and laws of the European.' But it

Figure 36

Mansions, coach-houses, and fine formal gardens. Shaughnessy Heights in the 1920s

was also engraved, metaphorically, in the landscape of the city, where the downtown core set two worlds apart. To the west and south the population was predominantly British and overwhelmingly European; to the east, on low-lying land between the Hastings sawmill and the mud flats of False Creek, clustered most of the city's 6,500 Chinese (and a substantial proportion of its 4,000 Japanese) residents.

First brought to British Columbia by the gold rushes of the 1850s and the construction of the CPR in the early 1880s, Chinese immigrants had settled on Burrard Inlet and the banks of the lower Fraser in considerable numbers before the incorporation of Vancouver. In 1884 there were over 1,600 in New Westminster and slightly more than 100 in the vicinity of the Hastings sawmill. By the end of the decade there were 29 Chinese businesses in Vancouver, all but a handful of them concentrated in the area known as Chinatown, near the intersection of Carrall and Dupont (Pender) streets. As the city expanded, so did its Chinese population. Although Chinese immigrants had been required to pay a tax of $50 on entering Canada since 1885, and that levy was doubled then raised to $500 early in the new century, between 2,000 and 4,000 Chinese entered Canada in most years of the 1890s, and annual arrivals

exceeded 6,000 in the boom years before the First World War. Soon there were more Chinese people in Vancouver than in New Westminster; by 1901 the city and its neighbouring municipalities had 2,840 Chinese residents, and Vancouver's Chinatown quickly supplanted Victoria's as the main centre of Chinese settlement in British Columbia.

Turn-of-the-century Chinatown was an overwhelmingly male place (Figure 37). In a population of 2,100, women and children were outnumbered forty to one. Family life was the privilege of a few merchants and a small handful of labourers. Most men lived in tiny quarters in narrow wooden tenement buildings; in 1900 a hundred or so boarding houses accommodated 1,500 men in the Dupont Street area, and several more were built as immigration swelled the population of the district. By 1910, the dense mix of commercial and residential land uses marking the confined territory of Chinatown encompassed four city blocks between Hastings and Keefer streets, Westminster Avenue (Main Street), and the narrow lanes to the west of Carrall Street known as Canton and Shanghai alleys. Here, wrote a visiting Chinese statesman in 1903, the majority of his immigrant countrymen were in 'distressed and cramped' circumstances, their situations 'pitiable beyond description.' Perhaps one in four worked in the city's lumber and shingle mills; others toiled in the city's hotels and the homes of its wealthy citizens as cooks, servants, and cleaners; substantial numbers were employed in Chinese laundries and tailoring firms; and some were itinerant peddlers of vegetables to Vancouver homes. Most lived on the economic margins. A 'small room and bunk, a table, a stove and long working hours, interrupted by gatherings with their countrymen in the lodging house common room' – these, according to one recent study, were the marks of daily life for most residents of turn-of-the-century Chinatown.

Circumstances were very different for a small group of Chinese merchants, however. Visiting one of their homes in 1910, a reporter described a 'beautifully neat' house with stained glass doors that opened onto a 'sheltered balcony,' 'carved wooden cabinets' and tasteful furnishings exhibiting a 'combination of Oriental and European ideas,' and a large print of Confucius. In 1907, four businesses – Sam Kee, Gim Lee Yuen, Lee Yuen, and Hip Tuck Luck – earned between $150,000 and $180,000 a year from the import and export trades, real estate investments, labour contracting, and other ventures. Another five returned some $66,000 to $85,000 annually, and there were 16 businesses with earnings of $31,000 to $55,000 each year. Some 36 smaller firms, yielding annual returns in excess of $1,000, made up the base of the small pyramid of relatively prosperous enterprises that provided a wide range of goods and services to Vancouver's Chinese.

Early in the 1920s, approximately 6,000 people lived in Chinatown. New buildings with deeply recessed balconies and other distinctive stylistic features – many of which remain today – lined Pender Street between Canton Alley and Gore Street (Figure 38). Chinese theatres and schools, a hospital, a library, and the premises of several clan, county, and benevolent associations were interspersed among the restaurants and dwellings of this largely self-contained community. Another thousand or so Chinese lived in Point Grey and South Vancouver, and others were located in Fairview, Mount Pleasant, and Grandview, where they worked as domestic servants, cultivated market gardens, or operated laundries and grocery stores.

Resistance to their presence was growing, however. In 1919, Grandview merchants agitated for a bylaw to prohibit the granting of business licences to Orientals; the already hefty $50 licence fee demanded of peddlers was raised to $100 (when retail stores paid $10); and unwritten agreements effectively excluded Chinese people from land ownership in Point Grey, Shaughnessy, and parts of east Vancouver. (Later, similar practices were followed in the University Endowment Lands and the North Shore's British Properties.) In 1923 the federal government responded to the strong lobby of British Columbian MP's by approving a new Chinese Immigration Act that effectively precluded further Chinese immigration. Through the late 1920s and 1930s an aging and, after 1931, shrinking Chinese community turned inward. In the harsh

Figure 38

Chinatown in the early twentieth century. Recessed balconies, commercial properties, meeting rooms, residences

years of the depression at least 175 customers of the city's Chinese soup kitchens died, and evidence that deaths were twice as common as births among the Chinese led the *Province* to observe, in October 1941, that the 'City Chinese Face[d] Racial Extinction.' Rejuvenation came only with the repeal of the Immigration Act in 1947 and a liberalization of the regulations governing the entry of Chinese people to Canada in the 1950s.

Through all of this Chinatown had several faces. For its residents, it offered familiarity, support, and shelter in a strange and often hostile environment. According to Won Alexander Cumyow, a Canadian-born court-interpreter, the chief reason for the concentration of Chinese in one part of the city was 'companionship,' because they knew that 'the white people have had no friendly feeling towards them for a number of years.' So marked was this antagonism that the spatial concentration of Vancouver's Chinese population might legitimately be seen as a result of necessity rather than preference. 'Their tendency to congregate' observed one Supreme Court justice, 'is directly owing to the fact that as foreigners, held in dangerous disesteem by an active section of whites, they naturally cling together for protection and support.'

However its residents regarded it, Chinatown held a very different place in the minds of most Vancouverites into the 1920s. For them it was a 'counteridea,' as much an intellectual construct as a physical reality, something to be contrasted with their own (ideal) society in order to illuminate and reinforce favoured norms of propriety and conduct. Reflecting a widespread conviction that Caucasians had moved further along an evolutionary path from barbarism to civilization than all other races, and the corollary belief that 'higher' races were vulnerable to contamination by the 'lower,' many residents of early Vancouver regarded the Chinese (inaccurately, but nonetheless anxiously) as an inferior and only partly civilized people, whose distinctive quarter of the city was 'an eyesore to civilization,' an unsanitary and immoral blight on the landscape. 'Conditions prevailing in the cities of China are familiar topics of the returned missionary, who will dwell at length upon the awful conditions of the slums, the armies of the unwashed, and the prevalence of vice in the shape of opium smoking and gambling,' claimed the *World* newspaper of 10 February 1912, before asking, rhetorically, 'Would you believe that the same condition of affairs is in existence in the city of Vancouver...?'

For several decades after 1886, Chinatown and its inhabitants were subject to the consequences of these perceptions. Time and again, medical health officers lamented the 'increasingly dilapidated and filthy' character of Dupont Street dwellings, described them as 'dangerous to the health of the city,' and successfully sought their demolition. Protests that pointed to worse conditions in some areas occupied by white residents were ignored, and officials continued to reinforce the exotic, undesirable image of Chinatown by fining its residents for the 'fowls, refuse, filth, dead dogs, and offal' in its streets, despite the ongoing efforts of several merchants to improve on the refuse collection service afforded the area by the city.

Moral anxieties about Chinatown were even more pointed. Early in the century, the location of the city's red light district in the vicinity of Chinatown gave issue to claims that the Chinese conducted a regular traffic in women, and that they were 'the most persistent criminals against the person of any woman of any class in this country.' Many years later, in the 1930s, hardline officials intent on 'cleaning up' Chinatown forced several Chinese restaurants to dismiss their white waitresses who, they said, were being 'induced to prostitute themselves with the Chinese.' Protests from women thrown out of work by these actions and indignant at the defamation of their reputations went for naught. Licences could be cancelled without the heavy burden of proof required for criminal prosecution, argued one advisor of the city police. 'A great many things are perfectly obvious ... [but] difficult to paint in Court,' he wrote. 'Suppose...the Licence Inspector makes a personal visit to one of these places and finds loose conduct, such as, a white waitress sitting down with a Chinese. No outward crime is being committed but the chances are that procuring may well be underway.'

By current accounts, Chinatown was also an opium den, the centre of a traffic in habit-forming drugs promoted at frequent 'snow parties.' According to the secretary of the city's Moral Reform Association in 1921, Vancouver had seen 'almost innumerable cases of clean, decent, respectable young women from some of the best homes dragged down ... very, very largely through the ... opium dens in the Chinese quarter.' Gambling was another vice associated in the moral reformers' minds with Chinatown, although it was not restricted to that area or the Chinese, and was opposed by significant numbers within the Chinese community. For many Vancouverites, Chinatown was a place where '"Chuck-a-luk" and other gambling games, were played behind mysterious doors without handles' in rooms that had hidden exits to underground tunnels. Police raids and multiple arrests were common, despite community protests about harassment and claims that many innocent bystanders were being arrested. 'We are subject to indignities and discriminating treatment' beyond the limits of legality, argued angry, 'law-abiding' members of the Chinese Board of Trade in 1905, who maintained that members of the Vancouver Police Force were 'in the habit of going into our stores and rooms where our families live,' although they showed 'no warrant whatsoever' nor claimed any business with them.

Little softened the harsh edges of this racial stereotyping and the oppression and coercion that accompanied it until the depression of the 1930s. Then, attitudes toward the Chinese began to change. In 1931 Professor Henry Angus, a UBC economist interested in the plight of underprivileged Canadians, pointed out that the economic disadvantages faced by minority groups were largely attributable to prejudice and discrimination in the host society, and he argued for the extension of the franchise to all men and women born in Canada as well as more open, tolerant policies toward the Chinese and others. At the same time, organized labour began to recognize that all workers would be served by eliminating the bigotry that made cheap labour of the Chinese. Such changes in the intellectual tide characteristically produce strong undertows, and this was no exception. Vehement opposition to greater acceptance of the Chinese was voiced by civic officials and others who saw the 'Foundation wedge of Communism' in arguments for 'the brotherhood of man.' By 1935, when a federal election was held, left and right were sharply divided over the question of enfranchisement: 'C.C.F. Party Stands PLEDGED to Give THEM the Vote...Insure Vancouver's Future by Voting Liberal' urged one advertisement in the *Province* of 7 October 1935.

In this climate, Chinatown began to assume a new meaning for Vancouverites. No longer expanding and, according to the *Province,* increasingly 'modernized' and 'westernized' in appearance, it was inherently less threatening to all but the most zealous of its opponents. To mark the jubilee of the city, in 1936, city council approved a proposal from the Chinese Benevolent Association for the erection of a temporary 'Chinese village' on Pender Street. Chinese merchants promised visitors a place 'most artistically and becomingly decorated with Chinese lanterns and hundreds of Oriental splendours,' and the local press reported favourably on this quaint place where 'the East lurks in windows crowded with strange merchandise ... herbs and bark and healing root, the properties of which were known when the Dragon Kingdom was young.' A bamboo arch, a pagoda, a 'Mandarin house' (including 'carved and jewelled' furnishings), and a 'Buddhist temple never before exhibited to the Occidental eye' were among the attractions that (according to the *Sun* of 18 July 1936) made city residents 'proud of the energy, initiative and citizenship of our fellow Canadians of Chinese birth or extraction.'

More than this, the jubilee village had called the attention of entrepreneurial Vancouverites to the potential of 'the Chinese community as a permanent tourist attraction.' By 1938, there were regular bus tours through Chinatown, and one authorized guide was hiring residents to 're-enact' the Tong Wars by running through the streets holding rubber daggers covered with ketchup. Two years later the *Province* claimed that 'no other city in Canada possesses a Chinatown which has retained the glamour, fascination and customs of the parent country to the same extent as this little corner of the Far East.' Here tourism flourished in the 1940s. Stores offering curios and other Chinese imports proliferated, along with meat, fish, and grocery outlets. To increase 'the glint of the Orient' in this 'occidental setting,' merchants added neon signs to their premises. By 1947, restaurants once the preserve of Chinese were crowded with others. So swiftly and thoroughly had the popular image of Chinatown been transformed that an account of the area in *Maclean's* magazine of 15 January 1949 bore the ironic title: 'What, no opium dens?'

In this new guise, Chinatown was far more fully and affectionately incorporated into popular conceptions of the city than ever before. Now it was quaint, less a threat than a titillating spectacle, a locale where West and East could meet, albeit essentially on Western terms. But the 'rare fascination' that Chinatown induced in its Caucasian visitors owed more to long-held European views of the exotic Orient than it did to any appreciation of the cultural and material circumstances of everyday life in this particular place. Its appeal depended on a sense of difference and lay in things 'symbolic of customs and traditions that were old when the oldest firs on the coast were seedlings.' Chinatown, enthused the *Province* of 6 July 1940, was 'China itself, age old, mysterious, inscrutable.' It mattered little what lay behind this façade of colourful parades, people who carried 'the look of the East' even when dressed in Western clothes, dazzling neon lights that had no counter-part in the cities of China, and mock Tong wars. Chinatown was a commodity, packaged for European consumption. Its unfamiliarity was its appeal. But beyond the bright play of appearances, its residents remained alien and apart. Although the image of their physical place in the city had been remade, their place in the wider society of the city had altered little. Through the first half of the century, few Vancouverites saw the Chinese community of their city in other than simple, monolithic terms; few appreciated its strength and diversity; few acknowledged the hopes and aspirations of its members. In the symbolic order of a predominantly 'European' city, they were outsiders, their circumstances largely defined (and confined) by the social, cultural, and political hegemony of the host society.

Insiders and outsiders, affluent establishment figures and impoverished oriental immigrants, these were the end points on the spectrum of human experience in Vancouver. Between them – between the man-sions of Davie Street and the bunkrooms of Canton Alley, between the West End and Chinatown – lay the social, economic, ethnic, and physical complexity of the city. Although few individuals and groups in this varied, vigorous, and fast-developing place experienced Vancouver in quite the same way as others, all shared something of the reality defined by these extremes.

For the Japanese, the Sikhs, and the East European Jews who came to the city before 1914, the circumstances of life in early Vancouver were not fundamentally different from those known by the Chinese. Most sought to establish homes and to gain reasonably secure livings. They brought with them distinctive habits and attitudes that influenced their choices of food, work, and recreational activity. They were subject to the biases and animosities of the predominantly Anglo-Saxon majority. And they occupied distinct and discrete quarters of the city, the Japanese on Powell Street between Main and Jackson, Sikhs in Kitsilano near First and Burrard, and Jews in Strathcona (Figure 39). If none of these districts had the conspicuous identity of, or developed along precisely the same trajectory as, Chinatown, each provided support and shelter for group members, allowed other Vancouverites to conceptual-ize the unfamiliar from afar, and gave territorial expression to the intellectual habit of setting minority groups apart from the majority, 'them' apart from 'us' (Figure 40). Each was also seen from beyond in rather simple, stereotypical terms that ignored the social, economic, institutional, and political complexities of the community and rein-forced common racial ideologies in ways not unknown to the present.

For working men and women who built small homes amid the stumps of Grandview or South Vancouver, and others who erected larger dwellings in more expensive areas of Point Grey, circumstances in sub-urban Vancouver differed in many respects from those experienced by

Figure 39

The Rubinowitz & Co. Department
Store, Abbott Street, Vancouver

Figure 40

The Sikh Temple and congregation,
1866 West 2nd Avenue, 1936

residents of Shaughnessy and the West End. Houses were smaller and their settings generally less spectacular; elaborate 'at homes' and garden parties were unknown; few indeed belonged to private clubs or sent their children to private schools. But friends and neighbours helped out when they could. Churches and lodges and ratepayers associations provided opportunities for social interaction and helped to foster a sense of community. If the picnic spots of Stanley Park were relatively far away, they could be reached by streetcar, and children played unsupervised in the long grass of King Edward Avenue and the vacant lots found in most parts of the city into the 1940s.

To be sure, most of those who lived in these 'ordinary neighbourhoods' probably knew more of hard work and financial uncertainty than of leisure and comfort. But however modest the dwellings of these new suburbs, however ragged their settings, however limited the financial resources of their occupants, they were (with their upscale counterparts in the West End and Shaughnessy) elements of a new urban order in which peripheral expansion and land-use specialization gave powerful play to the dream (fostered by the British Garden City movement and the rhetoric of Vancouver real estate promoters) of a single-family home in a semi-rural setting. And this was important, because the suburban cabins, Californian bungalows, and Voyseyesque dwellings scattered across city space were their occupants' castles. Humble though they may have been by comparison with the mansions of the élite, they were tangible signs of success in the general struggle for security and competence. They were, argued one 1911 article, 'the heart, the life, and the index of a city' in which it was possible to believe that property was the key to social and economic advancement and that diligence and frugality would bring their rewards in a piece of suburban land. Thus, these modest houses in still inchoate suburbs helped to sustain a loose partnership between the city's rich and powerful élite and its larger, less-favoured population, whose common aspirations for economic improvement translated into a 'shared enthusiasm for urban expansion and material progress' that would run largely unchecked into the 1960s.

The story of Vancouver and its region through these years of impressive growth cannot, then, be understood solely in terms of the material realities of population growth, land development, transportation improvements, areal expansion, wage rates, prices, and labour market fluctuations. Ideas – whether borrowed, quite consciously, from afar, or growing almost unremarked in the rich soil of local circumstances from the seeds of taken-for-granted traditions – played a pivotal role in shaping the fabric of the city. This was nowhere more evident than in the social realm, where representations of difference (between insiders and outsiders, 'us' and 'them') gave many residents a firmer sense of their identity and place in this dynamic environment, and where widespread acceptance of the rhetoric of opportunity helped to sustain the social contract.

And this is worth remark. For if there is a lesson in all of this, it is not that drawn by UBC historian Walter Sage, who took stock of the past and cast an eye to the future of his city in 1946 to argue that Vancouverites who looked back on 'the sixty years of progress since 1886' might anticipate the future with 'calmness and confidence.' Rather, it is that people shape places (if not always quite as they would wish), and that it is incumbent upon all who would understand this process, or who would exercise the privilege of citizenship intelligently, to weigh the discourse of time and place with care.

5

Primordial to prim order: A century of environmental change

T.R. Oke

M. North

O. Slaymaker

*T*he immensity of the changes wrought in little more than a century by the development of Vancouver are suggested by Figure 1. The main landforms of the setting remain essentially unaltered, but the surficial elements of the coastal forest ecosystem have been all but eliminated in the area now occupied by the city. Verdant forest has been transformed into urban 'jungle', and the hush of the forest has been replaced by the roar of traffic. Bubbling streams are now buried sewers, the forest canopy and haze are now urban canyons and smaze, and the haunts of the cougar and eagle are now the home of the cat and budgie.

Here and there, remnants of an earlier environment, such as the massive stumps found in several city parks, remind citizens of the past. But few people are aware of just how much natural and human forces have changed the setting of Vancouver. This is not to say that most Vancouverites are unaware of, or welcome all of the changes to, the physical setting of their city. Most revel in showing off the sheer magnificence of the landscape to visitors, but there is also growing evidence that Vancouverites are concerned about the city's environment. This concern is reflected in surveys, such as one conducted in 1990 by geographer Walter Hardwick and the Greater Vancouver Regional District (GVRD) which showed that Vancouverites ranked the environment ahead of economic, social, and lifestyle issues among their public worries.

This chapter focuses on the environmental changes that have occurred in the Vancouver region and the processes responsible for the transformation of the original ecosystem. Most of the changes are due to human activity. Materials and organisms have been removed or introduced, and natural processes have been interrupted or supplemented by human activities which collectively fall under the euphemism 'development.'

The biophysical and cultural environments of a city interact. The natural attributes of the Vancouver region were central to its choice as a site for settlement and continue to shape its form through time. On the other hand, as the contrasting images in Figure 1 so dramatically attest, settlers have fundamentally changed the environment. This chapter addresses

Figure 1

Top: depiction by the artist Jim McKenzie of the Vancouver area in 1792. With the addition of a couple of sawmills, houses for workers, and the surrounding forest clearings, this was how the landscape looked in the 1880s. *Bottom:* the same view today

both sides of this interaction but focuses mainly on the second: the human impact. The process of change will be particularly noted, although change usually occurs piecemeal and at small scales, so that the unwanted cumulative effects often 'creep up' on the inhabitants.

The impact of urban development on the environment

A series of 'snapshot' views of the built-up limits of Vancouver over time show how urban development evolved from the original nodes of settlement (Figure 2). The reasons why the pattern of development assumes odd shapes are many. Some are political (e.g., annexation of lands or designation of reserves), but many are the result of physical constraints. The most important of these for Vancouver is purely geometric. Given its coastal location, there are only 18,000 square kilometres of land within 50 kilometres of the city centre (compared with 40,000 for Toronto and 80,000 kilometres for Montreal), and much of that is not suitable for urban use. Generally, cities grow in those directions that present the least geologic or topographic resistance. Starting from its Gastown core, and initially with only a passenger ferry link to the North Shore, it was easier for Vancouver to expand into the southern half of its potential circle for growth. Except for some swampy stream courses, the land formed a good base for building. Soon after the forests were dissected by roads and interurban tram lines, suburban development filled much of the intervening areas. The resistance to expansion presented by Burrard Inlet was decreased by the opening of Second Narrows Bridge in 1925 and, especially, Lions Gate Bridge in 1938. Similar surges in suburban growth on deltaic lands to the south accompanied the construction of larger capacity crossings on the arms of the Fraser River. The earlier agricultural settlement of Richmond also depended on improvement of dyking along the Fraser and the provision of a drainage network.

Although the relationship between land use and terrain is complex, it is clear that largely uncontrolled development has had deleterious consequences, which include deforestation, the pollution of soil and water,

Figure 2

The built-up area of the Vancouver region at different stages of its growth

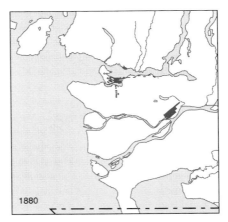

1880

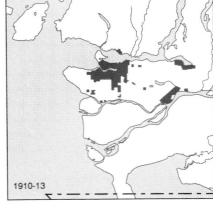

1910-13

erosion, and the release of sediment to adjacent watercourses. Systematic efforts to map land capability for specific uses date only from the 1960s, and resistance to restrictions on land development according to short-term profitability criteria (ease of clearance and construction) remains high, despite evidence that such a strategy frequently carries with it surprising long-term costs as well as environmental decline.

In most respects, the region is only now approaching the physical limits of its growth. To the north, further development up the North Shore slopes must incur increasingly high costs and decreasing attractiveness in terms of more cloudiness, rain, and snowfall. Clearly, the Howe Sound and Indian Arm fjords cannot provide much more land for residential or other growth – nor can the Seymour or Coquitlam watersheds, without compromising their role in water supply for the urban area. To the south, Burns Bog is essentially unsuitable for residential or industrial development, and much of the rest of the lower valley consists of valuable wetland or is held in agricultural reserve. The rate of urban expansion in the GVRD is declining. Some 26 square kilometres of land were converted from rural to urban uses each year in the late 1960s, compared to about 7 square kilometres currently; and barely 100 square kilometres is available for future urban growth. Any increase in population, therefore, must be accommodated on agricultural land, by building further up the lower slopes of the North Shore, or by increasing land-use densities in existing built-up areas.

Environmental change in microcosm

Except for the construction of roads and large commercial and industrial buildings, almost all of the biophysical impacts of urban development can be observed at the scale of the individual house lot or subdivision. Individual decisions affecting small areas accumulate to modify the urban ecosystem.

In the early days of Vancouver, development usually started with the total removal of the forest and ground cover, and with it, a complete

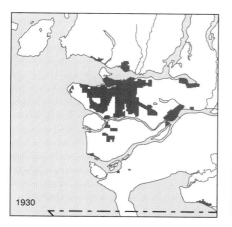

1930

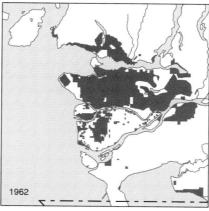

1962

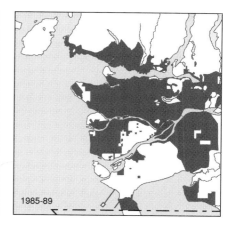

1985-89

Figure 3 (left)

The nature of the forest in 1895. This rich ecosystem teemed with wildlife, including bear, cougar, and deer. The view looks north of what is now the junction of Granville Street and 37th Avenue. This road provided a vital link between the delta farmlands and the developing settlement of Vancouver. Here, the Royal Mail Stage is on its way to Steveston, and a load of hay is trundling in to Vancouver.

Figure 4 (right)

The clearing of the Shaughnessy Heights residential area at Oak Street and 30th Avenue in 1912. This huge pile of stumps was one of many as the area from 16th Avenue to 41st Avenue was totally denuded of forest. This photograph was taken only about one kilometre from the site in Figure 3.

ecosystem. Just how brutal and comprehensive this process was can be appreciated by comparing Figures 3 and 4. The clearing of this land totally altered the biophysical characteristics of the site. The native vegetation was removed or burned, and most of the wildlife fled or died. Devoid of the buffer bestowed by the forest canopy, the ground was exposed to a completely new set of climatic and soil moisture conditions. Before the construction of a dwelling, the tree stumps were removed (Figure 4); the site was usually broken by plough and levelled if necessary; and, if a basement was involved, foundations were dug. These actions overturned the upper soil layers and loosened the soil itself, making it highly vulnerable to erosion and enabling heavy rains to wash the fine sediments into streamcourses.

Before the turn of the century, all the sidewalks and many of the roads were made of wood. Thereafter, they were mainly constructed of concrete with a base of stone (much of it from quarries on Little Mountain). The standard pattern established a grid of city blocks, with houses on opposite sides separated by a common, unpaved back lane. This adherence to geometric, rather than natural, order ensured the disruption of the natural drainage network and its replacement by a complex of storm and sanitary sewers.

To replace the vegetation destroyed by development, trees were planted along many city streets. In the West End, the early plantings were of native species (often dug up in Stanley Park), including vine and big-leaf maple and dogwood. They proved unsuited to urban life: they grew too fast and their root systems buckled the sidewalks. In 1928 Harland Bartholomew & Associates (who were hired to produce a city plan) recommended the creation of boulevards planted with trees of appropriate species. The result is that Vancouver now has about 70 trees per kilometre of road, with the street trees more prevalent on the west-side than they are on the east-side of the newer suburbs. The species planted are also overwhelmingly exotic, with over one-third being cherry or plum trees.

Houses were constructed of local materials, mainly wood but also stone. Generally, residential lots were sufficiently large for gardens to be set out; however, given the displacement of the soil and the destruction of vegetation on these lots, it was necessary to import soil to establish a rich upper layer as well as vegetation to please the largely European residents. Thus, an amazing variety of exotic plants were introduced to the ecosystem which, in addition, created a very different set of habitats for wildlife. Further, the settlers brought domesticated animals and pets to the region and unintentionally introduced new pests and diseases.

The carefully tended garden has become one of the most distinctive features of the Vancouver landscape. Indeed, the Bartholomew Plan even specified the type of planting in front yards in suburban neighbourhoods. The influence of this can still be seen in the less prestigious neighbourhoods: the front path leads up to a bungalow with a centrally located front door flanked by two tapering conifers, which are usually not of indigenous species but are ornamentals such as golden cypress. The more exclusive areas, such as Shaughnessy, were planted with non-native hardwoods prior to the sale of lots, and individual properties were hedged or walled off from their neighbours. Large gardens here and elsewhere showed a clear preference for English garden plants and plans; thus, grass lawns with specimen trees, herbaceous borders, and shrubbery became the standard replacements for the native forest flora. Expensively maintained by gardeners who were predominantly Asian and accustomed to a different gardening tradition, these 'English' gardens took on a neat and orderly appearance, somewhat removed from their original inspiration.

Over time, gardening operations have intensified. The demands of keeping the 'ideal' garden often led to excessive use of fertilizers, herbicides, insecticides, and irrigation. On a summer day, a typical suburban garden receives about 10 millimetres of water from sprinkling. In some locations, the cumulative impact of individual waterings is sufficient to cause water boils to form where the underground flow comes to the surface. Increased use of residential lots for parking also covered a large fraction of the area with pavement or concrete.

As immigrant groups moved into the city, they introduced familiar plants from their homelands. So Italians and Greeks, for example, tended to plant vines, figs, and tomatoes. In the past decade, 'native' gardens, in which indigenous species of trees and ground covers are retained or replanted, have become more common. Although these gardens reflect greater recognition of the natural conditions and heritage of Vancouver, they are hardly a return to earlier times because such gardens did not exist before.

Recently, as a result of the aging housing stock and growing pressure to increase the density of housing on higher valued lots, there has been considerable redevelopment of older residential districts. This process is somewhat different the second time around. An existing structure is usually destroyed, leaving considerable inorganic detritus and rubble to be incorporated into the soil. Modern-day equipment is then brought in, the considerable weight of which compacts the soil. Again, topsoil is easily washed off the site or carried away on vehicle wheels or tracks during the construction phase. Infilling by replacing a house with a much larger one, or with two houses (each built almost to its property lines) tends to eliminate the former backyards and often involves removal of mature trees. This is further exacerbated by a program of paving over back lanes, usually at the request of the residents, whose motives are to make lanes more accessible, neat, clean, and safe.

The net effect of all of these changes, however, leads to impoverishment of the ecosystem. Biological diversity is reduced by the elimination of sources of food and shelter for most of the indigenous animal populations as well as sites for the growth of vegetation. In particular, back lanes provide important corridors for the movement and migration of animals and plants. Trees are host to many insects, mammals, and birds. Moreover, the replacement of natural sites originally covered by soil, water, or vegetation with roads, sidewalks, houses, driveways, gravel yards, and parking bays has many undesirable side effects. Such surfaces are usually fairly impervious, thereby reducing recharge of soil moisture by rainfall. This causes rapid runoff and reduces evaporation, producing a microclimate that is considerably hotter and drier than before. These changes, in turn, lead to the need for irrigation, increase the need for expensive culverts, raise the risk of flood, and decrease the benefits of cooling by evapotranspiration. The new surfaces are also less effective at muffling sounds and reducing glare.

Urban impacts on the biophysical system

Human impacts on the biophysical environment of Vancouver have been diverse and substantial, affecting everything from the configuration of the coastline and the courses of rivers and streams to the character of vegetation and wildlife and the climate and quality of the atmosphere. Figure 5 shows the modern waterline of Vancouver in comparison with the pre-settlement map. Some of the changes would have occurred without human intervention: cliff erosion, longshore drift, channel shifting, and estuarine and deltaic sedimentation are continuous and relentless. But others are clearly human-induced. Dykes built on the Fraser River have had a profound effect. Before their construction, large areas of the low-lying land in Delta and Richmond were inundated every year. This may well have contributed to the difficulties the early explorers had in mapping the area. Dredging operations in the Fraser River started in the early 1900s and still continue to move very

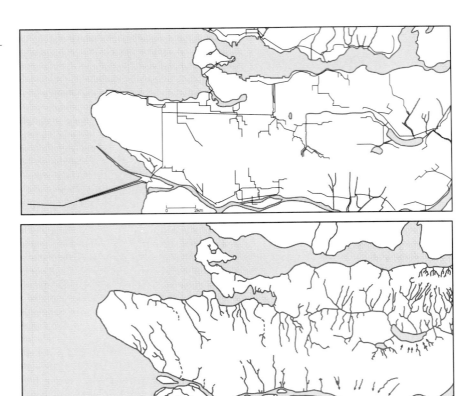

large amounts of sediment. This deepening of channels and harbours altered the natural dynamics of the river's evolution. Approximately 20 million tonnes of sediment, equal to a cube with sides 235 metres long, are carried every year by the Fraser River as it enters the delta. Almost 10 to 25 per cent of this, mainly sand, is dredged from the riverbed downstream of New Westminster to maintain navigable channels. The remainder is deposited on the delta front and in the Strait of Georgia. But because the river is dyked and regularly dredged, it can no longer shift its course on the delta in the way it did before the turn of the century (Figure 6). As a result of the dykes and jetties at the mouth of the Fraser River, most of the delta surface no longer receives sediment and has become inactive. Indeed, parts of the delta front which have been starved of sediment may start to retreat, and the dyked surface of the delta may become more susceptible to severe drainage and flooding problems, especially if there should be a rise in sea level.

The coastline of the Vancouver area has been substantially altered in several other ways. Stretches of the natural shoreline with tidal flats and reed beds were often perceived as a hindrance to water access and became convenient receptacles for dredged material, rubble, and wastes of all kinds. Sites in Burrard Inlet and along the Fraser River suffered this fate in a bid to provide good berthing for boats and to find simple, even 'positive,' solutions to the problem of disposing of large amounts

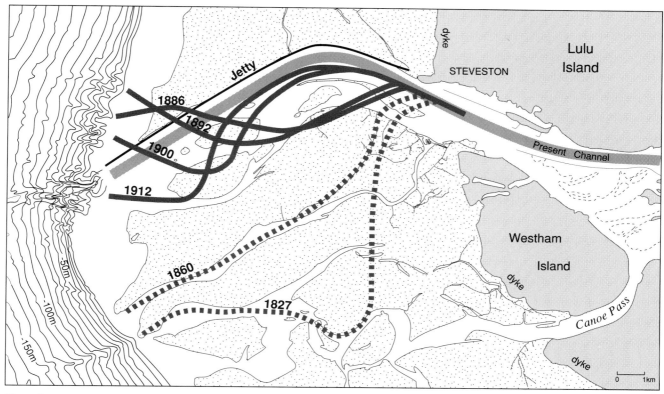

Figure 6

Shifts in position of the Fraser River distributionary channels during the period from 1827 to 1912 and subsequent stabilization of the channel as a result of dyking and jetty-building. Tidal flats and sand bars are stippled.

of solid waste. Dumping has also been used to fill in coastal embayments, build beaches, and fill bogs for industrial sites. The changes in the outline of False Creek are particularly striking (see Figure 12).

Many of the present-day sandy recreation beaches along the coastline from Point Grey to Stanley Park are largely human creations. The natural foreshore is an expanse of boulders, fronted only at low tide by sands and silts. Only three (Jericho, Kitsilano, and English Bay) naturally had pockets of sand between the upper and lower tide levels. Even these, which derived their sand from streams, were eroded and diminished when their drainage channels were filled. The present characteristics of these beaches were largely established through active intervention, beginning as early as 1913, and especially in the late 1950s and 1960s. Large quantities of sand dredged offshore were pumped onto the foreshore, and groynes were built to keep it in place. Still, the sand has to be replenished regularly to keep pace with the rate of removal. All sandy beaches are now continuously tended: the amount of sand is topped up regularly by pumping, mussel beds are scraped off, sand bars are shaved to gentle slopes, and selected drift logs are lined along the beach. The result is an artificial and carefully groomed environment.

Construction of the city required a huge volume of building materials, most of which came from the local region. The abundance of wood made it the prime material for houses in Vancouver, but sand, gravel, and limestone for cement, and rock of all kinds were also needed. Sand, gravel, cobbles, boulders, and mud were extracted from various stretches of the Fraser and other rivers by dredging and from elevated glacier and river deposits in Coquitlam Valley and elsewhere. This required pits and quarries to be dug in many areas of the Lower Mainland as well as on Vancouver Island and the Gulf Islands. Among the most important in the early twentieth century was the quarry on Little Mountain. Between 1906 and 1911 it was a source of shale for road surfacing and basalt for streetcar tracks. Then, after further excavation, the Water Commission converted some of the pits into reservoirs to supply south Vancouver with water from the Capilano Reservoir. In 1928 Little Mountain officially became a park and was increasingly manicured after 1948. The quarry reservoirs were roofed over in 1965 and other pits were transformed into the show gardens of what is now Queen Elizabeth Park.

Countless other excavations and dumpings have taken place. Building foundations, especially for large structures, generate large volumes of surplus subsoil, as do tunnels and road and railway cuttings. Early in the century this material, as well as household garbage and other waste, was dumped in ravines, bogs, swamps, and tidal wetlands that were commonly deemed worthless and unattractive. These sites were also the preferred routes for sewer lines.

Human activity has also triggered landslides. On the North Shore, Port Moody, and Coquitlam, buildings on steep slopes have overloaded the underlying materials, which have collapsed when saturated by heavy rainfall, irrigation, or road drainage outfalls.

The natural network of streams and rivers in Vancouver was also a victim of urban development. Land fill, drainage, and sewer construction have reduced the original 120 kilometres of perennial streams (see Figure 12, Chapter 2) to less than 20 kilometres today. In its place there are 125 kilometres of artificial channels (Figure 5).

As the forest (which holds precipitation for later, gradual release) was removed, and ever larger areas of the city were paved over, the rate and volume of urban runoff increased. Handling it required extensive and costly engineering works such as storm sewers. It also provided a headache for the North Shore. Sewers and culverts built on the lower slopes early in the century now have had to handle increasingly large flows as areas further up-slope, where precipitation is greater, were developed. On many occasions, the system has been unable to cope and water has cascaded down streets, damaging property below.

Changes in the Brunette River basin well illustrate the disruptions of urbanization. In 1900 only the southern part of this 60 square kilometre basin, adjacent to Kingsway, had been urbanized. By the 1930s, there were significant residential developments in Burnaby South, Burnaby North, and Edmonds. During this period, parts of the river

Figure 7

A perfect example of river engineering. The caption to this pair of photographs in the 1949 Annual Report of the Vancouver and District Joint Sewerage and Drainage Board needs no elaboration to illustrate the mind-set involved: 'The upper photograph, taken in 1927, shows a portion of Still Creek west of Burnaby Lake in its natural condition, while the lower shows the creek after improvements were completed in 1935. Such improvements increase the carrying capacity of a natural watercourse manyfold.'

network were 'improved' by straightening and deepening channels to carry runoff more efficiently (Figure 7). After the Second World War, the central part of the basin was filled with warehousing, trucking, and manufacturing plants, many of which located on the banks of Still Creek (the main tributary entering Burnaby Lake), and the population of Burnaby North soared. By 1975, little more than a quarter of the basin remained in forest and open space. Of the rest, over a third was residential; streets and roads accounted for 15 per cent of the area; 12 per cent was in commercial use; 5 per cent was occupied by industry; and 4 per cent was recreational land.

The surface and subsurface hydrology of the basin was totally transformed. Indeed, some reaches of Still Creek are covered over and the 'river' has been classified as a storm sewer by the GVRD. Although shallow, Burnaby and Deer lakes dampen extreme fluctuations in the flow of the rivers, and a dam at the outlet of Burnaby Lake regulates lake levels. Despite this dampening, the flow of the Brunette varies over a huge range: from 0.0028 to 83 cubic metres per second. This is a consequence of surface 'waterproofing' and the construction of storm sewers in the basin.

Runoff from roads and effluent discharges from industrial sources have also badly polluted the Brunette River. In general, concentrations of organic pollutants (hydrocarbons, phenols, and surfactants) in the wastewater are lower than in most cities, but trace metal (cadmium, chromium, copper, mercury, nickel, lead, and zinc) concentrations in the sediments of the Brunette River and Still Creek are exceptionally high and much higher in reaches of the river adjacent to industrial areas than near residential and green space areas. Indeed, the Brunette basin is more heavily contaminated with chromium, copper, lead, and zinc than are most urban drainage basins in North America. And this degradation has led to many fish kills.

With the expansion of Vancouver, much of the native vegetation and, with it, the habitats for most of the region's indigenous wildlife, have been eliminated from the city. There are few remnants of the Douglas fir forest and drier western hemlock that covered the area prior to 1880. The larger mammals, such as bear, cougar, and deer that were once common in the area, are now rarely seen. Many other species of animals, birds, fish, insects, and amphibia have had their essential habitat diminished. Species that are no longer found include the snowshoe hare, Roosevelt elk, wolf, yellow-billed cuckoo, purple marten, western bluebird, horned lark, and burrowing owl. Those threatened include the black bear, sandhill crane, barn owl, and yellow-headed blackbird. The wetland ecosystems, such as salt marsh, seasonal wet meadow, and bog, were almost eliminated. So, too, urbanization has sorely affected the habitats of salmon and trout once found in at least eighteen streams within the City of Vancouver. As long-time resident Thomas Dutton said to city archivist J.S. Mathews in 1955: 'No one would ever dream that during the salmon run, English Bay would be so full of fish that one could figuratively almost cross to the south shore by stepping from fish to fish and still harder to believe that I caught a salmon at the corner of Maple Street and 3rd Avenue.'

Yet Nature did not succumb. Soon after development, certain native species found a niche in the newly developing urban ecosystem. These included the seeds of indigenous vegetation and those animals able to lie low, re-invade the new landscape, or take advantage of abundant food sources. Some species even prospered, such as the ubiquitous rat, mouse, and herring gull. Perhaps, most surprisingly, certain relatively large mammals, such as racoons, and even the coyote, managed to survive the pressures of the city as long as they had vegetation cover to retreat to in the daytime and corridors through the city, such as lanes and rights of way, for nocturnal forays.

To these indigenous species, settlers introduced an amazing array of exotic plants and animals. These were important to the newcomers in economic terms (such as domesticated animals and crops) or in aesthetic and sentimental ways (such as pets, favourite wild animals, and ornamental plants to remind them of home). In addition, other species arrived as unintentional stowaways on carts, ships and trains, and in cargoes of plant materials, furniture, and foodstuffs. Further, although the construction of the city destroyed many habitats, it also created new ones: in and on buildings, and in culverts and gardens. The importation of exotic garden plants, often according to the dictates of changing fashion, provided a cornucopia of species from all parts of the world. Among the most striking examples of this phenomenon was the sudden appearance of monkey puzzle trees in the 1930s.

This wholesale, and largely uncontrolled, introduction of new species caused unexpected impacts on native populations and, often, colonization of much wider regions. A classic example is the European starling. There were none in North America, let alone Vancouver, in 1880. There are now 200 million in North America, and the starling is one of the most common birds in the city. Sixty birds were introduced in Central Park, New York, in 1890. They reached Canada at Niagara Falls by 1915, migrated to Manitoba in 1935, and to British Columbia in 1947. When they arrived in Vancouver, they greatly diminished the population of the less aggressive Japanese starling (crested mynah) from about 20,000 in the 1920s to about 2,000 birds today. These birds were part of the only resident colony in North America – established by crested mynahs that had escaped from the cages of early Chinese migrants in 1897. Squirrels provide another example. The olive brown Douglas squirrel is the only species native to British Columbia. In 1914 the grey squirrel was introduced to Stanley Park as a gift of the New York City Zoo. It has since spread throughout the city and has tended to oust the Douglas squirrel. The European earwig became a pest in Vancouver in the 1920s and 1930s, until the deliberate introduction of a small fly (a parasite on the earwig) reduced its numbers. In recent years, the region's only two native species of slugs and two accidentally imported European species have struggled for dominance. Many subtropical insects, such as silverfish, cockroaches, bed bugs, and weevils arrived in foodstuffs and furniture and found suitably warm homes in buildings. Domestic animals also escaped or were abandoned and became feral species. Most notable are the pigeon, dog, and cat.

The spatial pattern of ecosystems that has emerged in and around Vancouver is extremely varied. It includes a few relatively unmodified remnants of the original forest system in Lighthouse Park and Lynn Canyon Park, and small elements of some of the original coastal, estuarine, wetland, and bog environments. But the largest components are the created systems. These include the near vegetative 'desert' of the commercial core, managed parks with near monocultures of grass, and the rich and varied flora of the gardens of many suburban neighbourhoods. In general terms, the most built-over areas have relatively few species but surprisingly large populations of wildlife, such as starlings, sparrows, gulls, rats, and mice, which live off a wide range of foodstuffs found as waste or in warehouses or after spillage in transport. On the other hand, the garden districts have a greater diversity of species than was present before development, especially along the suburban-rural fringe where the range of native and exotic species and the choice of habitat are amazingly rich.

Each removal of the surface cover of Vancouver, and its replacement with built features, produced a new microclimate; and when sufficient area was developed, it produced a new urban climate. From the standpoint of the atmosphere, the most important change was the replacement of the relatively uniform tree canopy by an extremely heterogeneous surface of buildings, roads, gardens, and parks. This altered the roughness for airflow, the absorptivity for sunlight, the moisture availability for evaporation, and the ability of the system to heat up and cool down. Further, it added new sources of heat, particles, gases, and noise to the air. In general, built-up urban areas produce gustier, warmer, less humid, more polluted, and noisier areas. The effects are subtle and almost impossible to discern in the instrumental records of climate, air quality, and sound from year to year, but the cumulative impact has been considerable.

Although there is a climatic record for Vancouver dating from the turn of the century, data for it were collected at several different sites in the harbour, the downtown district, and (after 1937) at the airport on Sea Island. It is therefore of limited value for determining the climatic impact of urban development. Larger scale climate changes also mask local effects. One way of estimating the effects of urbanization is to look at modern-day differences between the city and its surrounding

Figure 8

The near surface air temperature distribution (°C) in a portion of the Vancouver region on the evening of 22 March 1973. These data are from thermometers at a height of 1.5 metres. Note the way the isotherms follow the density of building with warm spots in the city core, New Westminster, UBC, Marpole, and along major commercial arteries such as Kingsway, Broadway, and 41st Avenue. Cool spots occur in large parks (e.g., Stanley, Pacific Spirit, Queen Elizabeth, and Central) and lakes (e.g., Deer and Burnaby).

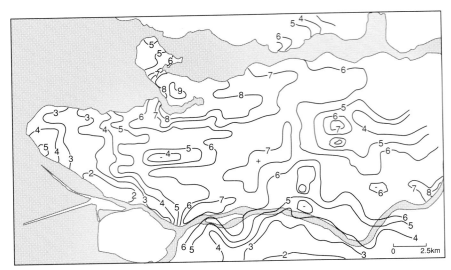

area. This assumes that only urban development is new; and it is only an approximation, because the rural areas around Vancouver have been developed for agriculture. Still, the air temperature pattern shown in Figure 8 shows that the built-up area is relatively warm. This is common to all cities and is called the urban heat island. Satellite images of the thermal pattern of the GVRD show that the city is always warmer than its surroundings. The maximum urban-rural temperature difference occurs on summer evenings during calm and cloudless weather. Based on the geometry and building materials in the city core and the population, it seems probable that the maximum temperature difference between Vancouver and its environs may have reached 9°C by the 1930s. The largest heat island recorded in Vancouver by the mid-1970s was 10.2°C and, in the short distance between Stanley Park and the downtown core, the difference reached 8°C. The surge of high-rise construction in the downtown area in the past twenty years has probably pushed the maximum value even higher. At the time that the data in Figure 8 were collected, conditions were not ideally conducive to heat island development, and its intensity was about 7°C; the Stanley Park-downtown difference was about 4°C. Averaged over all weather conditions and times, the annual heat island is probably of the order of 1.0°-1.5°C, which may not seem very much, but it does mean that the city has about fifty fewer days with frost than does the Fraser delta.

In the early years, Vancouver was a smoky place, especially in winter. Smoke, soot, and ash from wood- and sawdust-burning stoves in houses often hung in the air, which was further sullied by emissions arising from lot clearance, ships, trains, and the lumber milling and drying operations concentrated around Burrard Inlet, False Creek, Marpole, and the banks of the Fraser River. The growth of all types of manufacturing and processing industries in False Creek, and the lack of significant regulation before the 1930s, led to that area becoming extremely heavily polluted with particulates of all kinds.

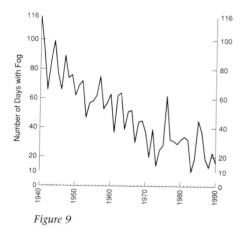

Figure 9

The number of days with fog at Vancouver International Airport, 1940-90

Although there are no instrumental records of early atmospheric pollution, the sediments of Deer Lake provide a buried record of air and water pollution in the region. Recent sediments show the rising number of automobiles by the increase in concentration of lead from exhausts since 1947. By the late 1940s, however, the industrial base of False Creek was in decline and tepee burners for kiln-drying wood were banned in the city. Steam locomotives had given way to diesel, and air quality regulation increased; domestic furnaces were converted to oil or gas, and oil refineries and the Burrard Thermal Generating Plant were constructed at the eastern end of Burrard Inlet. All these changes led to the steady decline of particulate and smoke pollution but to an increase in gaseous pollutants such as carbon monoxide, nitrogen oxides, and secondary photochemical pollutants such as ozone. Ozone has now become the major air quality concern in the region, and although Vancouver is primarily reponsible for this problem, the brunt of the effects are experienced to the east of the GVRD and further up the Fraser Valley as winds move pollutants up-valley in summer.

On the face of it, the elevation of temperature in the city and the reduction of smoke pollution may have contributed to the obvious, relatively steady decline in the number of days with fog in Vancouver (Figure 9). The increasing heat island effect could have helped keep temperatures above the threshold for condensation on some occasions, and the change from particulate to gaseous air pollutants may have reduced the number of nuclei available for condensation. Both explanations are logical, but the full story of the decline is probably more complex. Similar reductions in the incidence of fog are evident at other stations along the BC coast, including places where urban or industrial development is only minor. At least part of this trend is probably due to changes in the frequency of weather conditions conducive to fog formation, especially of high pressure ridges in the fall months. As always, identifying the relative impact of human and natural factors on the environment is not easy.

Case studies of individual urban sites

False Creek

The rapidity and scale of environmental change in False Creek have been arresting. It is difficult to find another example of a site which has moved, in the space of forty years, from an almost pristine forested coastal inlet, rich with wildlife, to a disgusting industrial mess, almost devoid of nature and posing a risk to human health; then, within twenty years, its fortune has swung around and it has emerged, after a period of stagnation and indecision, as a sought-after residential, recreation, and exhibition area.

The creek at False Creek is indeed false; it is an erosional trough (perhaps a former arm of the Fraser River) invaded by the sea. In 1880, except for the small Indian village of Snauq, it was a quiet inlet with

Figure 10

Looking north across False Creek from the forest on the Fairview slope in 1886. This is now close to the site of Granville Street and 4th Avenue, and the sand bar which became Granville Island is visible in the creek. The far shore is near the present line of Richards Street, and the cleared area extends to the skyline at about Drake Street. This picture was taken just after the Great Fire of the same year.

tidal flats and sand bars, fed by several creeks, and ringed by massive forest (Figure 10) with marshy shores. It supported an amazing diversity and abundance of wildlife, including deer, elk, beaver, muskrat, geese, wildfowl, salmon, trout, perch, flounder, and sturgeon. The inlet extended 5.5 kilometres inland to the present position of Clark Drive. At low tide, the eastern half was mostly exposed tidal flats, but at high tide it was connected to Burrard Inlet.

The Main Street Bridge was constructed in 1872 and selective logging occurred in the next decade; but the transformation of the area really began in 1885-6 when the CPR clearcut 480 acres on the north shore of the creek over the entire area of what is now downtown. The arrival of the CPR line in 1887 and the leasing out of the land along the north shore produced a surge of industrial construction in the subsequent decade. On the south shore, clearing and development also proceeded near the Main, Cambie, Granville, and Burrard bridges (Figure 11), especially after 1900. Industrial operations ringing the basin included sawmills, planing mills, sash and door manufacturers, shingle mills, lime kilns, brickyards, cement works, building suppliers, breweries, tanning works, ironworks, foundries, metal works, machine shops, a cooperage, slaughter houses, a crematorium, and (later) shipyards and a gas works.

By 1920 the size, shape, and character of the basin had been transformed (Figure 12). The need for deep-water access led to dredging. The dredged material was dumped back onto the land to fill in large areas of slough and marshy land and to extend land holdings. Edges were filled in by industrial dumping along the shore, which choked off streams,

Figure 11

False Creek and Vancouver as seen from Westminster Avenue (now Main Street) and 7th Avenue in 1889 (top) and 1898 (bottom). The area to the right (east) of the bridge is now completely filled.

Figure 12

Changes in the shape of False Creek in the past century

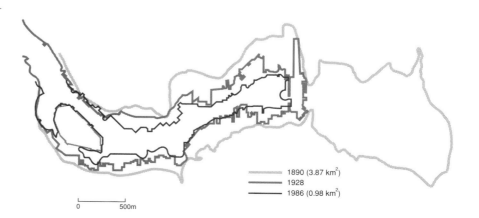

1890 (3.87 km²)
1928
1986 (0.98 km²)

0   500m

Figure 13

The industrial operation of the south shore of False Creek west of Cambie Street in the 1930s

covered marshes, and dramatically changed the shape of the creek. The eastern tidal flats were viewed as a nuisance, an inconvenience to road transport, and of no use as a harbour. They continued to support wild-fowl but were also used as a dump for garbage and domestic sewerage. The flats became a rat-infested, unsanitary swamp and an obvious site for reclamation; by 1918 railway companies had filled in 221 acres to construct their terminals. Another 34 acres were reclaimed by building up the sand and mud bars near the creek entrance to form Granville Island, which was quickly leased to new industry.

By the 1930s the basin was in a dreadful state: a quarter of its foreshore was strewn with dilapidated woodwork, rusting metal, and dumps of every kind of industrial waste. Its soils were contaminated with heavy metals, wood wastes, tannins, lignins, tars, wood preservatives, paints, solvents, ash, slag, manure, and rubble. The waters were a stinking slop pool of industrial effluent and flows from combined sanitary and storm sewers. This was especially true of the eastern half, where tidal flushing was poor, because a raised sill kept the bottom water almost stagnant. The bottom sediments were contaminated with extremely high concentrations of toxic substances, including mercury, lead, cadmium, arsenic, petroleum hydrocarbons, PCB's, and organic debris. The diversity and abundance of marine and benthic organisms plummeted. The air was often foul-smelling, and over the whole basin hovered a murky pall of smoke containing the output of every conceivable combustion and processing operation (Figure 13).

The basin had become a civic horror. Some suggested that it be filled in completely. This had been recommended as early as 1897 by CPR general manager, W.C. Van Horne, and the idea kept re-surfacing through the first half of the century. In the 1950s some remedial action on sewer outfalls and control of air emissions improved conditions slightly, but in the long term the economic difficulties of False Creek industries, the inadequacies of the site, and competition from new facilities elsewhere were more effective agents of change. For a decade or so the area declined. Then in the late 1960s the idea of ousting industry and turning the basin into an attractive area for residential and recreational use took hold, and redevelopment of the south side of the creek began. Later, the north shore and east end became the site of Expo '86. Old industrial uses were cleared away and the shoreline was 'neatened up' by rip-rapping and sculpting. Now, there is no semblance of a natural shore anywhere in the basin, and its total area has been reduced to barely 25 per cent of its pre-development size (Figure 12). There was further dredging to deepen the channel and clean out polluted sediments from the western part of the creek. Some polluted sediments in the eastern part had to be left undisturbed. Several species of fish and wildfowl now frequent the basin. Some attempts to cleanse soil were carried out on the south side, but the rest of the area presents a difficult challenge that will be cripplingly expensive (certainly several tens of millions of dollars) to solve. Cessation of the noxious emissions means that air quality is at least as good as in most parts of the region. The attractiveness and access to park space has made the area a favourite with visitors and residents alike. There is no comparison with the original environment of the basin, but at least many of the worst aspects have been patched over.

Figure 14

Lost Lagoon, about 1870, in what is now Stanley Park. This was a site occupied by a Squamish Indian settler.

Stanley Park

Stanley Park is generally regarded as Vancouver's environmental jewel. Time and again, people have marvelled at the foresight and wisdom of preserving these 405 hectares when the city was only a few years old. The *News-Herald* set the tone at the establishment of the park in 1889 with the comment: 'A city that has been carved out of the forest should maintain somewhere in its boundaries evidence of what it once was, and so long as Stanley Park remains unspoiled that testimony to the giant trees which occupied the site of Vancouver in former days will remain.' But closer inspection reveals that the gem is flawed and shows signs of the piecemeal degradation that is the fate of many urban parks.

The park was selectively logged between 1860 and 1880, and since 1889 its Douglas fir, cedar, and hemlock forest cover (Figure 14) has been substantially altered. New facilities and routes have continued to reduce and divide the once extensive and contiguous forest. In 1890 the first civic sports field was opened at Brockton Point, and the first road and bridge were built to link it to the city. A collection of caged animals was present in the park as early as 1889, and trees were cleared for a zoo in 1912. Further projects throughout the 1920s added bowling and

Figure 15

An isobel (lines of equal decibel level) map of Stanley Park based on daytime readings on summer weekdays. Notice the way the pattern of access breaks up the forest hush, especially the road to the Lions Gate Bridge.

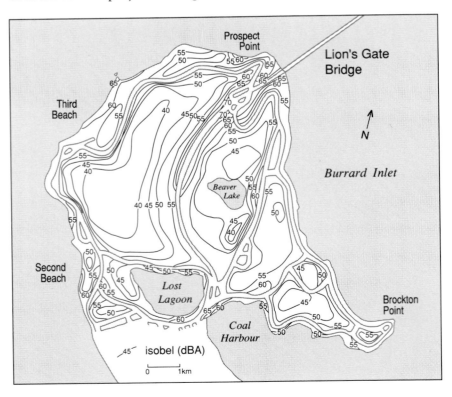

putting greens, a rose garden, and a children's playground. Many of these schemes involved removing trees and, in 1935, a severe February storm removed more. Undoubtedly, the most environmentally damaging decision was to allow the construction of the First Narrows (Lions Gate) Bridge and the road through the park in 1938. The steady deterioration of the trees along the roadside and the constant noise of vehicles (Figure 15) have been an ongoing source of concern. The process continued with the addition of a children's zoo in 1950, the aquarium in 1956, and a petting zoo in 1982, use of which continues to increase. The park has also become the home of unwanted household pets which have become feral.

But as the population of the city increases and the forested area of the park is whittled away, opposition to the loss of vegetation grows. The strong attachment of many Vancouverites to the natural heritage of Stanley Park became evident when the forest-products company MacMillan Bloedel offered to replant it with conifers to establish a typical mature native coastal forest. The plan, which would have cost $1.5 million, involved the removal of a large number of mature deciduous trees, which are only found in disturbed successional forests. Park users especially value these trees for the colourful variation of their foliage and the habitat they provide bird species not found in densely shaded, mature coniferous forests. Opposition to this attempt by professional foresters to 'improve' the park by returning it closer to its original mix of species led to the establishment of Friends of Stanley Park, a citizen's group seeking to formulate a management plan for the park that avoids further erosion of the forested area, yet meets the diverse wishes of all users.

Hastings Park

When Hastings Park was opened by Mayor David Oppenheimer, he envisaged its development as 'a pleasant pleasure resort in the near future, having many equal advantages with Stanley Park. This park must eventually become a constant resort for all lovers of romantic and woodland scenery and lovely groves.' It was not to be; the council allowed it to be developed as an amusement and sports park, the site of an annual fair and horse racing.

On the opening day of the first fair in 1910, the *Vancouver News* noted that 'where a short while ago there was nothing but the densest of woods defying the hand of man to work a change in their primaeval condition, now [stand] commodious buildings.' Soon many more buildings were erected, the fair became a great attraction, and the site more densely developed; in 1948 the site was renamed Exhibition Park. The Agrodome, the racetrack grandstand, Empire Stadium, Playland, the Coliseum, and their associated parking lots covered almost the entire area with buildings and pavement. Today, virtually all of the 162 acres are built over, there is hardly a mature tree on the site, no natural drainage, and the summer microclimate is probably the most oppressive in the city. Recently, the vision of a green park has reappeared in a proposal to convert 102 acres of Exhibition Park to grass and trees and to establish 'natural' streams and lagoons. If it comes to pass, the new landscape will provide another example of an over-developed urban site being returned to a less artificial, but not natural, state.

Camosun Bog

Camosun Bog is small compared to most other bogs in the Lower Mainland; it covers approximately 15 hectares on the eastern edge of Pacific Spirit Park. It originated as a lake in glacial deposits set down about 12,000 years ago on what is now the Burrard Peninsula. Over time the lake filled in, and approximately 5,000 years ago a swamp of sedges and grasses replaced the open water. Death and decay of swamp plants gradually depleted the oxygen in the standing water, and the acidification of the accumulating organic debris created conditions favourable to invasion by sphagnum moss. By 2,000 years ago, a sphagnum bog was established and, with its associated shrub communities, remained the dominant vegetation type until fifty years ago when hemlock started to invade the area. In the normal process of succession, bog surfaces gradually become drier as they build up. In Camosun Bog this change has been accelerated by human activity. The area surrounding the bog was logged in the early 1900s and a major fire swept over it in 1919. Now residential streets and a school border the bog on its eastern side, and drains were installed in 1929 and extended in 1959. One of the pressures to drain the bog was the desire to eliminate mosquitoes and frogs, which were perceived as pests by nearby residents from the 1930s to the 1950s. In addition, construction fill was dumped on the western side.

Improved drainage and the addition of nutrient-rich soil water from the forested western slopes and dumps destroyed the high water table and acid conditions essential for bog plants. Species from the adjacent forest colonized the drier site, and the low growing shrubs, herbs, and mosses of the bog flora were shaded out.

To preserve the bog, it is necessary to re-establish the year-round high water table, to ensure that the water is nutrient-poor, and to remove the invading species, especially hemlock, which tend to dry the area by transpiration. The bog has been instrumented to give information on groundwater fluctuation and nutrient chemistry that is essential for proper management. Surface ditches on the periphery have been redesigned with control gates to maintain a high water level, and mature hemlocks have been removed. A hand-weeding program has been initiated to remove invading forest species, and permanent plots have been established to monitor the effectiveness of these combined management techniques. Both community opinion and public funds have supported this manipulation of the environment, but the bog's need for water during drought may become an issue, because this use will compete directly with growing human demands.

Future options and prospects

After reviewing the human impacts on the natural environment of the Vancouver region one might ask, what remains? Within the city, the answer is, very little. Outside the city, three areas of high resistance to development stand out: the steeply sloping lands of the North Shore Mountains, the waterlogged soils of the delta flatlands, and the tidal marshes at the leading edge of the Fraser Delta. So far, they have been saved by the relative difficulty of developing them and, to a degree, by their designation as parks, Indian reserves, or protected watersheds for the city's drinking water supply.

Even these sites are under pressure, however. North Vancouver is growing rapidly and there have been clashes over proposals to build housing in Lynn Canyon Park. A proposal to build a gas pipeline through the Coquitlam Watershed may threaten the sanctity of the Water District, and Natives have leased large tracts for commercial development at the northern end of the Lions Gate Bridge. On the delta, the bogs are used for cranberry and blueberry farming, peat cutting, and sanitary landfill. A recent proposal calls for a deep-water dock and industrial zone where Burns Bog meets the Fraser River. The status of the tidal flats as vital habitat for migrating waterfowl and salmon (via the detrital food chain) has served to protect them; but plans to extend the runway system at the International Airport pose new concerns about the loss of wetlands.

As producers of environmental change, residents of Vancouver are in a curious, and often confusing, position with respect to these and other management problems. In Camosun Bog, efforts are directed to returning the environment to something like its original state by culling invaders and removing human influences. In areas such as Shaughnessy, efforts to save existing trees seek to protect non-native trees planted by the first residents. In Stanley Park, the two approaches come up against each other: should the forest be returned to its original state if this means removing beautiful trees that are a direct result of human interference with the natural world? Or, to take another example, should wildflowers growing on parts of the former Shaughnessy golf course be protected? This area was cleared, intensely managed, and subsequently invaded by a mixture of indigenous and exotic species. It is by no means natural but it is regarded as 'wild' and worth saving. Vancouverites also need to ask themselves whether they wish to encourage the mind-set that lies behind the search for 'prim order' in the urban landscape as illustrated by manicured beaches, neat paved back lanes, and straightened river courses.

The environmental metamorphosis sketched in this chapter, and the questions it provokes, continue. The process is essentially the same as that occurring in cities everywhere, and the broader view suggests that Vancouver may become a significantly different, perhaps less attractive, environment in the future. Change is bound to occur; the challenge is to maintain the natural systems that are vital to the environmental quality for which this area has become renowned. It is both important and encouraging to realize that we are at the same time the actors, directors, and audience of the drama being played out. We have roles and responsibilities to exercise, and we need to be informed, aware, active, and creative in our roles as joint stewards. The phoenix-like revival of False Creek and, perhaps, Hastings Park are enormously encouraging; but they also point out the incredible cost of environmental remediation. Urban physical geography is an active and important part of every person's daily life. We ignore it at our collective peril, cost, and shame.

6

Vancouver, the province, and the Pacific Rim

Trevor J. Barnes
David W. Edgington
Kenneth G. Denike
Terry G. McGee

Geographers have long distinguished between site and situation. They use the former to refer to the attributes of a particular locale and the latter to describe the relationship between a place and its surroundings. Site and situation are closely linked, however. Nowhere, perhaps, is this more evident than in cities, where outside connections clearly and profoundly affect the character of the place. These connections, whether they be economic, social, or cultural, are the subject of this chapter, which examines the impact on Vancouver of some of the fundamental changes that have occurred in British Columbia, Canada, and Pacific Rim countries during the last forty years or so.

Between 1945 and the early 1970s, both the Canadian and global economy were dominated by the United States and the multinational corporations it spawned. Large firms producing standard, mass-produced goods were at the very core of this production system, which came to be known as 'Fordism.' Because an uninterrupted supply of raw materials was essential to the continued growth of Fordism, multi-national corporations (the largest of which were located in the U.S.) invested in Third World countries. In doing so, they maintained and strengthened a global resource periphery first created during colonialism.

Canada's role in these developments was ambiguous. Traditionally an exporter of staple resources, politically stable, and with strong historical ties to the U.S., Canada was an ideal source of many of the raw materials required by American mass producers. This was formally acknowledged in a report produced for President Truman in 1952 that identified Canada as a prime source of twenty-two key resources and which, subsequently, led to direct investment in this country by American corporations. In mining and smelting, for example, U.S. ownership increased from 38 per cent to over 70 per cent between 1946 and 1957, and Canada quickly became part of a system labelled 'continental resource capitalism.' Yet Canada also possessed considerable manufacturing capacity, particularly in southern Ontario and, unlike many developing countries, offered an affluent market for American mass consumer products. For these reasons, Canada was more than a simple resource periphery, and over time its manufacturing sector and market were increasingly absorbed within a wider continental Fordist system. Thus, the Auto Pact, signed in 1965, effectively integrated the Canadian car industry (by far the most important manufacturing sector) with its counterpart based in Detroit. Likewise, several trade treaties enabled approximately 80 per cent of goods traded between Canada and the U.S. to cross the border duty-free and quota-free.

Early in the 1970s, new developments began to erode the foundations on which Fordism rested: the Bretton Woods currency agreement ended in 1971, making international trade more volatile; the Organization of Petroleum Exporting Countries (OPEC) sharply increased oil prices, creating global inflation, debt, and unstable flows of 'hot money'; the markets for traditional mass-produced consumer goods declined; productivity in traditional Fordist industries fell because of higher wages, labour unrest, and technical limits on the very nature of the Fordist production technology itself; and the dominance of American multinational corporations was increasingly challenged by European and Japanese firms. In short, the U.S. economy lost its pre-eminence, and as it did, the Canadian economy also encountered difficulties. The global economy lurched from crisis to crisis, culminating, in the early 1980s, in the second most severe depression of the century.

As the Canadian economy recovered in the mid-1980s, traditional Fordist production systems were becoming less important. The telecommunications revolution and the increasing importance of highly mobile financial capital opened up the world to the multinational corporation. Falling transport costs and improved production and communication technologies allowed multinational corporations to locate routinized tasks in low-wage countries almost anywhere in the world. At the same time, a number of Pacific Rim countries capitalized on the new opportunities available to them by establishing Free Trade or Export Zones. Among these countries, Hong Kong, Singapore, South Korea, and Taiwan were especially successful in attracting manufacturing investment and were soon identified as Newly Industrialized Economies that, in turn, formed the basis of a New International Division of Labour.

Such developments hastened a process of deindustrialization in many traditional West European and North American manufacturing regions: in Europe steel production and allied industries such as shipbuilding declined, and in North America the manufacturing belt became the 'Rust Belt.' At the same time, novel types of manufacturing activities began to emerge in newly created industrial spaces (e.g., the electronics and aerospace industries in the American Sunbelt). More broadly, the complex of new kinds of industries, new (often computerized) production methods, and new working practices has been called post-Fordism or flexible specialization. However imprecise these terms are, they denote a move away from the mass production of the standardized goods of Fordism toward the provision of more specialized products and services designed to meet niche consumer markets.

The effects of these changes on Canada are multifaceted. The country's location, particularly that of British Columbia, offers the potential to capitalize on the rise of the Pacific Rim nations. But at the same time, by signing the Free Trade Agreement in 1989, Canada further integrated its economy with that of the U.S. Moreover, service industries have grown at a faster rate in Canada than anywhere else in the world, and this ostensibly marks the country's transition to a post-Fordist economy. But staple exports continue to be the most important source of external revenue. Ironically, despite the changes, the foundation of the Canadian economy still rests on those who hew wood, extract ore, and harvest crops.

To describe and analyze Vancouver's geographical situation between 1950 and 1990, this chapter is divided into three main sections: the first examines the relationship between Vancouver and British Columbia as a whole; the second discusses the links between Vancouver and Pacific Rim countries; and the third explores possible futures for Vancouver given the likelihood of continuing changes in external relations. The main argument is that Vancouver has evolved from an important local metropole, concerned with economically managing and accumulating capital from the rest of the province in the 1950s and 1960s, to a 'world city' with a global role and reach in the 1970s and 1980s.

Vancouver: Provincial centre, 1950-90

Fordism in British Columbia, 1945-73

The success of Fordism after 1945 rested upon twin foundations. The first was a distinctive set of production and organizational characteristics. Foremost among these was the mass manufacture of standardized products, such as automobiles, domestic appliances, and heavy machinery, that yielded considerable economies of scale. This, in turn, required a distinct industrial structure, dominated by large, oligopolistic capital-intensive corporations that often combined all stages of production at a single place. Associated with this vertical integration were labour practices based upon principles of 'scientific management,' which entailed production-line work and the repetition of limited tasks. Typically, labour in such settings was unionized, and there were strict rules about job demarcation and order of seniority.

The second reason for Fordism's success was a sustained demand for large consumer durables. Mass production necessitated mass consumption, and just such a congruence occurred with the emergence of a consumer society in North America in the 1950s and 1960s. With unprecedented levels of disposable income, and widespread suburbanization requiring the purchase of automobiles and consumer durables, the American Dream became a reality, and, with it, the 'golden age' of Fordism was assured.

In retrospect, it seems clear that these developments were fostered by an informal coalition of three groups: private corporations, unions, and the state. By seeking steady growth in investments, corporations earned adequate returns for their shareholders and improved both productivity and the general standard of living. For their part, unions largely ceased to contest changes in labour practices (for example, the de-skilling that mass production technologies entailed) in exchange for steadily rising wages. In this way, there could be both rising output brought about by productivity improvements in the technology of mass production and sufficient effective demand because of concomitant increases in wages. Finally, the state assisted corporate investment by providing the necessary infrastructure and suitably trained labour as well as by managing aggregate demand through such measures as monetary policy, direct intervention in wage negotiations, and the provision of a social welfare system. In these ways, the state regulated both the spheres of production and consumption. Of course, there were strains and fractures among corporations, unions, and the state along the way, but, in general, the coalition held and produced an unparalleled long boom within industrial capitalism.

This account of Fordism best pertains to mature industrial regions, such as the manufacturing belt in the United States or the 'Golden Horseshoe' in Southern Ontario. In British Columbia, by contrast, there was relatively little manufacturing of end products, and the economy was dominated by the production of staple goods – crude and semi-fabricated materials. In 1970, for example, manufactured end products represented about 17 per cent and staple goods 83 per cent by value of the province's goods producing sectors. In British Columbia, Fordism was necessarily modified by its orientation toward staples production.

Between 1945 and 1975, British Columbia's economy was dominated by the export of forest products and minerals (Table 1). During the early 1970s the forest products sector accounted for 50 cents of every dollar made in the province. In this, at least, forestry's role in the provincial economy is directly analogous to the function that more traditional Fordist manufacturing played in 'core' economies. Furthermore, the general nature of the product, and the processes by which it was produced, were strikingly similar to those of archetypal Fordist industries. Three main commodities – lumber, pulp, and newsprint – accounted for between 80 and 85 per cent of the total value of production in the forest products sector during this period. At least until the mid-1970s, all three were standardized products that could be mass-produced and exported. Even lumber, which could be manufactured according to countless specifications, was produced mainly as a single construction grade and was used in both the Canadian and U.S. housing industries.

Table 1:

Foreign exports of selected major BC products, 1960-88 ($ millions)

Products	1960	1965	1970	1975	1980	1985	1988
Forest products	493.1	752.3	1,176.8	2,057.6	5587.1	6135.1	9,706.7
lumber	252.7	365.0	534.1	780.6	2,461.1	2,847.9	3,915.0
pulp	99.0	181.0	356.1	869.5	1,898.4	1,594.5	3,368.0
newsprint	100.6	133.8	181.0	269.0	530.0	831.7	943.0
Minerals	152.3	221.4	364.9	531.7	1,472.2	1,835.4	2,224.2
copper	9.3	31.1	121.8	201.0	428.8	380.7	743.0
aluminum	58.5	55.7	85.5	97.1	441.7	341.3	534.0
molybdenum	NIL	11.3	41.7	61.0	297.4	44.8	81.9
zinc	46.0	47.4	54.3	95.5	144.8	310.7	436.0
lead	20.0	42.3	32.4	33.2	75.5	41.9	114.4
Chemical products	30.1	29.2	39.5	50.1	60.7	209.6	192.5
Energy products	13.6	33.4	92.1	563.5	1,087.1	2,441.0	1,940.8
coal	3.4	4.2	32.5	334.7	515.3	1,617.8	1,458.0
Food products	36.2	67.1	81.9	141.4	311.8	658.8	459.4
Manufactured end products	N/A	0.4	18.5	12.6	57.8	290.9	811.8
Other	92.6	115.1	258.4	515.6	1,077.1	743.7	2,041.9
Total	817.9	1,218.9	2,020.3	3,872.5	9,653.8	12,314.5	17,377.3

Source: *Economic and Financial Review* (various years)

The milling of lumber also provides a good illustration of the Fordist production methods used (Figure 1). First, logs were sorted and graded on a moving conveyor belt – the 'green chain.' Work on the chain was based on a strict order of seniority, where the last employee hired was the first sorter and grader, and the most senior employee was responsible for final sorting and grading. After grading, the logs were sawn and planed using 'dedicated machines' set to standardized dimensions. Here, labour was more skilled. Both sawyers and planers needed to know how to 'read' the wood in order to maximize the proportion of finished lumber extracted from each log. Finally, the lumber was stacked to await shipment to markets. During the 1960s and 1970s, many technological improvements were made in each stage of production. All such changes were intended to increase production capacity and to allow even greater economies of scale.

The corporate organization of the forest products industry similarly reflected Fordist tendencies. Before the war, the industry was dominated by small companies, many of them family-owned. This changed quickly during the 1950s. This was partly the result of American firms, such as Weyerhaeuser and Crown Zellerbach, moving into the province, but it was also because of the large number of mergers and takeovers among BC and Canadian firms. For example, MacMillan Bloedel, the largest

Figure 1

A sawmill on the Fraser River (New Westminster)

forest products firm in the province throughout the postwar period, was created by several mergers among BC-controlled companies. Increased corporate concentration and changes in provincial forestry policy meant that a large and growing proportion of harvesting rights fell into the hands of a few corporations. By 1974 eight firms controlled 82 per cent of harvesting rights in the province; in 1940 the top fifty-eight companies had held only 52 per cent of these rights. The postwar forest products corporations were also essentially Fordist in their high levels of capital investment, except in the actual process of logging itself (although even that changed during the 1970s). Unlike the quintessential Fordist firm, however, the forest products sector was not always vertically integrated. Indeed, the increasing relative importance of pulp, which more than doubled its share in forest products exports to over 40 per cent between 1960 and 1988, reflected the fact that production of the end product – paper – was carried out elsewhere. In this respect, the BC forest products sector exhibited the characteristics of a truncated branch-plant structure often associated with multinational corporations, in which production is carried out far from the head offices of the firm.

Another typically Fordist characteristic of the forest products industry was the high rate of unionization among workers, principally represented by either the Canadian Paperworkers Union or the International

Woodworkers of America (IWA). The latter was the largest union in the province in the 1960s, with between 40,000 and 50,000 members. Wages of IWA members were estimated to be 5 per cent higher than those of comparable workers elsewhere in Canada, and they had good benefits, well-defined rules about career advancement based upon seniority, and clearly demarcated job tasks. Moreover, there was an implicit coalition between organized labour and the forest products companies. In the interests of both higher wages and an enhanced role for the union itself, the IWA backed the oligopolistic corporate structure of the industry. As one IWA union official reflected:

> We have dealt for decades with 'independents' who compete not by innovation, better equipment, better management, etc.; but by cutting corners on legitimate costs such as safety, wage and bene-fits, reforestation, etc. The popular appeal of allowing 'the small guy' into the forest should not disguise the fact that to ride out bad markets, bear legitimate forestry, safety and labour costs, etc., in basic industry, requires considerable size.

Corporate and labour interests were in large part reconciled through the policies of the third member of the coalition – the provincial govern-ment. Between 1952 and 1972, the provincial Social Credit government under W.A.C. Bennett benefitted corporate forestry interests by invest-ing in large-scale infrastructure projects and passing favourable resource legislation. Thus, roads were developed in the Interior; rail links were extended to the resource periphery (including the construction of the Pacific Great Eastern Railway – now the British Columbia Railway – to Fort St. John and Dawson Creek and beyond to Fort Nelson); and an extensive electric grid was built, based upon massive hydro-electric projects on the Columbia and Peace rivers. In addition, the government, which owned nearly all timberland in the province, undertook to auction harvesting rights to the private sector according to a 'sustained yield policy.' This required management, restoration, and protection of the resource, conditions that only the large, oligopolistic firms operat-ing with long-term resource horizons could meet. But having met such conditions, firms were then able to charge themselves very low prices for lumber. Because of this undervaluing, corporate forest products companies extracted handsome profits. Furthermore, provincial rev-enues from stumpage fees (based on the value of logs) were paltry. At the same time, the provincial government won favour among the electorate, including organized labour, by upholding the social welfare system – one that both provided a safety net for a volatile economy and guaranteed the corporate forestry sector a stable, but elastic, supply of labour.

The features that characterized the forest products industry were even more evident in mining and smelting. This sector was very volatile, with mines opening, closing, and reopening in accordance with fluctua-tions in world commodity prices. Low value-added, unprocessed miner-als made up the bulk of production and exports, with lead and zinc most important by value during the 1950s and 1960s and copper and molybdenum more important during the 1970s (Table 1). In the 1960s a growing share of production came from capital-intensive, open-pit mines that yielded lower grade ore, such as at Tasu (iron), at Cassiar (asbestos), and in the Highland Valley (copper). Given the long-term, high volume contracts associated with such operations, Fordist mass production techniques were generally employed to maximize economies of scale. Similar techniques were also evident in the Alcan smelter in Kitimat (which provided a consistent source of export income through its manufacture of aluminum ingots) and at Cominco's integrated lead and zinc smelting and refining complex in Trail. In mines and smelters labour was heavily unionized. Corporate control lay largely beyond British Columbia in the hands of multinational corporations, such as Cominco, Noranda, and Rio-Algom mines. Finally, as in forestry, the provincial government assisted corporate investment. Examples of this include: government mega-projects, such as construction of the Kenney Dam and associated power tunnels beginning in 1953, which provided hydroelectric power; the undervaluation of resources, which resulted in corporations appropriating large resource rents (by one estimate the multinational Kaiser Resources garnered $42 million a year in resource rents during the 1970s); and the Instant Towns Act (1965), which effectively transferred the costs of resource town development from the company to the municipality and, ultimately, back to the province itself.

Although staples production in British Columbia between 1945 and 1973 mirrored important characteristics of Fordism in terms of production technology, industrial organization, and the relationship among corporations, organized labour, and the state, it also differed from the archetypal pattern in at least four important ways.

First, the dispersed spatial distribution of industry in the province is not typical of Fordism (Figure 2). Usually Fordist industries are urban-based and concentrated in a few specific regions. Rather than the metropolis, however, the home of the staple industry is the single-resource community. In fact, with over a hundred such communities, British Columbia led the nation in the ratio of single-resource towns to total settlements. Such dispersion, and its fluctuating pattern, is, of course, a direct result of the nature of staples production itself, for non-renewable resources are location-specific and are subject to depletion. So, the overcutting of coastal timber stocks was partly responsible for the demise of such communities as Ocean Falls, and the recent systematic exploitation of interior forests explains the rise of new communities such as Mackenzie.

Second, and related, the hallmark of staples production is price and output volatility. A number of studies conclude that British Columbia is the most economically volatile of all provinces, experiencing the highest highs and the lowest lows. British Columbia also tends to be out of

Figure 2

Mills and mines of British Columbia

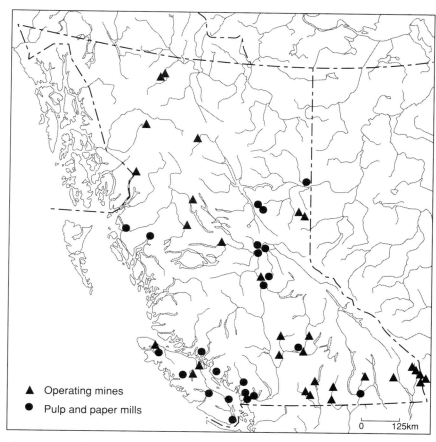

▲ Operating mines
● Pulp and paper mills

0 ――― 125km

phase with the economy of the country as a whole, experiencing both recession and recovery later than other regions. Both features are attributed to the province's specialization in the export of staples to competitive markets characterized by high price elasticity. Even making generous assumptions about the definition of manufactured end products, nearly 90 per cent of merchandise exports from British Columbia in 1970 were staple exports. The economic volatility that is a consequence of such dependence is manifested directly in the landscape, with single-industry towns literally appearing and disappearing with the vicissitudes of export prices and demand. By contrast, classic Fordist industries are renowned for the stability associated with the long-time accumulation of capital in places such as Toronto, Hamilton, and Windsor.

Third, traditional Fordist industries generally demanded a less publicly funded infrastructure than did those on the resource periphery. The need for power, transportation, and settlement in remote areas entailed much greater government expenditures in hinterlands such as British Columbia. Moreover, provincial ownership of resources implicated the government in development and exploitation from the outset, and the volatility and limited range of employment in resource communities enhanced the importance of the social welfare system and counter-cyclical public investment projects.

Finally, although resource firms share with more traditional Fordist industries the characteristics of large size and capital intensity, they are not usually vertically integrated or strongly linked to other local sectors. For example, the increasing importance of kraft pulp for export reflects the efforts of some firms to remove the most profitable stages of production from British Columbia. Similarly, most of the equipment used in the forestry and mining sectors is imported. A measure of the paucity of such backward linkages is revealed by statistics showing that metal fabricating, machinery production, and the manufacture of transportation equipment in the province accounted for barely one per cent of all employment in 1970. In part, these characteristics are attributable to the important position of multinational corporations in the economy, but they also owe something to the collective support of the staples economy by the three partners in the Fordist coalition: capital, labour, and the state.

At the hub of this resource-dependent provincial economy and dispersed urban network, Vancouver lacked traditional Fordist industries and did not develop into a classic industrial city. Such manufacturing employment as existed was either in the processing of staples or the production of goods for local consumption. But even consumer industries were proportionately underrepresented because of high labour costs in the city – a result of the 'shadow' effect of elevated wages in the wood products sector. The city's central roles, therefore, were that of distribution and control: from Vancouver, staple products from British Columbia and beyond (such as Prairie wheat) were sent around the world, and the activities of scattered resource sector operations were co-ordinated. Several consequences followed. First, service sector employment was the fastest growing sector in the city between 1945 and 1973. In the 1960s alone, employment in transport, communications, and trade increased by over 40 per cent; in finance, insurance, and real estate the growth was approximately 75 per cent; and in community, business, and personal services it exceeded 80 per cent. Second, the city's importance as a distribution centre brought in a good deal of investment in transportation projects (Figure 3). The most spectacular was the Roberts Bank bulk loading terminal, built primarily for East Kootenay coal exports supplied by Kaiser Resources. In addition, there were improvements to the port facilities in the GVRD, including a container terminal, and, in the mid-1960s, Vancouver's airport was expanded. Such developments complemented Vancouver's existing warehouse districts on Burrard Inlet and along the North Arm of the Fraser River in Richmond. Third, as local resource firms merged and outside companies entered the province, Vancouver was the best site for the control and organization of such operations. To the marine insurance offices, shipping agencies, and banks associated with the city's long-established entrepôt function,

Figure 3

Vancouver as a distribution centre
(North Vancouver)

the Fordist boom added the headquarters of both private corporations and public institutions. As such offices expanded, they promoted the growth of ancillary business functions – accountancy firms, law practices, hotels, and restaurants. Employment in business services increased by over 130 per cent between 1961 and 1971. In the same decade, the city's complement of managerial, professional, and technical workers also rose by over 30 per cent. In addition, Vancouver's stock exchange (established in 1907) grew significantly, specializing in the penny stocks that raised venture capital for small-time resource prospectors in western Canada. To accommodate these expanding service functions, there was an office building boom in the city core. Between 1957 and 1980, the amount of gross office space almost tripled to reach 15.7 million square feet.

Because of its greater range of economic activity, Vancouver was less subject to external perturbations than were the smaller resource communities that it effectively administered. More generally, the relationship between city and province during the period seemed to conform to the classic core-periphery pattern associated with staples production. By such indices as unemployment rates, per-capita income, and the level of publicly funded services, Vancouver generally fared better than did the rest of the province. Yet the core-periphery relationship was not a simple one. Not all parts of the periphery were equally subordinate nor was the core always dominant. Specifically, because control of many of BC's major firms lay outside the province, the city often played only an intermediary role and was itself greatly affected by decisions made in other metropoles. In this sense, Vancouver as a core was itself part of an even larger global periphery. This reality became even more marked once the 'long boom' turned into the 'long bust.'

Recession, economic restructuring, and post-Fordism, 1973-90

Looking back, commentators have viewed the early 1970s as a time when the economy was on the cusp, a time when the old verities of Fordism began to crumble and new ones developed. First, there were changes in production technology. Through computer-assisted design and computer-assisted manufacturing (CAD-CAM), firms began producing to customer-specifications and serving niche markets. Such flexible specialization is often associated with the rise of small, 'vertically-disintegrated' producers, where each firm is responsible for only one part of a product and is connected to others through a web of input and output linkages. Of course, CAD-CAM techniques are also used by large corporations that formerly employed traditional Fordist methods. Everywhere the new technology has brought major changes in labour practices; production-line work and the rigid demarcation of tasks has given way to co-operative 'work-teams' responsible for a number of different job assignments, and the use of part-time workers has increased substantially. The latter constitute, in effect, a secondary workforce hired and fired, as needs dictate, to supplement the work of a core group of full-time employees.

Second, the post-Fordist period has been marked by greater international economic linkages and the emergence of a clear hierarchy of 'world cities.' Defined by their role in directing global flows of commodities and finance, world cities are a direct consequence of the telecommunications revolution. Downtowns are now the electronic crossroads of information, money, and power. In particular, control over finance capital, the most flexible of all commodities, is assuming a pivotal place in the major cities of the world. Producer services have rapidly expanded in tandem, and, in turn, this has brought significant change to the physical spatial structures of downtown areas (such as the gentrification of residential areas, the 'boutiquing' of traditional retail areas, and various displays of spectacle).

Finally, there has been a change in the political climate in many advanced industrial nations, which, in its most extreme form, has manifested itself as neoconservatism or the rise of the 'new right.' The advocates of this new political thinking called for a reduction in the role of government and the promotion of the ethos of private enterprise. In line with such beliefs, neoconservative governments have sought to dismantle aspects of the welfare state and to privatize sectors that hitherto were publicly controlled. More broadly, this has had the effect of undermining the former coalition among corporations, labour, and the state, with organized labour, in particular, suffering a series of defeats.

In British Columbia the emerging post-Fordist regime has been laid upon the foundation of the staples producing economy and its associated Fordist practices, creating a distinctive economic and social environment. To investigate it further we need to consider in more detail the development of the provincial economy between 1973 and 1990.

When OPEC imposed the first of its major oil price hikes in 1973, the ripples of recession spread through the world economy. Because the BC forest sector is tied so closely to the U.S. (lumber to the housing market and pulp and paper to the general level of business activity), provincial forest products exports took an immediate downturn. By 1976, however, output levels had returned to pre-recession levels, and by the late 1970s both production and export price levels reached record highs. A similar story holds for mining, although the details of recovery varied by commodity type. In 1972 the volatility induced by OPEC was compounded by the election of a New Democratic Party provincial government intent on diversifying the BC economy. The revenues necessary to achieve this policy were sought from increased royalties and new taxes imposed upon the mining sector. The passage of the Mineral Royalties Act (1974), however, provoked vehement opposition. Mining companies quickly mobilized support from other industries, the press, the Opposition, and their own associations. In the face of such pressure, the government backed down. At the same time, the NDP government sought to establish a 'window' on the forest industry by creating the BC Resources Corporation, which took over Crown Zellerbach's Ocean Falls' mill and the properties of Columbia Cellulose, including a suphite pulp mill in Prince Rupert. In little more than three years the NDP government was defeated in a campaign based on the issue of 'Free enterprise versus socialism.' Yet the return of the Social Credit party could not obscure fundamental changes occurring in the province. The unsettling effects of a two-year recession, combined with the NDP's flirtation with diversification (including proposed higher charges on resource extraction) had generated increasing strain within the Fordist coalition and thrown into question the relationship between Fordism and resource production.

Such strains increased further in 1979, when another round of OPEC oil price rises triggered a second recession that quickly turned into a deep global depression. As in the past, British Columbia was slow to feel the downturn, but when it did, in late 1980, the effect was particularly severe, with staples-producing industries not recovering until 1986. Because of its greater diversity of activities, Vancouver fared better than most of the province, but unemployment rates still peaked at over 14 per cent. (In the single-industry towns, these rates were two to three times as high.) At the bottom of the depression in 1982, over 50 per cent of all loggers, between 25 per cent and 30 per cent of all workers in sawmills, and 60 per cent of shakes and shingles workers were unemployed. Every important company in the forest-products sector incurred financial losses, and some of them for three successive years. In 1979 profits in the forest-products sector amounted to $500 million. Two years later, the industry recorded a $500 million loss. And in the mining sector, a number of pits were abandoned.

One immediate effect was corporate restructuring. Between 1975 and 1987, seven American corporations sold their BC holdings, including Rayonier in 1980 and Crown Zellerbach in 1982. In addition, MacMillan Bloedel, by far the largest forest products corporation operating in the province, was taken over by the eastern Canada-based Noranda in 1981, which, in turn, was bought out in the same year by Brascan (owned partly by the Bronfman family). Even so, continued cash flow problems within MacMillan Bloedel precipitated a number of changes: a 35 per cent decrease in its labour force, the sale of its controlling interest in a number of firms worldwide, and the sale of its head-office building in downtown Vancouver. Even management and research and development staff were affected: over 700 head-office employees were laid off between 1979 and 1984. Furthermore, analysts began to suggest that this depression was more than just an exaggerated version of the boom-and-bust cycle that had characterized the forest products sector since the war. They pointed to the province's poor reforestation record, the increased demand for hardwoods (that were more available and cheaper than BC's softwoods), the levelling out of the U.S. housing market, the increased use of wood substitutes, trade sanctions imposed by the Americans (and later the Free Trade Agreement itself), and the difficulties of raising productivity using Fordist techniques. In short, they argued that the forest industry had moved into a new era.

Emerging from this turmoil of the early 1980s was quite a different economy. Computerization fundamentally changed the nature of work and livelihood for many types of workers. The intangible commodity of information became the most valuable commodity of all and was now bought and sold around the world and around the clock. And the always uneasy coalition among corporations, organized labour, and government was rocked by strong new political winds.

Technological change was widespread. New tasks, new skills, and new kinds of workers were required in this 'brave new world' of the microchip. Such changes are well exemplified by what happened after the introduction of new sawmilling technology in MacMillan Bloedel's plant at Chemainus. Touted as the largest mill on the coast at its completion in 1923, the Chemainus mill was acquired in 1950 by the MacMillan Export Company, which, one year later, became the MacMillan Bloedel Corporation. Through the next three decades, the mill employed typical Fordist techniques and specialized in the production of construction-grade lumber. With the massive restructuring of MacMillan Bloedel in the 1980s, the Chemainus mill was closed, and in 1982, 654 workers were laid off. A year later the company announced plans for a new state-of-the-art facility to be built on the old plant site.

When the new mill re-opened in 1985, working practices were quite different. Many operations were now controlled by computers. The 'green chain' was replaced by the automated 'J' bar, planing was subcontracted out to non-unionized local shops, and there was no need for sawyers 'to read' the raw lumber, because laser-sensing devices automatically measured the length, width, and height of each log and assessed the best cut to maximize recovery and revenue. In addition, the mill also instituted work teams in which small groups of employees were responsible for a large number of different job tasks. All workers now needed to learn 'the manual,' implying that every employee on the shop floor should be able to undertake any assignment within the mill. In this sense, job tasks were recomposed rather than made more specialized (as under Fordism). The broader effect of such changes was greater flexibility. Literally, by the push of a button, the mill's machinery could be reconfigured to produce any lumber dimension. Niche markets and particular customer specifications could thus be easily met.

In many ways, these changes were driven by MacMillan Bloedel's attempt to reorientate production toward new, and more profitable, international markets. In this sense, flexible specialization was initiated in response to the traditional export concerns of staple production. In part, MacMillan Bloedel wanted to move away from the standard products of Fordism (and from its reliance on the United States and Canadian housing construction) and move toward the production of more specialized and valuable timber products for other markets, particularly in Japan and the European Economic Community. This goal was realized in Chemainus by setting up a flexible manufacturing system; this increased the value of the lumber produced from $45 million ($25 million exported) in 1981 to $68.5 million ($65 million exported) in 1986.

These changes affected Chemainus severely. All workers were laid-off for two years, and when the new mill opened only 145 people were hired. Because of the length of closure, the seniority provisions of the union contract did not apply, and many long-time employees were left unemployed. Many resorted to informal work and suffered drastic reductions in income. Spouses forced to find employment often did so as low-paid workers in the local tourist industry, which was centred around a series of giant murals celebrating the town's past. In this way Chemainus mirrors yet another aspect of post-Fordism – work and leisure based on the organization of spectacle.

Changes in the Vancouver office sector paralleled those on the Chemainus shop floor. Many offices were automated during the early 1980s, and the number of clerical workers in the city fell by 14 per cent between 1981 and 1985. In addition, there was also organizational change with the development of a producer services complex in downtown Vancouver (Figure 4). Consisting of a tightly interwoven set of often small, vertically disintegrated business services (consultants,

accountants, lawyers, architects, advertisers, and so on), the complex principally serves the local community and has few input linkages beyond the city. Employment in this sector has grown at an annual rate of 6 per cent, and the businesses that comprise it exhibit a high degree of spatial concentration. Over 70 per cent of all producer services within the Greater Vancouver Regional District are located in the city of Vancouver. Thus, even during the depression of the 1980s, the number of managerial and professional jobs in the city increased by 40,000. By contrast, other occupations declined by over 90,000. Between 1971 and 1984, the number of law firms in Greater Vancouver doubled to 2,108, and 300 of those firms were clustered in just one downtown block. This pattern reflects the fact that producer services deal in information and that face-to-face communication is still the best, and the most complete, means of conveying knowledge.

Figure 4 (opposite)

Home of Vancouver's producer services complex (downtown Vancouver)

More generally, the task of the producer services complex is one of buttressing the control and organizational role of the head offices that administer operations in other parts of the province (and, increasingly, in different parts of the world). In this sense, Vancouver takes its place within the hierarchy of world cities. According to one commentator it is on the third rung, behind first-order cities such as New York and second-order cities such as Toronto. Consonant with this position, Vancouver is enmeshed in a global network whose currency is information. Vancouver's incipient role as an international financial centre is also indicative of its world city status. Benefiting from recent federal legislation giving the city special tax advantages for foreign financial investors, the city now has its own International Financial Centre that attempts to lure 'hot money' to Vancouver (currently there are about forty registered companies at the centre). This growth in a quaternary sector specializing in knowledge-based industries has brought social as well as economic changes, including the rise of a 'new class' of often young, highly educated, and affluent workers. Gentrification of such inner city neighbourhoods as Kitsilano and Fairview Slopes, both close to downtown, are perhaps the most visible evidence of this process, but its effects are also seen in the boutiquing of retail areas and the growth of eating, accommodation, and entertainment establishments in the downtown area.

In broader terms, both British Columbia and Vancouver have experienced continued rapid growth in the service sector during the 1970s and 1980s (Table 2). In 1989 three-quarters of all employment in the province was made up of service-producing industries, reflecting a 19 per cent increase over the decade. By contrast, employment in goods-producing industries fell by 8.7 per cent over the same period. Producer services, which grew by over 90 per cent in Vancouver between 1971 and 1986, were the major gainers, but more menial services also expanded. Employment in consumer services, for example, increased by

Table 2:

Labour force growth in Vancouver Census Metropolitan Area by industry, 1961–86

Industrial group	1961	1971	1981	1986
Agriculture	3,806	5,430	7,735	10,220
Forestry	2,518	3,355	4,175	3,890
Fishing and trapping	1,836	1,575	4,175	3,890
Mining	1,581	3,370	3,325	2,920
Manufacturing	57,485	78,770	96,315	91,065
Construction	19,897	32,034	44,575	46,155
Transportation and communication	34,934	49,980	69,625	73,005
Trade	59,899	85,700	126,485	138,815
Finance, insurance, and real estate	15,918	28,180	47,345	53,560
Community, business, and personal services	70,380	127,775	216,715	N/A
business	7,898	18,180	N/A	49,885
education	12,113	27,530	41,015	41,940
health and welfare	18,828	30,280	52,625	60,925
accommodation and food	12,905	22,336	42,630	58,725
Public administration and defense	18,003	22,330	37,885	39,350
Total	274,759	474,560	690,955	757,520

Source: Census of Canada: 1961, 1971, 1981, 1986

over 160 per cent in Vancouver between 1971 and 1986. Furthermore, there are clear differences within the service sector in terms of pay, benefits, working conditions, regularity of employment, and so on. This had led some to argue that a dualistic labour market is emerging under post-Fordism. Highly-educated, well-paid, professional employees form the core group, and low-paid, unskilled workers, many of whom have part-time or temporary work, comprise a peripheral group. The proportion of full-time employees in British Columbia fell from 87 to 81 per cent of the workforce between 1975 and 1989, while, during this same period, part-time workers increased their share from 12 to 18 per cent. In British Columbia, and Vancouver, this peripheral group is exemplified by workers in the tourist industry. This sector has grown remarkably in the last two decades. Over the 1980s it ran neck-and-neck with mining as the province's second largest industry and brought in $3.8 billion in revenue in 1989, including much valuable foreign exchange. But in Vancouver, average weekly wages in accommodation and food services were just over a third of those in business services, and in amusement and recreational services they were just over half.

The triumph of neoconservative politics in British Columbia was also modified by the importance of the staples economy to the province. Not until July 1983 were policies of 'restraint,' and the wider politics of the New Right, initiated with vigour. Then the government embarked on the strategies of public sector down-sizing (the civil service was to be reduced by 25 per cent); the privatization and contracting out of existing government services (for example, highway maintenance); elimination or reduction in welfare services; and changes in the Labour Code intended to make it more difficult for unions to operate. The ostensible reason for 'restraint' was the danger of a provincial fiscal crisis, but the government's own figures threw doubt on this justification. In part, the response was ideological, an espousal of doctrine made popular by Ronald Reagan in the United States and Margaret Thatcher in the United Kingdom. It was also a response to economic changes. With internationalization, and technological and organizational change, there was no need for the state to prop up the alliance between labour and capital by offering the former a welfare safety net. The new motif was flexibility, and, if labour was neither sufficiently cheap nor pliant, multinational corporations would simply leave. This said, British Columbia remained a staples periphery, and so the necessity of state-funded mega-projects continued: some were to assist in the process of exploiting the province's resources, including their recreational value (e.g., the North East coal project at Tumbler Ridge built specifically to meet the coal demands of the Japanese, or the construction of the Coquihalla Highway that further opened up the Interior's tourist industry); others were justified by the greater international visibility they would give to British Columbia, such as enhancing tourist potential (e.g., Expo '86); and yet others were to assist in the development of Vancouver as a 'world city' (e.g., the construction of the light rapid transit system). Since the change of Social Credit leadership in 1986, neoconservative financial policies have been less evident (partly a result of economic recovery), but many of the New Right's articles of faith are upheld, including privatization and the continued attack on organized labour.

Vancouver's role in British Columbia between 1973 and 1990 both continued and broke with the earlier established patterns. The continuing importance of staples production in the province ensured that Vancouver's major role as a distribution centre was maintained. Indeed, the shift toward Pacific Rim markets made the port even more vital and sheltered the city to some degree during the 1980s recession. The emergence of a producer services complex also strengthened the control function of the city, thereby reinforcing the classic core-periphery relationship between city and province. For some, in fact, there are now two distinct economies in British Columbia: the service economy of the core Georgia Strait region and the staple hinterland of the rest of the province. In this view, it is not surprising that an assessment of the implications of major investments in Prince Rupert estimated that between 80 and 90 per cent of the resulting retail trade, construction, wholesale trade and finance, and insurance and real estate business would leak from that city, and that most of it would be captured by Vancouver. But Vancouver is not only a local control centre – it is taking its place within an integrated hierarchy of world cities.

Vancouver and the Pacific Rim

As Vancouver's relationship to the province has changed, so, too, has its international position. Long considered Canada's back door, the city is now seen as the country's gateway to the fast-growing Asia-Pacific region. Federal and provincial politicians, as well as business interests, recognize that by the year 2000, 60 per cent of the world's population, 50 per cent of global production, and 40 per cent of all consumption will be Asian. They are also well aware of the financial power of Japan and the economic vitality of the so-called New Industrial Economies of Hong Kong, Singapore, Taiwan, and South Korea, which have brought about a new environment of global competition (and encouraged the flexibility, re-orientation, and specialization associated with post-Fordism in North America). In this context, many in British Columbia, Washington, Oregon, and California have begun to claim their place on the Pacific Rim.

The potential of this 'region' (which at its most extensive includes Australia, New Zealand, Indonesia, Malaysia, Chile, Peru, and Colombia, as well as the countries flanking the North Pacific) seems almost boundless. Indeed, advocates of its importance point out that seven of the world's ten largest ports are on the Pacific, that trans-Pacific trade recently surpassed trans-Atlantic trade, and that the value of this trade is expected to increase from U.S.$300 billion to U.S.$1 trillion in the next decade.

However, it must be stressed that the notion of a Pacific region is an abstraction. Few cultural, environmental, or political ties bind its parts and its people. The Pacific Rim is a functional area, held together by the circulation of goods, people, and ideas. Yet there is every prospect of expansion in these exchanges, and the consequences for Vancouver will be profound; indeed, the repercussions of these quickening movements are already clearly evident in the city.

Movement of goods

In recent years, trading links between Vancouver and the Pacific Rim have expanded considerably, and it is clear that any continuation of this trend will consolidate the city's position within this dynamic and growing region and further shift its trading and commercial orientation from the rest of Canada. Of course, Vancouver has long shipped goods across the Pacific. But the expansion of this interaction after 1945 was a decisive factor in the rise of the Port of Vancouver to pre-eminence in Canada and its position as the largest port (by tonnage) on the Pacific coast of North America (Figure 5). Initiated by Japan, which began to consume increasing quantities of Canadian raw materials after the Second World War, this trade has more recently taken large quantities of western Canadian wheat, potash, sulphur, coal, and wood pulp to China, South Korea, and Taiwan. Since 1986, West Pacific countries have accounted for nearly three-quarters of all shipping tonnage passing through the Port of Vancouver. Air cargo traffic with the Pacific Rim has also accounted for a significant proportion of the recent 20 per cent per year expansion in freight handled at Vancouver International Airport.

Perhaps the most important feature of the dramatic shift in British Columbia trade from the U.S. and Europe to the Pacific Rim is the rise of Japan as a trading partner. This began in the 1960s, with the interest of Japanese trading companies (*sōgō shōsha*), and their banks, in Vancouver-based corporations associated with the exploitation of western Canadian resources. Typically, this interest involved minority investment in, or loans to, Canadian mining and other resource projects. Japanese trading companies, and the steel mills and power utilities that they represented, sought secure access to raw materials and, in contrast to U.S. corporations, which characteristically sought acquisi-

Figure 5

Vancouver and the Pacific

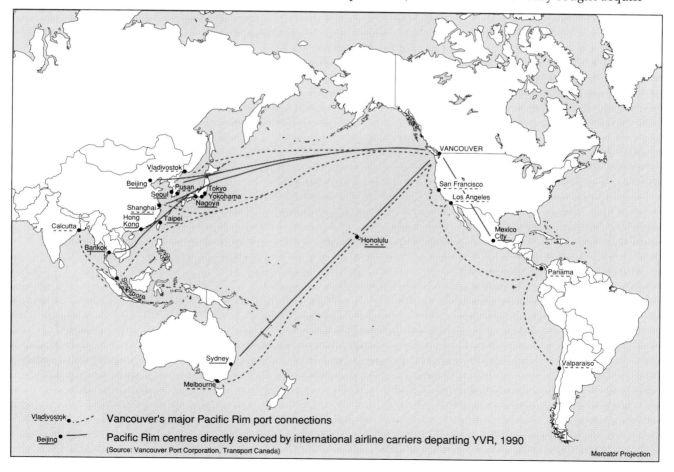

Vladivostok •---- **Vancouver's major Pacific Rim port connections**

Beijing • ——— **Pacific Rim centres directly serviced by international airline carriers departing YVR, 1990**
(Source: Vancouver Port Corporation, Transport Canada)

Mercator Projection

Figure 6

Toyota's import and distribution centre, Surrey Docks, Fraser River

tion or equity control of Canadian enterprises, they were willing to leave organizational control in other hands. Japanese control of the Canadian resources trade was maintained by different means, such as monopsonistic bargaining power and the threat of quickly diversifying to other suppliers (e.g., Australia in the case of coal). In contrast to the outflow of primary products, Vancouver also provided a gateway function for a rapid increase in Japanese imports to Canada after 1970. Fully built-up automobiles (Toyota and Honda) are carried by special vessels to unloading docks in the Lower Fraser River and, thence, to the major markets of eastern Canada by rail and truck (Figure 6).

By the late 1960s, the value of Canada's trade with Japan exceeded that with the United Kingdom, and through the last quarter century only the U.S. has been a more important trade partner. Japanese trade with Canada expanded by 370 per cent over the 1960s and 1970s, and the

Canadian-based branches of the Japanese sōgō sōsha generated most of the business. Almost two-thirds of this trade passed through Vancouver, and between 1977 and 1987 British Columbia's imports from Japan grew more than twice as quickly as did those from the U.S. By 1988 they accounted for 37.5 per cent of the province's imports (compared with 37.1 per cent from the U.S.). In the same year, Japan took 27 per cent of British Columbia's exports.

Between 1955 and 1985, five of the eleven major Japanese general trading companies that established subsidiaries in Canada chose Vancouver as their local headquarters. Others, such as Mitsui, which located their head offices in central Canada, established substantial branch offices in Vancouver, with financial and other decisionmaking autonomy. Among the four largest private companies registered in British Columbia, three are sōgō sōsha. Only the Jim Pattison Group reported 1988 revenues greater than Mitsubishi Canada ($1.22 billion), Sumitomo Canada ($1.2 billion) and C. Itoh and Company ($1.1 billion). Indeed, the Canadian subsidiaries of each of these trading houses ranked among the top forty corporations (both public and private) with headquarters in British Columbia. For all their importance in the trading economy, however, they employ surprisingly few people. For instance, in broad terms, Mitsubishi handled trade valued at over $1 billion in 1988, but only 120 persons were on its payroll.

As the focus of Japan's economic development has shifted away from processing industries requiring large quantities of industrial raw materials, so the original role of the trading companies – securing agricultural, mineral, and timber-related resource supplies for Japan – has declined in importance. With the rapid expansion of the Ontario economy in the 1980s, and the Canada-U.S. Free Trade Agreement of 1989, many sōgō sōsha have looked to central Canada for expansion. So Mitsubishi Canada (established in Vancouver in 1956) moved to Toronto in 1989, where the headquarters of their arch-rival Mitsui Corporation was located. Even so, Mitsubishi claimed that its Vancouver office handled four times as much trade as did its Toronto counterpart. Still, relative growth was stronger in Ontario, despite the rapid expansion of Japanese demand for Canadian forest products and food products after 1987.

The movement of people

There may be no better measure of the falling costs of movement (in terms of time and money) that lie behind the growing integration of Pacific Rim countries than the swelling tide of people who cross the ocean for business and pleasure. Prominent among them are women engaged in low-order service jobs. Filipino and other Asian nannies working in Vancouver are a prime example. Living in Canada under labour agreements that limit their freedom to change jobs and remain in the country on a permanent basis, most of these women return home after a few years in the city. They are part of a growing number of people (mainly women) who move temporarily from the Third to the First World for meagre wages and, perhaps, experience. Others of their number in Vancouver are employed as chefs, in more menial restaurant jobs, and as illegal factory workers.

Near the other end of the economic spectrum are the successful Asian entrepreneurs who have established businesses on both sides of the Pacific and who make frequent visits to Vancouver while continuing to live abroad. They, of course, are matched in number by British Columbians departing Vancouver to conduct or pursue business opportunities in Asia and other countries of the Pacific Rim. In addition to these developments, many wealthy residents of Hong Kong and Taiwan have acquired real estate or other enterprises, often run by younger family members in Vancouver. Such strategies have strengthened the bonds between Vancouver and Asia and made the Asian presence in the city more visible, although the numerical impact of these people on the population of Vancouver is still relatively negligible.

A similar pattern has been established with the expansion of tourism, which is now British Columbia's second major industry and the mainstay of the burgeoning Vancouver service sector. In the decade after 1978, when the number of foreign travellers entering the province rose by 118 per cent, the number of Asian visitors grew by 170 per cent. This has had a marked impact on air passenger travel patterns. Vancouver International Airport is Canada's gateway airport to the Pacific (Figure 5), and the second busiest airport in Canada, carrying almost twice as many international passengers as does neighbouring Seattle and rapidly approaching the level of international traffic at San Francisco. On the Vancouver-Hong Kong air route, passenger traffic increased 125 per cent between 1983 and 1987; by contrast, traffic on the Vancouver-Toronto air route increased by only 40 per cent.

Figure 7

Vancouver's 'Ginza.' Corner of Burrard and Alberni streets

Figure 8

The Pan Pacific Hotel and Canada Place, Vancouver, built and operated by Japan's Tokyu Corporation

Another telling indicator of the changing orientation of Vancouver toward Asia is the fact that Japan has now replaced the United Kingdom as the prime source of overseas tourists reaching British Columbia. Currently, the Japanese dominate the Asian tourist presence in Vancouver, and Japanese shop signs are now commonplace in the downtown core (Figure 7). The Japanese have also begun to invest in tourism in the province. Major downtown hotels are owned by Japanese investors, including, in 1990, the Pan-Pacific, Westin, and Coast hotels (Figure 8). Investors from Malaysia, Macau, and the People's Republic of China have also acquired Vancouver hotels in recent years. There is also a smaller but growing counterflow of tourists from Vancouver to Asia. Asian tourism is the major growth area of the Vancouver overseas retail tourism market; and together, the West Pacific countries have replaced Europe as the second most common destination (after the U.S.) of those leaving Vancouver.

Movement of ideas and information

In an effort to enhance the province's competitiveness abroad and to ameliorate racial tensions at home, British Columbia's Ministry of Education has committed itself to increasing the awareness of Asia among provincial residents. Asian languages and cultures are taught in some primary and secondary schools. Each of the four provincial universities has an institute of Asian research where advanced level studies and academic exchanges are conducted, and the federally chartered Asia Pacific Foundation is also located in Vancouver. A large and rapidly increasing number of students from Asia also attend schools and universities in Vancouver. By one estimate there are some 20,000 graduates of Vancouver educational institutes in Hong Kong alone.

Figure 9

Entrance to the Dr. Sun Yat-sen Classical Chinese Garden, Vancouver

The growth in educational exchanges has been matched by a general rise in interest for 'things Asian.' This is especially evident in the growing popularity of Japanese sushi and other Asian food as well as in the range and quality of the city's Japanese and Chinese restaurants. In 1970, despite an earlier resident Japanese population in Vancouver, there were only a small handful of Japanese restaurants; twenty years later there were at least thirty-five. There have also been changes in the range of Chinese food available: 'supermarket' restaurants have been established outside Chinatown; and there has been a proliferation of Asian restaurants offering cuisines from Vietnam, India, Thailand, and elsewhere.

Festivals, featuring Asian films, and dragon-boat races that bring crews from Hong Kong to row in False Creek, are other signs of the cultural exchanges that occur at many levels. The cultural imprint of Asia is also evident in the Japanese and Chinese formal gardens within the city (Figure 9) and in the Asian-oriented cultural events and/or projects sponsored by the city's many Asia-Canada business organizations, one of which (the Hong Kong-Canada Business Association) has 800 members and is second in size only to the Vancouver Board of Trade.

Investments in the technologies of information flow, which allow long-range control of corporate activities, have done much to foster the emergence of the Pacific Rim as a dynamic economic entity. Computer networks, fax lines, public data bases, and the private control/communications systems of large corporations all ease the friction of transoceanic distance, although Vancouver has only a peripheral position in many of these networks, which remain centred in Los Angeles, Tokyo, and Hong Kong and incorporate important links to London and New York. This means, to use a refreshing example, that someone interested in the distribution of BC beer in Asia would need to contact the international offices of the breweries in New York for the relevant marketing information.

In an attempt to change this situation, all levels of government are attempting to improve Vancouver's centrality in the telecommunications-dominated financial world of the late twentieth century. The provincial government, for example, has established an International

Financial Centre, where members' transactions are exempt from certain provincial and federal tax laws and other regulations, with the hope that this could help Vancouver secure a North American 'west coast slot' in a global financial system. Assisting such endeavours are the more limited city and regional infrastructure programs of civic government which focus on social services, amenity, and transport infrastructure.

Vancouver benefits from its longitudinal position, which allows it to do business with both Asia and Europe on the same business day. But competition is tough, because Los Angeles, San Francisco, Portland, and Seattle all share this advantage. In addition, the most coherent Canadian-based in-house communications networks spanning the Pacific belong to the large Canadian banks centred in Toronto, bypassing Vancouver. Similarly, most Asian banks, other than the Hong Kong and Shanghai Banking Corporation, have established their Canadian head offices in Toronto. Somewhat paradoxically then, for all its close interpersonal family and business connections across the Pacific, Vancouver is poorly served by formal corporate information.

Movement of capital

Since the 1950s, Vancouver and Canada have attracted a good deal of foreign investment. On a national scale, the leading sources of direct foreign investment in Canada have traditionally been the United States, the United Kingdom, and Europe, and much of this investment has been concentrated in the resource industries. In the early 1960s Japan was a relatively minor player in this sector, but over the decade it moved from fourteenth to eighth place among countries investing in Canada. During these years the focus for Japanese investment was in the mining and forestry industries of British Columbia and Alberta, particularly copper-concentrating plants, pulp mills, and sawmills.

In the 1970s, Japanese trading companies were heavily involved in purchasing western Canadian coal and, in particular, the substantial Kaiser Resources deposits near Sparwood in the East Kootenay region of British Columbia. Partly as a consequence, the value of coal production in western Canada rose from $14 million to $103 million between 1969 and 1972. Further expansion was facilitated by the investment of $27.5 million by several Japanese steel makers and the Mitsubishi trading company in the acquisition of 27.3 per cent of Kaiser Resources in 1973. This expansion led directly to the construction of the purpose-built coal loader at Roberts Bank in the early 1970s.

A few years later, after the OPEC-driven escalation of oil prices, Japanese interests were associated with the Tumbler Ridge (or North East) Coal development. This enterprise, predicated on long-term fixed price contracts from Japanese utility companies, involved enormous public investment (estimated at $1 billion) in the development of mines, railroads, and other infrastructure, including the town of Tumbler Ridge. Recently, however, this project's viability was seriously undermined by

Japanese attempts to reduce both the price and quantity of coal contracted – the consequence of a real fall in the world market for coking coal since the early 1980s.

In the mid-1980s, the internationalization of Japanese paper companies (such as Oji Paper Company, Daishowa Paper Manufacturing Company, and Sanyo Kokusaku Pulp Company) generated a new wave of Japanese investment in pulp and paper mills in British Columbia and Alberta. Acquisitions in the Howe Sound project (Oji Paper), and the construction of new plants at Peace River (Daishowa) and Athabasca (Mitsubishi/Honshu Paper Company) in northern Alberta, gave the Japanese a larger presence in the industry, and Vancouver became the head office centre from which these investments are controlled. With this has come a demand for financial, insurance, real estate, and business services, which has contributed to the rise of the service economy in Vancouver. In addition, Japanese imports of Canadian pulp and paper products rose by 23.5 per cent between 1986 and 1987.

In contrast to raw materials, Japanese interest in the city's secondary manufacturing sector has been limited to date; the largest venture is a major aluminum wheel plant established by Toyota. Japanese interest in Vancouver as a banking and financial centre has also been restricted by both the small size of the provincial economy and Toronto's domination of the Canadian financial system. Still, major regional branches of half a dozen Japanese banks, which have their Canadian head offices in Toronto, are located in Vancouver, as is the Canadian head office of the Mitsubishi Bank. And in 1987, Daiwa Securities Company opened an office which competes for local pension funds. In the early 1990s, most Japanese banks and securities companies were unwilling to commit resources to developing Vancouver's financial markets. Strong sectoral competition limited their inclination to extend operations beyond Toronto, their original point of entry into Canadian financial markets. But if current rates of population and job growth continue to favour western Canada, and if the expanding Pacific economy causes a re-evaluation of Vancouver as a financial and business centre, others may follow the lead of the Mitsui Bank of Canada, which became the first Japanese bank to take advantage of Vancouver's designation as an international banking centre by registering as an international finance office under BC legislation.

Over the last thirty years, the Japanese connection with Vancouver has deepened and shifted in and out of sectors which mirror the rapid changes occurring in the Japanese economy. In the foreseeable future, and contingent upon a continuing Japanese tourism boom and expansion overseas of Japanese firms into this field, it is likely that BC ski resorts, hotels, and other property will follow copper, coal mines, and forests and be dominated by the most vigorous economy in the Pacific. Opportunities for a more diversified involvement in Vancouver's economy exist in the city's rapidly expanding software industry, which has the potential to complement Japanese strengths in computer hardware. Also, spin-off commercial entreprises from British Columbia research centres (e.g., the TRIUMF high energy physics project at UBC) may be attractive to Japan as it enters a more technologically focused stage of development.

Toward the future

In order to respond to changes brought about by the advance of flexible organization and the growing importance of Pacific connections, Vancouver City Council established an Economic Advisory Commission in the late 1970s. Its 1983 report stressed the importance of understanding Vancouver's role as a provider of retail, business, and public services to British Columbia and western Canada, and of promoting the city as a key link between Canada and the Pacific Rim.

Five years later, these objectives were redefined as a need to develop Vancouver as Canada's 'Pacific Rim Gateway' city – a centre for trade, finance, travel, and tourism. To further this end, the city is being publicized in a handful of important Pacific Rim cities as a progressive and outward-looking centre of two-way trade and investment. Complementing these initiatives, the Greater Vancouver Regional District's plans for the year 2000 address the regional economy and the need for an infrastructure to support future levels of residential and manufacturing activity in the area.

Together, these schemes envisage the rise of Vancouver to 'world city' status and as part of a 'mega' urban region encompassing both the lower sections of the Fraser Valley and Vancouver Island. Even more extreme are projections that the Vancouver area will become part of a 'New Pacific metropolis' stretching from Portland to Burrard Inlet and linked by a high speed train. This larger region already includes some ten million people, has a substantial industrial base, and has several fine ports that are 1.5 days closer to Asia (by ship) than are Los Angeles and San Francisco – cities which stand as competitors for pre-eminence in the developing circulation between North America and Asia.

Yet development of this sort is bound to be controversial. It would mean many changes for Vancouver. Further interest by Asia-Pacific investors may contribute to rising real estate prices as Vancouver's market is considerably 'undervalued' in comparison with those of Hong Kong and Tokyo. The city may have to sacrifice autonomy to attract capital (by lowering regulatory restrictions, for example). It might also find it necessary to use scarce resources to enhance local amenities, especially those emphasizing livability and consumption, in order to entice the new global wealth. Moreover, as it is unlikely that even a 'New Pacific metropolis' will join the first rank of world cities, Vancouver might well become a centre of small dependent enterprises spawned by the process of vertical disintegration, in which large corporations subcontract facets of their business to cut risk and increase flexibility. Because small firms need to be extremely flexible in meeting the demands of the quickly changing global marketplace and of the large corporations which purchase their products or services, they can be a dynamic, but very unstable, element in the economy. All of this is likely to lead to extremely high levels of social polarization – which has been the experience of other world cities.

Conclusion

Since the Second World War, changes in the relationship between site and situation have had a profound effect on Vancouver. In the 1950s and 1960s the city was principally a distribution and local control centre for the province's resource industries. Such industries, often part of larger foreign-owned corporations, extracted and processed primary resources throughout the rest of the province using large-scale production methods characteristic of Fordism. These resources were exported, especially to the large American market, through Vancouver. The city was clearly important but, primarily, as a facilitator, articulating between British Columbia's resource periphery and American buyers. More generally, there was a certain stability to this period, despite the periodic booms and busts of the economy. Residents of Vancouver knew what to expect with regard to wages and work, the nature of government and its expenditures, and life within the city. Vancouver was a cozy place, slightly in the backwaters, but one with a clear sense of itself, stemming, in part, from seemingly enduring relationships with the rest of the province and the outside world.

During the 'whirligig' years of the 1970s and early 1980s, the world and Vancouver seemed to remake themselves. Now seems quite different from then. With improvements in telecommunication and transportation equipment, and the pervasiveness of multinational corporations, there are no longer any backwaters. Rather, all places are hierarchically linked into an international economic order characterized by a constant global circulation of money, people, information, and investment. In this new era of post-Fordism, Vancouver is no longer simply a facilitator. It is now an active participant, particularly among the countries of the Pacific Rim. Of course, it maintains its former role as a local distribution and control centre, but it is a lot more besides. The consequence is that the coziness of Vancouver is lost; the 'world city' and the 'Pacific Rim' are now the twin planks of a new civic ideology. Yet this ideology often fails to recognize the stresses and strains that change brings: the creation of peripheral groups of service workers, tensions between old immigrants and new, inflation in the real estate market and consequent homelessness, and congestion and environmental degradation. Clearly, to opt out is not a viable option. But perhaps, with political will and energy, it will be possible to redress the now seeming imbalance and reassert the importance of Vancouver's site over the forces of its geographical situation.

7

Vancouver since the Second World War: An economic geography

Robert N. North
Walter G. Hardwick

Between the end of the Second World War and 1991 the population of what is now the Greater Vancouver Regional District nearly tripled, from half a million to over 1.4 million (Figures 1 and 2). The city rose to prominence both within the nation and within the emerging economic system of the Pacific Rim. There were a number of profound changes during this period. New economic activities appeared and old ones relocated. The city expanded spatially, and the functional links between the downtown and the suburbs changed. Old administrative arrangements proved inadequate, and their replacements were no longer of appropriate scale by the end of the period.

This chapter describes the major changes in Vancouver's economic landscape in the decades since the Second World War and examines the processes that contributed to these changes. Three phases of growth are recognized. The first phase runs from the end of the war, through the years of postwar expansion and into the late 1950s, when the city, along with Canada, experienced an economic downturn. The second phase extends from the early 1960s to the recession of 1981. And the third phase begins with the Expo '86 world fair and continues into the 1990s.

Postwar expansion, 1945–60

When the Second World War ended, Vancouver was still a provincial city where a myriad small firms directed the extraction of raw materials, imported processed goods, acted as shipping agents, and provided goods and services to the local population. To a prewar economy heavily dependent on the export of grain and lumber, the war had added shipbuilding and the manufacture of munitions. Indeed, manufacturing grew faster during the war than in any other Canadian city. But the peacetime legacy of this growth left Vancouver with more facilities than production and, in terms of economic importance, the city remained a distant third, behind Montreal and Toronto.

In the late 1940s, there were few high-rise buildings downtown, and these were more likely to be occupied by doctors, dentists, insurance

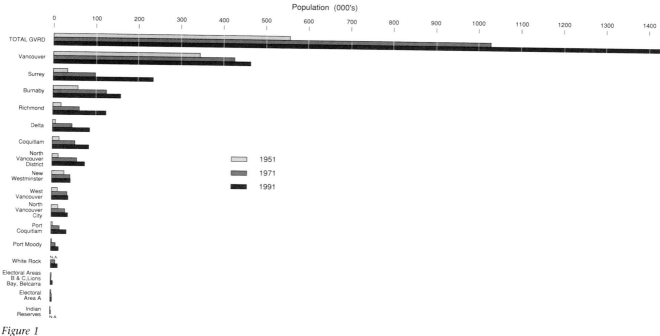

Figure 1

Greater Vancouver: population, 1951-91

agents, brokers, bailiffs, and accountants than by captains of finance and industry. In the ring of warehouses and industry (sawmills, ship-yards, metalworking firms, and fish plants) that encompassed downtown on the Burrard Inlet and False Creek waterfronts and adjacent streets, mills and plants characteristically employed fewer than fifty persons and occupied lot- or block-sized sites rather than the acreage common in cities with more industrial employment. Both specialized and mass consumer shopping were concentrated downtown, while shops on streetcar ribbons provided day-to-day foodstuffs and other convenience goods for shoppers from largely single-family residential suburbs. There were more sawmills, fish canneries, and boat yards along the North Arm of the Fraser River. The original Eburne sawmill at Marpole (where the Canadian Pacific Railway's Lulu Island branch line crossed the river) became part of a Canadian Forest Products complex in 1940. Industry extended eastward from there to beyond Mitchell Island (Figure 3).

Settlements outside the Burrard Peninsula were loosely connected to Vancouver. Two bridges across Burrard Inlet linked West and North Vancouver to the city: at First Narrows a suspension bridge for road traf-fic had been built shortly before the war, and the Second Narrows was crossed by a double-decked road and railway bridge. There was a similar bridge over the Fraser River at New Westminster, and the nearby Pattullo road bridge had been opened in 1937; the roads to Vancouver, however, were simply arterial streets.The North Arm of the Fraser was spanned at Marpole and Fraser Street by two low-capacity bridges with opening swing spans for river traffic (Figure 4).

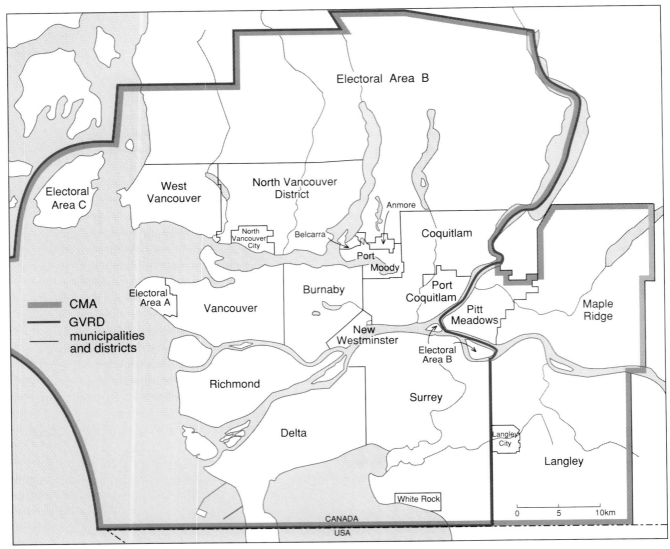

Figure 2

Boundaries of the Greater Vancouver Regional District, Census Metropolitan Area, municipalities, and districts

The two North Shore communities differed radically (Figures 1, 3, and 5). West Vancouver had eschewed industry, but in North Vancouver approximately a thousand workers were employed in two dozen forest products firms and in the principal shipyard. In north Burnaby and Port Moody a minor industrial concentration generally replicated the activities of Vancouver's downtown waterfront. An exception to this was the main regional concentration of petroleum terminals and processing facilities located east of the Second Narrows Bridge.

A separate belt of industry stretched along the North Arm of the Fraser River from southeast Burnaby through New Westminster to Fraser Mills. New Westminster continued to attract industry, and many of the region's largest sawmills and plywood and paper plants were located there. The size of these operations is indicated by the fact that they employed five times as many people in approximately the same number

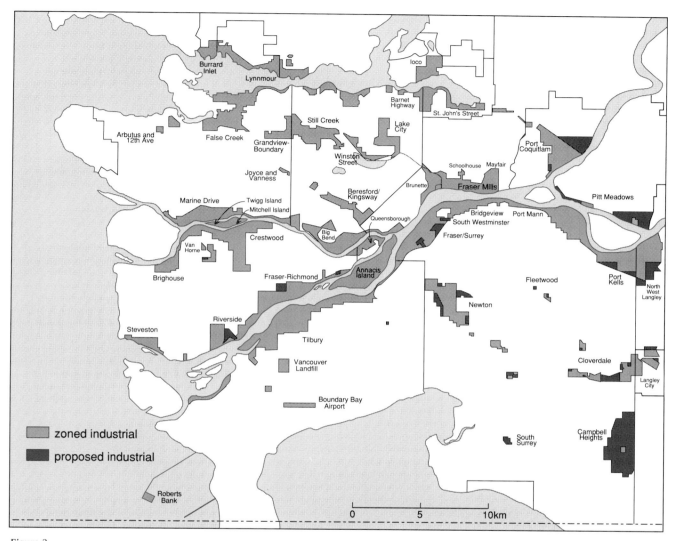

Figure 3

Greater Vancouver: industrial areas, 1990

of plants as there were in North Vancouver. Another 2,500 workers were employed in a much wider range of industries than on the North Shore, particularly in transport and, seasonally, in agricultural processing. Another dozen lumber companies operated on the Surrey side of the river, but such other industrial activity as there was south of the Fraser River was scattered and heavily dependent on agriculture, fishing, and peat-cutting, and, therefore, of a much more seasonal nature than that north of the river.

On the national scene, in the decade after 1945, the government of Canada promoted a laissez-faire market economy, allowing for the free flow of capital, labour, and energy across the country. This policy stimulated economic growth in Montreal and Toronto at the expense of the Maritimes, rural Quebec, and parts of the Prairies. Vancouver also benefited from expansion in the national and international economies, which increased the demand for raw materials and brought both

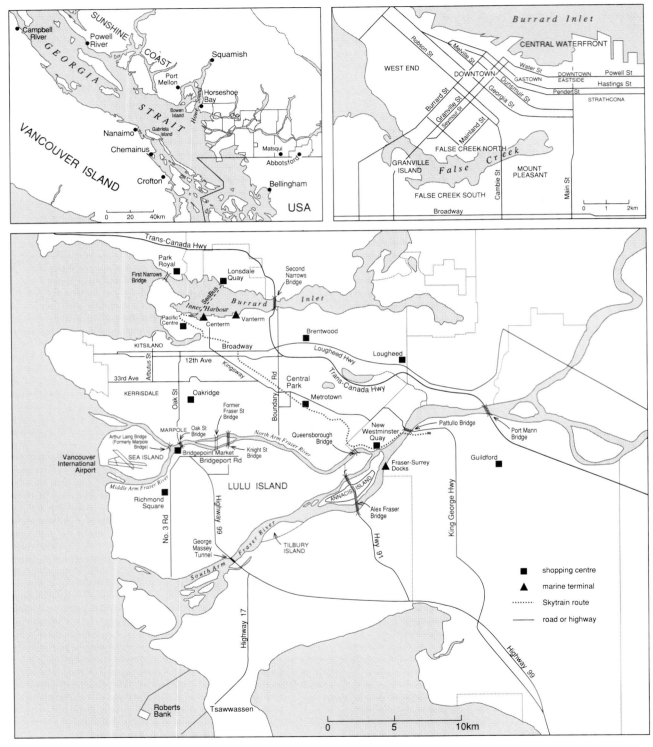

Figure 4

Places referred to in the text

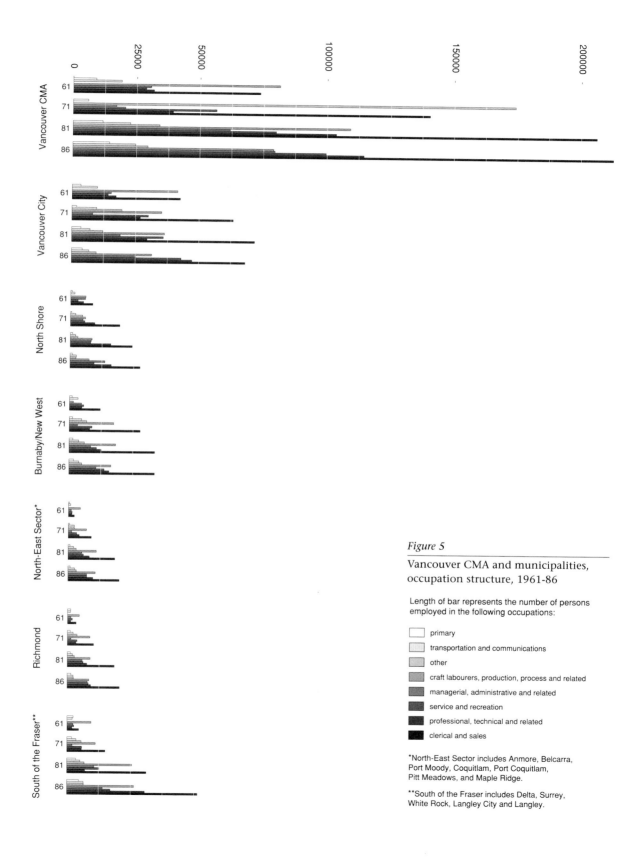

Figure 5

Vancouver CMA and municipalities, occupation structure, 1961-86

Length of bar represents the number of persons employed in the following occupations:

- primary
- transportation and communications
- other
- craft labourers, production, process and related
- managerial, administrative and related
- service and recreation
- professional, technical and related
- clerical and sales

*North-East Sector includes Anmore, Belcarra, Port Moody, Coquitlam, Port Coquitlam, Pitt Meadows, and Maple Ridge.

**South of the Fraser includes Delta, Surrey, White Rock, Langley City and Langley.

Canadians and immigrants into the city. But the scale of expansion paled when compared to those of the eastern centres. Local entrepreneurs sought to meet the challenge of the times through modernization and by enlarging their scales of operation, which led to vertical and horizontal integration. Companies, such as the forestry giant MacMillan Bloedel, emerged through a series of mergers and acquisitions to encompass every aspect of the forest business from logging to the overseas marketing of lumber, newsprint, and pulp. In the fisheries, two companies – BC Packers and the Canadian Fishing Company – became dominant. And in mining, national firms such as Noranda began to play as important a role as did the small speculators financed by the penny stocks of the Vancouver Stock Exchange.

The process of industrial consolidation shaped the landscape of the city in two ways. First, natural resource companies began to build head offices downtown. Among the first purpose-designed office buildings erected since 1929, they included MacMillan Bloedel on Pender Street, BC Forest Products on Melville Street, the Utah Mining Company on Burrard Street, Rayonier on Georgia Street, and the BC Electric Company on Burrard (Figure 4). Second, small sawmills and woodworking establishments along the waterfronts of False Creek and Burrard Inlet were closed, and bigger mills were constructed or old ones were extended along the Fraser River and in the outports of the Strait of Georgia. Although the City of Vancouver had an industrial commissioner, whose job, in part, was to attract industry to the facilities built during the war, many shipyards and metal fabrication plants closed or moved out. Wholesale warehouses on Water Street (adjacent to the harbour rail lines) and along Mainland Street on False Creek were vacated. Their functions were transferred to larger sites along the railways and highways in the suburban municipalities, particularly in Burnaby and, later, Richmond. Some wholesalers moved into purpose-designed industrial parks such as Lake City in Burnaby or Annacis Island near New Westminster, which were both opened in the 1950s. Business migration was further encouraged by improvements in access from downtown; the Oak Street Bridge to Richmond, for example, was opened in 1957.

As the suburbs grew, retail facilities were established to compete with those downtown. Park Royal, opened in 1950 in West Vancouver, was the first automobile-oriented shopping centre in Canada (Figure 6). Seven years later the Oakridge centre was built halfway between False Creek and the North Arm of the Fraser River. Both competitors were within five kilometres of downtown, but the city core maintained the advantage of centrality. Until well into the 1950s, the streetcar system provided efficient access from the suburbs. Land in the vicinity of the Granville-Dunsmuir and Granville-Pender intersections was the most expensive in the region, and it was here, adjacent to the growing office and financial districts, that Vancouver's prime shopping facilities were

Figure 6

Park Royal Shopping Centre, West Vancouver, Sunday, 10 May 1959. The first planned shopping centre in Canada, built by Guinness interests to serve their nearby British Properties residential development. View looking east along Marine Drive to the Capilano River. The original shopping centre is on the left, and the White Spot drive-in restaurant is on the right of Marine Drive with cars parked outside. Within two years, most of the forest was cleared for the new South Mall.

concentrated. During the postwar years the range of services offered, both market and non-market, rose sharply. The growth of population in the city itself, the general growth of disposable income, the building of civic parking facilities, and shrewd merchandising helped maintain the dominance of the core. Although a few peripheral retail stores were abandoned, along Hastings and Seymour streets in particular, downtown Vancouver was a more lively and profitable place than was the centre of almost any city in North America.

As the North Shore municipalities, Burnaby, and Richmond became more urbanized and integrated with the city, the fragmentation of political power posed new and difficult problems. Throughout the 1950s the Vancouver city government had dominated regional policy and planning discussions, and the adjacent municipalities had played reactive roles. This was changing, however, and the latter were becoming more assertive. The municipalities were already linked formally through special-purpose districts, such as the Greater Vancouver Water District and the Joint Sewerage and Drainage Boards, and recognition of the need for a wider regional view came with the establishment of the Lower Mainland Regional Planning Board in 1949. Nonetheless, progress toward effective regional government was slow.

Toward the livable region, 1960–81

The city and the province suffered an economic downturn in the early 1960s. Blight marred the fringes of the urban core, and Vancouver City Council was haunted by the spectre of American-style urban decay, which it feared might imperil the regional centrality of the downtown area. So American planners and business consultants were brought in to advise on revitalizing the downtown economy. They saw the city through American eyes and urged publicly financed commercial revitalization, city-subsidized parking, and new cultural facilities. Following this strategy, the city expropriated the core shopping block on Granville Street and leased it to the Cemp/Eaton interests, on very favourable terms, to build a department store, hotel, and office buildings. This initiative was based on the assumption that downtown Vancouver was as vulnerable to abandonment as were the centres of American cities. The city provided parking and charged low ground rents. Yet new office construction downtown was under way as early as 1966 at the Bentall Centre on Burrard Street, without any civic subsidy. Shopping too was already being rejuvenated, particularly on Robson Street, as the West End high-rise buildings accommodated more and more people with inner-city interests and tastes.

To promote the continued regional dominance of the downtown area and to improve truck access to the central waterfront, the American consultants also recommended building an urban freeway. It was around this issue, however, that neighbourhood community feelings coalesced, becoming a political force. The freeway would have destroyed both Gastown (the early commercial and warehousing district) and Strathcona (the main residential area for the Chinese community). The scheme was eventually defeated, leaving Vancouver as the only major North American core city without an urban freeway.

By 1968 the city council had expanded its outlook and was promoting Vancouver as a business and financial centre for western Canada. There were efforts to attract federal agencies, to encourage banks and trust companies to increase their local presence, and to ensure that decision-making in the expanding resource, transportation, and energy sectors was focused in Vancouver. New head and regional offices prompted a rapid expansion of office space in the core.

From the debates on the future of Vancouver emerged a reform political party, The Electors' Action Movement (TEAM). TEAM members articulated the growing sense that Vancouver was different from American cities, that it was developing on its own trajectory, and that it needed made-in-Vancouver plans. They recognized that a post-industrial society was emerging and that city residents were less and less dependent on jobs in the port and in primary industry. They saw the future of the city centred around management and service activities, from producer services to finance, tourism, and the information industries. They

Figure 7

City of Vancouver: land suitability for
industry and rezoning prospects

☐ suitable for continued industrial use
and likely to remain in industrial use

▨ suitable for continued industrial use but
uncertain or unlikely to remain in industrial use

▨ not suitable for continued industrial use and
unlikely to remain in industrial use

■ former industrial land
rezoned for non-industrial use 1968-1988

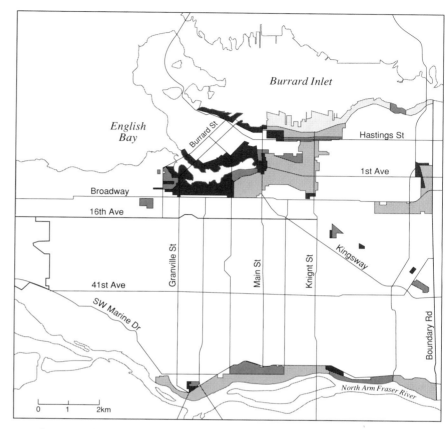

sought to make the city more livable for a population engaged in such activities. Universities, colleges, and science parks were expanded with the recognition that higher education, research, and development were important components of the new order. And the desirability of creating a high-quality public environment, with safe, attractive, and accessible neighbourhoods, was widely acknowledged. These designs had the support of the majority of city electors and provided the basis for a broad political consensus for more than a decade. The central waterfront and lands around False Creek were rezoned away from industry (Figure 7), and the south side of False Creek was completely redeveloped into medium-density housing (Figures 8 and 9). On Granville Island (a decayed industrial area that belonged to the federal government), old buildings were converted to theatres, shops, restaurants, a college of art, and a public market, which became a model for others both locally and internationally.

As the city was turning away from encouraging industrial growth, the southern municipalities sought industries to broaden their tax bases and offset the cost of servicing hitherto little-planned residential development. The aims of both these municipalities and Vancouver were promoted by the improvement of transport links southward from Vancouver. This process took place throughout the period. The Oak

Figure 8

Granville Island and False Creek in December 1961, before development. The north shore is dominated by the CPR yards and adjacent warehouses. On the water, log booms await utilization in the many sawmills. The main port on Burrard Inlet can be seen to the north, with working-class suburbs to the east. The large tank in the upper centre was part of a coal gas manufacturing plant, long gone with the adoption of natural gas but a source of modern-day soil contamination.

Figure 9

The south shore of False Creek in 1990 after redevelopment. The housing enclaves have a variety of tenure arrangements to accommodate households with a range of incomes, ages, and life-styles. The star-like marina is a live-aboard co-operative.

Street Bridge was followed by the Deas Island (now George Massey) Tunnel under the Main Arm of the Fraser, a freeway linking the two and continuing to the United States border, and a connecting highway to a new Vancouver Island ferry terminal at Tsawwassen (Figure 4). All of these were completed by 1962. Shortly afterward came the Port Mann Bridge at New Westminster, which carried the Trans-Canada Highway over the Fraser, and the Queensborough Bridge to Lulu Island. Then in the mid-1970s the Knight Street Bridge over the North Arm of the Fraser replaced the bridge at Fraser Street, and the Arthur Laing Bridge connected Marpole to Sea Island and the Vancouver International Airport.

With the completion of these facilities, the southern municipalities attracted planned industrial estates, the first of which was on Annacis Island. By the early 1970s Richmond had three of them – Brighouse, Van Horne, and Crestwood – each competing with similar parks elsewhere in the region to attract mainly light industry (Figure 3). The municipalities also co-operated to promote the development of Fraserport, covering the South Arm and main channel of the Fraser, which offered many sites suitable for heavy industry and deep-sea docks (Figure 10).

Other initiatives raised the status of road transport relative to all other modes in the region. Construction of the ferry terminal at Tsawwassen, to provide service to southern Vancouver Island and the Gulf Islands, hastened the replacement of privately operated coastal steamers by roll-on roll-off ferries. The latter, introduced by Black Ball Ferries, were soon

Figure 10

Fraserport: freight traffic, 1945-89

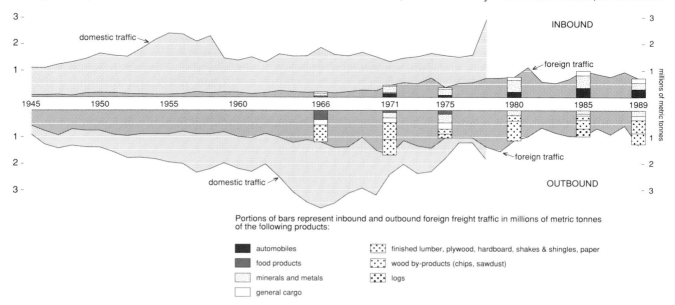

Portions of bars represent inbound and outbound foreign freight traffic in millions of metric tonnes of the following products:

- automobiles
- food products
- minerals and metals
- general cargo
- finished lumber, plywood, hardboard, shakes & shingles, paper
- wood by-products (chips, sawdust)
- logs

Domestic traffic figures not available after 1978.

taken over by the provincial government to link its highway systems. In the late 1950s the road and ferry system was further integrated by expansion of the Horseshoe Bay ferry terminal in West Vancouver and the development of a new high-speed section of the Trans-Canada Highway through West and North Vancouver to the Second Narrows Bridge.

Finally, in another government initiative, in 1967 the provincial government set up regional districts as co-ordinating and planning bodies for groups of municipalities. It then abolished the Lower Mainland Regional Planning Board. The area covered by the board, which extended 160 kilometres up the Fraser Valley to Hope, was divided into four regional districts. The Greater Vancouver Regional District (GVRD) included Surrey, Port Coquitlam, and eleven other jurisdictions westward to the Strait of Georgia (Figure 2).

In 1971 Vancouver was a core-ring city. It was made up of a radially organized, core-focused inner zone (the legacy of early twentieth-century expansion) comprising Vancouver, the North Shore, and parts of Burnaby, and a ring of recent suburbs. In this new, fast-growing outer area there was more mobility within and among suburbs than between most of them and the city core. Whereas commercial patterns in the core were linear along radial arterial streets, suburban commerce was far more nodal, reflecting the central place configuration of shopping centres. And whereas employment in the city was focused on the central business district, in the suburbs it was spread over dispersed industrial and office parks and along the Fraser River.

In 1975 a new regional plan was adopted by the GVRD and its constituent municipalities. Called the 'Livable Region Strategy,' the plan established goals for the next fifteen years, which were intended to preserve and enhance the livability of the region. The plan attempted to bring order to development that was now sprawling well to the south and southeast, the only directions in which population and industry could spread. Among the most important elements in the design were plans to decentralize work and residence, to make transit a more attractive alternative to commuting by car, to stimulate regional town centres for shopping and services, to encourage medium- and high-density residential development in the vicinity of these town centres, to expand public open space, and to start an environmental monitoring process. Both public and private sectors embraced the plan, and the rudiments of a new urban form emerged. Regional town centres were founded, and several key employers relocated in the suburbs. The BC Telephone Company built new headquarters near Central Park in Burnaby, and the federal Regional Taxation Centre moved to Surrey. Such relocations limited total aggregate travel. This was also achieved by improvements in the regional transit system that provided not only radial, but also crosstown, services. Express buses began to run from outer suburbs such as White Rock, and they were given priority access to bridges and transit-only lanes on arterial roads. But several initiatives that required provincial government action were stillborn, because the Social Credit government, in electoral power from 1976, was much less proactive than was its New Democratic predecessor.

Industry and many transport-related functions also moved to the suburbs during the period. In the 1970s the municipalities south of the Fraser River increased their share of the GVRD's industrial output from 20 to 40 per cent. Several factors were responsible for this trend. With growth in the scale of traditional processing, manufacturing, and freight-handling activities, these functions needed more space than they could afford and, in many cases, more than was physically available at most locations near the city centre.

Space needs were also increased by technological change, particularly affecting freight-handling and warehousing methods. The introduction of fork-lift and straddle trucks made it more economical to handle general and packaged freight on one level. Old-style, compact multi-storey warehouses fell out of favour. They were replaced by single-storey buildings or simply open areas for weatherproof freight. Enough space for such operations was not easy to find around Burrard Inlet and False Creek. The Fraser delta lands, by contrast, had ample space, and the lack of a firm subsoil in the area was no longer a great disadvantage, since the new style of warehouse placed relatively light loads on the ground. At the same time, high-speed transfer facilities – modern grain elevators and coal loaders, for example – were introduced for bulk freight, which was the main business of the Port of Vancouver. These new facilities placed great demands on the railways serving them. More space for trackage was needed to maintain an uninterrupted flow of freight cars, and both the CPR and the CNR moved their classification yards. The former moved from the north side of False Creek to Port Coquitlam, vacating land which became the site for Expo '86; the latter went to Port Mann on the south side of the Fraser, east of New Westminster. Still, these moves alone could not fully compensate for the lack of space adjacent to waterfront facilities on the south shore of Burrard Inlet, especially as new types of traffic – such as containers and automobiles – demanded huge areas for storage. All of this prompted the relocation of activities from Burrard Inlet to the Fraser Delta. In addition, some types of industry moved out of the city, or chose an initial location on the fringe, in order to distance themselves from neighbours who might object to their noise, smell, or other unpleasant characteristics. Examples include the Port Mellon pulp mill, built on Howe Sound in 1907 and converted after the Second World War from sulphite to sulphate, and the Dow Chemicals plant on Tilbury Island in the South Arm of the Fraser.

Technological changes in ocean shipping restricted the locational choices of many industries and imposed a high level of spatial sorting on freight-handlng activities. As economies of scale were pursued, extremely large vessels came into use for some purposes, including the transport of coal. Even Burrard Inlet could not handle the largest of these, so the Roberts Bank superport, near the Tsawwassen ferry terminal just north of the U.S. border, was built to accommodate them. Opened in 1970 and dredged to 75 feet (22.9 metres), it is now the largest coal terminal in North America. Vancouver's Inner Harbour, with berths dredged to 45 and 50 feet (13.7 and 15.2 metres), can accommodate smaller coal carriers and most other bulk freighters. The South Arm and main channel of the Fraser can take vessels drawing up to 31 feet (9.4 metres) as far as New Westminster, but, even for vessels of this relatively modest size, the Fraser is far from an ideal channel. Maintaining advertised depths in the river sometimes requires dredging every few days at Fraser-Surrey Docks, opposite New Westminster; larger vessels have   difficulty turning in the river at low water; and in certain conditions adequate depths across the shoals at the river mouth are available for less than an hour a day. Finally, the North Arm of the Fraser offers depths of only 14.8 feet (4.5 metres) – adequate for tugs, barges, log booms, fishing boats, and small coastal craft but not for deep-sea vessels.

As a consequence of all these factors – water depths, land costs, and space requirements – most bulk-handling facilities, including all the grain terminals, came to be located on Burrard Inlet. New facilities to handle the coal, sulphur, and potash exports, and phosphate rock and salt imports, that are an ever more prominent component of port traffic, were built by private interests, mainly in North Vancouver (Figures 11 and 12). There, waterfront land was less congested than in Vancouver, and the extension of the Pacific Great Eastern (now BC) Railway from Squamish provided a direct link to the northern interior of the province from the late 1950s. Container terminals, on the other hand, opened on the south side of the inlet, largely through the redevelopment of old finger pier sites on the initiative of the National Harbours Board and its successor, the Vancouver Port Corporation. In 1956 a large cargo-handling area, named Centerm, was developed on the Hastings Mill site, and container-handling facilities were added in 1970. Three years later another container terminal (known as VanTerm) opened a little farther east.

Forest-products and construction-materials industries continued to focus their activities on the North Arm of the Fraser. Raw materials were brought in by barge or boom, and finished products were shipped out by rail or road; overseas exports had first to be barged to docks with deeper water. Light industries situated back from the river insulated the heavy industries on the waterfront from commercial and residential uses. Some long-established sawmills also remained on the South Arm and main channel of the Fraser, but most new forest products plants were built farther afield, especially on Vancouver Island. Land costs there were low and raw materials could be brought in by water. The

Figure 11

North Vancouver waterfront, west of the Second Narrows. The Saskatchewan Wheat Pool grain elevator, completed in 1966, is one of the newest in British Columbia. Neptune Terminals, the only remaining coal facility on Burrard Inlet, also handles potash, rapeseed oil, and phosphate rock. Seaboard Terminal, at the bottom right of the photograph, was opened in 1971 on reclaimed land and is one of the largest packaged-lumber terminals in the world.

Figure 12

North Vancouver waterfront, east of the Second Narrows. Canadianoxy Chemical Ltd. imports salt from Mexico and manufactures chemicals for local pulp mills. Note the large area of land which can be made available for waterfront industry on the north shore of Burrard Inlet in comparison with the more restricted space on the south shore.

Figure 13

Annacis Island, looking downstream along the Fraser River. The island is entirely occupied by a privately owned industrial estate. It was developed by the British Grosvenor interests in the 1950s and reflects the suburban industrial park development strategy of that time. It attracted several companies which migrated from Vancouver in the 1970s and 1980s. Its locational advantages grew with the opening of the Alex Fraser Bridge (upper left) in 1989. Note the large terminal for imported Japanese and Korean cars (bottom right).

Harmac pulp and paper mill in Nanaimo and the Crofton mill to the south provide cases in point (Figure 4). By the late 1960s ocean freighters commonly assembled full cargoes of wood, pulp, and paper from several locations within the BC port system.

Old-established chemical industries also remained on Burrard Inlet, east of the Second Narrows Bridge, but new plants gravitated to relatively remote locations along the South Arm and main channel of the Fraser. So, too, did car-unloading terminals, attracted by low-cost land necessary for their extensive parking areas (Figure 13).

By the 1960s Vancouver International Airport was also an important focus of industry. Set up as a municipal airport in 1931, it was taken over by the federal government in 1961 and greatly expanded. Eventually it attracted over a hundred firms engaged in such activities as aircraft servicing, airline catering, the training of aircraft mechanics, and freight brokering. Many of them necessarily located on Sea Island; others occupied nearby industrial estates. In the late 1980s airport-related activities were far and away the largest employers in Richmond.

Other light industrial activity in the region can be divided into two broad groups. First, warehousing, distribution, and light manufacturing (for the local consumer market and, in the case of electronics and machinery especially, for North American and global markets) comprise a large and growing proportion of such activities. Land costs and road access to the regional market are the essential determinants of their location, because labour is a highly mobile and relatively unimportant input. As the centre of gravity of the regional population shifted south and east and new roads and bridges were opened, the attractiveness of locations changed. Industrial estates south of the river were most attractive immediately after the building of each new bridge, but increasing traffic eventually brought congestion and raised the relative attractiveness of estates in Burnaby and other municipalities to the east, as well as of the old industrial zones in Vancouver. A second group of light industries tended to locate closer to downtown. Some did so because they serviced downtown office or waterfront activities. Others were infant industries, attracted by relatively low rents in small old premises and the ready availability of services.

Some indication of the way in which these forces played on the fabric of the city is given by a recent study of fifty firms that gave up their False Creek locations during the redevelopments of the 1970s and 1980s. Sixteen, mostly in heavy industry, went out of business. Having failed to move from small lots and outdated facilities, most were at a competitive disadvantage and could not bear the cost of relocation. Of the rest, eight stayed in Vancouver. Five moved to the Marine Drive industrial area and three to other widely dispersed locations. Another eight moved to Burnaby, with those involved in wholesaling and distribution locating on industrial estates and heavier industries moving to the waterfront on the North Arm. In addition, four firms moved from False Creek to Annacis Island (engineering and road transport), four to North Vancouver (mainly marine-oriented), four to Surrey, and three to Richmond. The rest went variously to Port Coquitlam, Abbotsford, and Washington State. All firms reported that they were satisfied with their moves, which had given them more space and more efficient premises and had rarely hampered their access to suppliers or customers (although both might have changed since the move).

As the population of the Vancouver region grew and dispersed southeastward, new suburban shopping centres were built to meet its needs. Generally, they followed the common form of such places in an increasingly car- and consumer-oriented society. In physical layout and store mix, Brentwood, Lougheed, Richmond Square, and Guildford malls were strikingly similar to shopping centres elsewhere in the country. At the same time, despite the planning initiatives of the GVRD, classic North American commercial 'strips' with their land-extensive, neon-lighted, auto-oriented mix of car dealers, muffler shops, fastfood outlets, building-supply stores, and discount furniture warehouses spread along the region's arterial highways. Number 3 Road in Richmond, King George Highway in Surrey, and Kingsway in Burnaby are examples of such blighted landscapes. Shopping facilities in downtown Vancouver began to take on a dual character. Some stores, such as high-fashion outlets along Robson Street and ethnic and other specialty shops, catered to a regional clientele or the tourist market. Others, including the Pacific Centre Mall, offered little that was different from suburban malls. Their principal market was local – the permanent population of the West End high-rises and the daytime population of the downtown offices.

Offices also became suburbanized but not as quickly as did shopping and industry (Figures 14 and 15). By 1990 the downtown contained 56 per cent of all office space in the GVRD, 7 per cent less than five years before. Yet the city still claimed 80 per cent of the GVRD total. Among the regional town centres, only Metrotown in Burnaby attracted significant amounts of office space. Major business became ever more

Figure 14

Greater Vancouver: warehouse and light industrial floorspace, 1980 and 1990

1980
1990

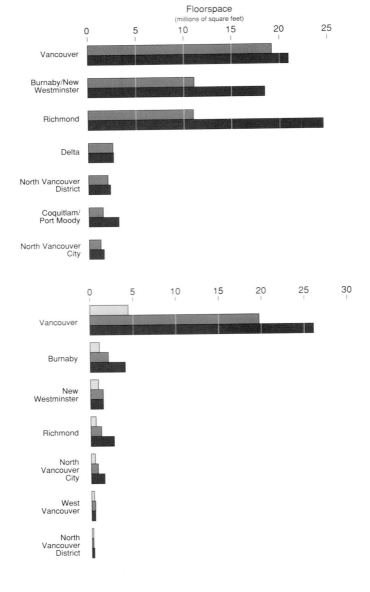

Figure 15

Greater Vancouver: office floorspace, 1976-90

1976
1981
1990

concentrated in the downtown. The stock exchange, the regional headquarters of all the major banks, and most other financial institutions and corporate headquarters were located there. New office towers housed up to ten times the number of workers that industry could have sustained in the same area, and the proportion of managerial staff among downtown employees expanded from 4.5 per cent in 1971 to 11.7 per cent in 1986. For all that, the entrepreneurial, decisionmaking leaders of several new industries chose to locate next to their manufacturing plants, even if these were in the suburbs. They included the computer-remote sensing firm MacDonald Dettwiler, Epic Data, and Mobile Data International (the largest mobile data manufacturer in Canada) in Richmond, and Nexus Engineering, a manufacturer of satellite telecommunications equipment, in Burnaby.

Among the most significant of all changes between 1961 and 1986 was the transformation of Vancouver's service sector. At the beginning of the period, most services catered to household consumers. These services typically included public schools, local health facilities, and leisure facilities, and their numbers and location reflected the distribution of households and income levels. By the late 1980s such services were far more numerous than they were in 1960, and they were offered in town centres, office parks, and strip developments throughout the region. Still, their share of the regional workforce had declined over twenty-five years. By contrast, marketed producer services, which produced and exchanged information and depended upon electronic communications networks, had risen to prominence and were heavily concentrated downtown. By the 1980s, such functions as accounting, management consulting, public relations, law, research, and engineering, which large companies once provided in-house, were widely offered by independent firms or partnerships that specialized in meeting the complex needs of decisionmaking élites in both the public and private sectors. They were an industry in their own right. Hotels, too, had left behind their ancillary status to become significant components of the downtown business complex due to the effective marketing of tourism and convention services.

Other expanding services had located outside but close to downtown. They included design services (attracted by lower rents) and centralized medical services. Indeed, the medical zone east of Oak Street and south of Broadway, including the Vancouver General , Childrens', Grace, University (Shaughnessy site), and St. Vincent's hospitals and associated facilities, has become and remains among the half-dozen largest in North America.

As the number of jobs downtown increased, high-density housing expanded, first in the West End and then in Kitsilano and Mount Pleasant. Zoning regulations and locally developed concrete high-rise technology meant that most such building occurred as multi-storey apartment blocks. With the transformation of False Creek South into a medium-density inner-city residential area, large numbers of young people entering the labour force were able to live close to downtown in amenity-rich neighbourhoods, some of which soon became identified as particularly fashionable ('yuppie') enclaves (Figure 16). Other neighbourhoods, such as Kerrisdale, Ambleside, and South Granville, attracted waves of empty-nesters – aging persons who traded large single-family homes for quality condominium suites and townhouses. In the early 1980s the redevelopment of Lonsdale Quay in North Vancouver and New Westminster Quay, in particular, provided attractive dwellings for empty-nesters from suburban municipalities and from the Prairies (Figure 17).

Figure 16

Granville Island and False Creek South. A view looking west from Granville Bridge to the Burrard Bridge. On Granville Island (bottom right), Bridges restaurant was formerly the headquarters of Arrow Transfer Company (now on Mitchell Island in the North Arm of the Fraser River). South of Granville Island is some of the denser apartment housing built recently at False Creek South; south of that are the light industrial areas stretching to Fourth Avenue, a product of urban renewal policies in the 1950s. This area is dominated by the Molson Brewery (top left).

Figure 17

The New Westminster Fraser River waterfront. Comprehensive redevelopment has replaced docks and waterfront industry by condominiums (bottom left), a hotel, and a covered public market (bottom centre-right). Note the proximity to a Skytrain station and to Douglas College (right, above the highrise block). Redevelopment has taken place alongside railway yards, which still support sawmills and industry further to the west.

Most families with children migrated to the outer suburbs, where land was cheaper and new subdivisions were being opened up with striking rapidity. Long-distance commuting increased, and this produced new challenges for the planners.

Expo and renewed growth, 1986-92

After the economic recession of the early 1980s, which stripped over 30,000 jobs from the Vancouver economy during a period of modest population increase, the provincial government sponsored Expo '86. It was hoped that the world's fair would focus international attention on the city and boost the provincial economy. While it was a great public relations success, the growth and prosperity that marked the Vancouver region in the late 1980s cannot be attributed to Expo alone. After 1985, commodity prices began to inch upward. Economizing measures and reorganization made industry more competitive than it had been before the recession: the forest products industry, for example, invested heavily in new technology, which both reduced labour costs and improved its ability to vary the type of output. And the provincial government invested in a series of major public works to stimulate economic growth, including the North East Coal Project at Tumbler Ridge, the Coquihalla Highway, and the light rail 'Skytrain' system between downtown Vancouver and New Westminster (all of them billion-dollar projects). Massive investments in the resort town of Whistler during the early 1980s made it a major destination for international tourism; immigration to the region increased; and, as in every other boom in the city's history, prosperity was propelled by the building of the city itself in terms of housing, offices, warehouses, transit lines, bridges, sewers, and roads.

The consequences of the boom for land use were first evident in the construction of even more offices, hotels, and residences in downtown Vancouver and the expansion of the latter to include the Broadway corridor, two kilometres to the south. In 1990, Vancouver, New York, and San Francisco had more downtown office space in proportion to their populations than any other city in North America. Hotel construction was stimulated by demonstrable growth in international travel. Its considerable impact on the city is manifest in the transformation of the federal Expo pavilion, itself a redevelopment of an old finger pier on the central waterfront, into a convention centre, along with a cruise-ship terminal (catering to the Alaska trade) and a hotel (Figure 18). As for residences, Vancouver is remarkable for having such a great variety of housing within 2-3 kilometres of the centre. Even the downtown eastside, where the old skid road was located, has, under indigenous leadership, improved its environment, housing, and community life more than have comparable areas in almost any other city. Moreover, in 1992, several new neighbourhoods are planned for, or are under construction, in the inner city. These include a high-density residential and commercial development on the central Burrard Inlet waterfront, which will

Figure 18 (opposite)

Downtown to the West End. A view looking west over the business district to the West End high-density residential area, English Bay, and, at the top right, Stanley Park. Note the Canada Harbour Place cruise-ship terminal and convention centre. It was once Pier B/C, a dock for trans-Pacific and coastal steamers. Then it was reconstructed as the federal pavilion at Expo '86 and, finally, converted to its present uses. Note also the open waterfront, lined with marinas, floatplane docks, and the railway freight (and formerly passenger) ferry terminal serving Vancouver Island. The area is scheduled for redevelopment, by Marathon Realty in the 1990s, into high-density commercial and residential uses.

displace boat-yards and industrial facilities, including the CPR's truck and railway ferry terminal serving Vancouver Island. High-rise residential construction by local developers and builders is also taking place on the north shore of False Creek, both east and west of the former Expo site. And a major development of the Expo site has been approved by city council. This project includes 7,500 dwellings and 2 million square feet of commercial space in a comprehensively planned community.

In the neighbouring municipalities some regional town centres are maturing, and this has made them more attractive as office locations. Helped also by the building of Skytrain, the introduction of the Seabus ferry across the Inner Harbour, and other improvements to the transit system, the suburbanization of offices is no longer confined to Metrotown in Burnaby. Places such as New Westminster Quay, Lonsdale Quay in North Vancouver, and Richmond Town Centre also partly satisfy the aspirations for decentralization held by the planners of the 1970s, although in detail they leave much to be desired. Several of these nodes were built by developers used to building suburban malls rather than integrated town centres. Metrotown consists of three poorly connected shopping malls, and the offices and residential towers are almost too far apart for pedestrian movement. However, municipal officials are aware of the deficiencies and are devising bylaws and incentives to encourage more coherent and compact development around the Metrotown focus.

New Westminster Quay has been particularly successful in this regard. It is a second-generation inner-city residential and commercial development (second-generation in the sense that it is modelled to some extent on False Creek South but incorporates lessons learned from both positive and negative experiences with False Creek) that provides an excellent living and working environment, particularly for empty-nesters. This achievement owes something to the improvements in the old city centre nearby, which has been enhanced by a new campus for Douglas College, new provincial buildings, and the Skytrain, which effectively links the community to Vancouver. New Westminster Quay typifies a growing trend to convert waterfront areas from industrial to commercial and residential uses. Industry has virtually disappeared from downtown New Westminster, pushed out by lack of space for expansion and the escalating prices offered for commercial and residential land. Lonsdale Quay has followed the same pattern, and both developments imitate the example of False Creek in one further respect: they both have public markets reminiscent of that on Granville Island. Richmond's Bridgepoint Market on the North Arm of the Fraser near the Oak Street Bridge is another example. Further waterfront conversion is in process on the Vancouver side of the North Arm, at the east end of the Marine Drive industrial area near the Knight Street Bridge, and toward the river mouth on land formerly occupied by the net lofts and fish boat repair yard of BC Packers (once the site of a major fish cannery).

In the City of Vancouver, small industrial areas that are not on the waterfront are also vulnerable to conversion. One such parcel, comprising a dozen hectares at 12th Avenue and Arbutus Street, once served by CPR's Lulu Island branch railway line, is scheduled for commercial and residential development early in the 1990s. Reflecting changing needs and priorities as well as pressure on space, Vancouver has also allowed large retail stores to locate in industrial zones, to the irritation of competitors who have assembled commercially zoned land at much greater cost. A similar trend is also evident in Richmond, where retailers along Bridge-port Road have been pushing very hard against the limits of space allowed for retail sales on lots zoned for wholesaling and light industry.

Space pressures have also spurred a vigorous debate over housing in the Vancouver region. In an attempt to cope with the heavy concentration of employment downtown and the associated problems of commuting imposed on many lower-paid white-collar workers (who form a large and growing proportion of the downtown labour force and cannot afford to live anywhere near their work), municipal planners have proposed that the density of residential neighbourhoods be increased. Such a policy would permit cheaper homes relatively near the downtown, and incidentally raise the city's tax income, but it is strongly opposed by many residents of single-family neighbourhoods who want to preserve the character and amenities of their districts. Some of those neighbourhoods, as it happens, are the home of blue-collar workers who can no longer find jobs in Vancouver and are commuting to the suburbs.

In the early 1990s, the core city contained about a third of the GVRD's people and the same share of its industrial floor space (Figures 1 and 14). But only 10 to 15 per cent of new industrial investment in the district occurs in the city, and the mix of activities has shifted away from manufacturing toward wholesaling and freight-handling. Industry also occupies less land than before and uses it more intensively. Pressures to convert industrial land in the city to commercial and residential zoning, which increases its value, are exacerbated by the extremely steep land cost gradient outward from the downtown (Figure 19). Since the recession of the early 1980s, however, there has been increasing concern expressed in city planning documents about the loss of industry, an attitude which was less evident in the 1970s. Partly this reflects doubt about the wisdom of narrowing the range of activities which forms the city's tax base and partly it reflects the sense that a shortage of available industrial land could make it hard to attract immigrant entrepreneurs, who would prefer premises reasonably near their homes. Conversion of waterfront land is of particular concern, since once the land is lost to industry it is unlikely to be returned to such uses. City planners regard all industrial land in Vancouver, except the Burrard Inlet waterfront, as threatened (Figure 7) and are attempting to ensure that some land is retained for industry.

Figure 19

GVRD: land cost gradients (in thousands of dollars per acre), 1970 and 1991

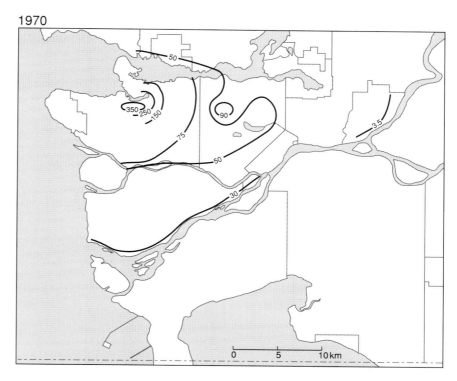

1970

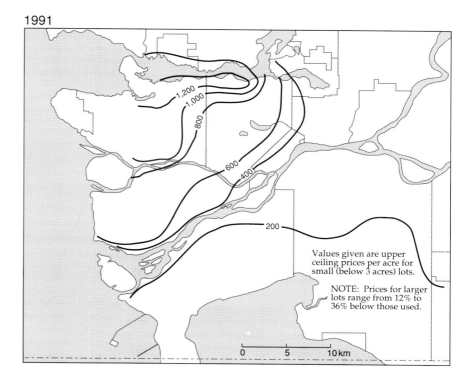

1991

Values given are upper ceiling prices per acre for small (below 3 acres) lots.

NOTE: Prices for larger lots range from 12% to 36% below those used.

However successful these initiatives may be, the processes of relocation and change will continue, as firms respond to evolving trends or adapt to shifts in the fabric of the city. So while clearance for housing and commercial development east of the Knight Street Bridge displaced the Point Grey Towing Company (including a small shipyard which serviced the tugs and the Harbour Commission patrol boat), these services are still needed and new sites need to be found nearby. The tugboat company has moved to Sea Island, near the airport; the marine ways from the shipyard have been moved to a Twigg Island boat yard (formerly located on False Creek); and servicing of the Harbour Commission vessel has been taken over by a shipyard near the mouth of the North Arm, recently set up by the Musqueam Indian band on premises formerly used by the BC Forest Service for servicing its patrol boats. For other activities, however, the problems of relocation may be difficult, especially since the suburban municipalities are less eager to attract some types of industry than they were in the past. Indeed, the suburbanization of industry is now spreading well beyond the GVRD. The Regional District still contains about half the manufacturing industry of British Columbia, but that share is declining, and a growing proportion of provincial manufacturing is located elsewhere in the Lower Mainland and on Vancouver Island.

With the growth of Vancouver's importance as a West Coast hub in the last twenty years, seaport and airport activities have also had a significant impact on the regional economy. A 1975 study suggested that the seaport accounted, directly and indirectly, for a tenth of the jobs in Greater Vancouver. A more recent study attributed 20,000 Lower Mainland jobs to port activity and estimated labour income at $550 million. Industrial activity relying on the port, such as fish processing, shipbuilding, and marinas, accounted for another 2,500 jobs and a payroll of $72 million. As for airport activity, in 1986 (the busiest year because of Expo), Vancouver International Airport handled over 8,385,000 passengers and nearly 100,000 tons of freight. Estimates suggest that it generates 29,000 jobs in the Lower Mainland, and that it contributes as much to provincial GDP as do the coal- or metal-mining industries.

The division of functions among the seaports of Greater Vancouver changed little in the late 1980s (Figures 10, 20, and 21). Burrard Inlet and Roberts Bank handle over 60 million tons a year. This makes the Port of Vancouver the largest port by volume on the west coast of the Americas and, in most years, second or third in North America. The North Arm handles some 16 million tons, most of it wood and construction materials coming in by barge or log boom. Fraserport accounts for only about 6 million tons, but its traffic in vehicles and chemicals makes it more prominent in terms of value. Similarly, containers – a very small proportion of traffic by weight in the Port of Vancouver – account for over 40 per cent of its value and help achieve a near bal-

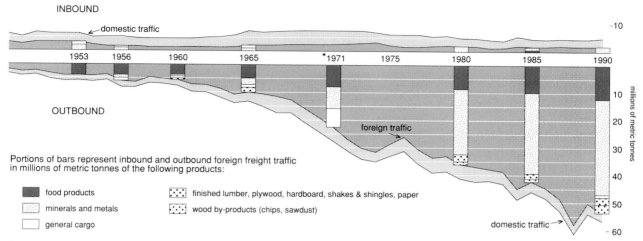

INBOUND

domestic traffic

1953 1956 1960 1965 *1971 1975 1980 1985 1990

OUTBOUND

foreign traffic

Portions of bars represent inbound and outbound foreign freight traffic
in millions of metric tonnes of the following products:

- food products
- minerals and metals
- general cargo
- finished lumber, plywood, hardboard, shakes & shingles, paper
- wood by-products (chips, sawdust)

domestic traffic

millions of metric tonnes

*Data for 1971 exports are incomplete. Total tonnage for exports is uncertain.

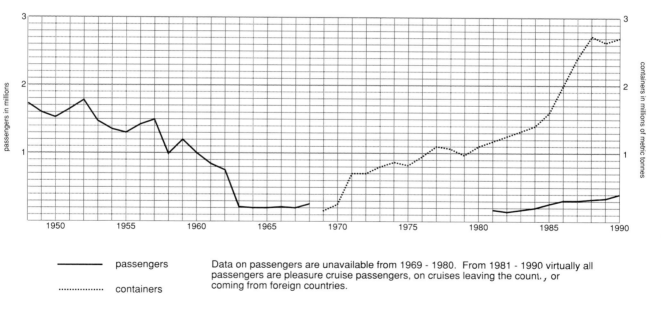

passengers in millions

containers in millions of metric tonnes

——— passengers

··········· containers

Data on passengers are unavailable from 1969 - 1980. From 1981 - 1990 virtually all passengers are pleasure cruise passengers, on cruises leaving the count., or coming from foreign countries.

Breakdown of passengers:

		Landed	Embarked
Foreign	1948	2,035	2,014
	1958	5,574	4,859
	1968	8,652	9,052
Domestic	1948	852,277	874,818
	1958	476,066	494,199
	1968	119,185	123,692

Breakdown of containers (metric tonnes):

		1976	1990
Foreign	inward	43,000	995,166
	outward	29,600	1,622,788
Domestic	inward	11,200	68,502
	outward	12,100	21,705
Total Harbour		95,900	2,708,161

Figure 20

Port of Vancouver: freight and passenger
traffic, 1948-90

Figure 21

North Fraser Harbour: freight traffic, 1955-90

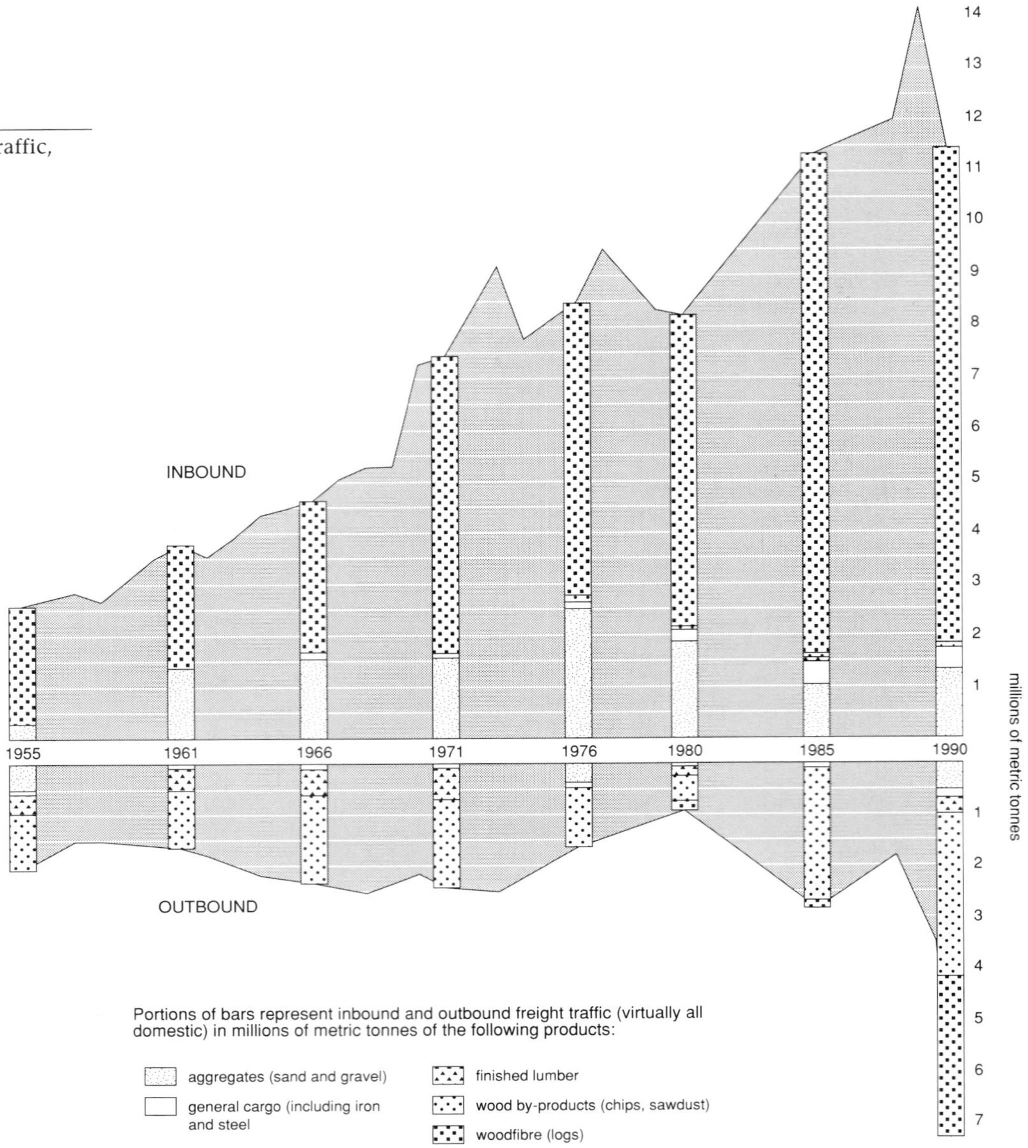
INBOUND
OUTBOUND
1955
1961
1966
1971
1976
1980
1985
1990
14
13
12
11
10
9
8
7
6
5
4
3
2
1
1
2
3
4
5
6
7
millions of metric tonnes
Portions of bars represent inbound and outbound freight traffic (virtually all domestic) in millions of metric tonnes of the following products:
aggregates (sand and gravel)
general cargo (including iron and steel
finished lumber
wood by-products (chips, sawdust)
woodfibre (logs)

ance in the values of imports and exports, although 85 per cent of traffic by weight is outbound.

Container-handling, however, has been a problem for the Port of Vancouver for many years because of a restrictive labour agreement which discouraged shippers from using the port. Half the containers crossing the Pacific bound for Vancouver were shipped to Seattle and carried by road. Even containers bound for central Canada went to Halifax or Saint John instead of Vancouver. Following an improved labour agreement in the late 1980s, container traffic is now increasing, although Seattle still handles far more.

Passenger traffic has also increased. Almost all of it is in the Alaska cruise trade, which benefitted from a new Burrard Inlet terminal and from terrorist incidents in the Mediterranean that persuaded some of the major cruise lines to switch to North America. In 1989, nine lines carried over 300,000 passengers.

The changes described so far in this section are not only a reflection of purely local pressures but also reflect changes in the wider world. Economic decisionmakers in the region have always been sensitive to events in the wider world. In the past, this was particularly true of events affecting markets for raw materials and opportunities to serve them. The local leadership has relied on finding ways and means to take advantage of new conditions such as the opening of the Panama Canal and the postwar expansion of the Asian Pacific economies. Such sensitivity remains important today, even if the mix of wares for sale has changed. Indeed, the export of management and producer services has no less potential as a generator of regional income than does the export of raw and semi-processed materials.

Within Canada, Vancouver is increasingly assuming the role of the second city in the management of enterprise for English-speaking Canada. During the past two decades, with Montreal focusing ever more on francophone business, the English-speaking business community has looked to Vancouver as an alternative to Toronto for investment and employment opportunities. Until 1980, investors seemed uncertain about whether to prefer Calgary or Vancouver among western cities, but the oil crisis of the 1980s demonstrated that Vancouver has a more diverse and stable economy than its Alberta rivals. The need to avoid excessive concentration of urban investment (if the national economy is to maintain relative stability) augurs well for Vancouver to assume an even more prominent role in Canada's economic affairs.

Internationally, too, Vancouver has been able to expand its management and service role. The high-order knowledge and expertise of the producer services firms finds a ready market internationally as well as locally and nationally. Engineering consultants, such as H.A. Simons and Sandwell-Swan Wooster, provide design and construction services to five continents; financial services reach specialized boutique markets; and business services are sold throughout the Pacific Basin. Real estate development firms, which have cut their teeth in the competitive and highly regulated market of the Lower Mainland, have expanded into the United States and around the Pacific, and along with them have flowed the services of architects, contract lawyers, accountants, and a host of others on the 'soft costs' side of the business.

Other services also bring in foreign income. Tourism is a major activity to which the city's hotels and convention centre, its bus and shipping companies, and its adventure and conventional tour operators all contribute. Although the primary role of universities and colleges is to create an educated and trained population, they also contribute to foreign income through such things as conventions and English-language courses. And international research is becoming a more important stimulant to economic change than discoveries of new mines. Consider, for example, the impact of the TRIUMF facility and the proposed Kaon factory at UBC.

The Vancouver region is also benefitting from structural change in the North American economy. In particular, contracting out by large corporations, which were formerly but are no longer integrated, is enabling small businesses to expand. The variety of activities is considerable and includes location shooting in the motion picture industry, wheel manufacture for Toyota, and electronic equipment for aircraft control systems. The value added in the motion picture industry alone totalled $180 million in 1990. The industry is attracted to Vancouver not only for its environment but also for its skilled pool of technical and administrative personnel.

One consequence of this changing scope of activities is that Vancouver's fortunes are no longer linked as closely to those of the province's resource industries as they once were. And, given the direction of so many of the external links, these fortunes are no longer closely linked to those of the rest of Canada or the United States. As it gradually distances itself from the boom-and-bust cycles of the rest of the province, which is still dependent on world commodity markets and the United States construction industry, Vancouver's needs and priorities are likely to diverge from the very activities which built it up – a potential source of tension in the future.

This portrait of Vancouver as a city of expanding importance in the wider world must be tempered in one respect, however. During the 1970s the number of companies headquartered in Vancouver grew rapidly, but, thereafter, several of the resource companies became subsidiaries of transnationals. The acquisition of MacMillan Bloedel by Noranda is one example. Some of the producer services firms, too, have been affiliated with others in central Canada. The implication of this is that the decisionmaking power of these firms has moved elsewhere.

In the early 1980s it became fashionable to talk of Vancouver as a 'post-industrial' city. For much of the century the city had been seen as the interface between the raw material frontiers of the province and western Canada and the urban industrial markets reached by sea. That is not to say that Vancouver was merely a port through which raw materials flowed. Value was added to some of the materials through manufacturing to produce such items as toilet tissue, millwork, sashes, and doors. Mills operated in False Creek, and the pall of smoke and fly ash often penetrated the residential and shopping streets of the core. But in the 1980s, heavy industry was largely absent from the core city and even from the inner ring of suburbs. In social surveys it became difficult to identify neighbourhoods that were 'working class' in the British sense.

Where industry had expanded it was based on a new kind of commodity – knowledge and skills. Telecommunications, marine technologies, computer software, motion pictures, and television films were of increasing importance. And the production of goods was secondary to the production of services. The expansion of producer services changed the complexion of the employment pool. Managers were one of the fastest growing groups, followed by those with skills based upon the physical and social sciences, and with clerical work expanding as well. Each of these categories more than doubled in the 1970s, and many of them were up sharply in the 1980s.

The old factors of location that gave the city its importance in the early twentieth century have gradually been superseded. Although the port and the Fraser Valley transport corridor are by no means insignificant for growth of the city in the 1990s, quality-of-life issues have come to be recognized as prime factors. A quality environment for living, which includes safe neighbourhoods, a myriad of discretionary leisure activities, and ready mobility are all higher on the list of concerns than is relative location. Clearly, the factors of location are necessary conditions for the economic viability of the city, but they are no longer sufficient. The intangible factors of livability now provide the impetus for the growth of the city. As the regional population grows, even the outlying municipalities are experiencing this shift in growth factors and attitudes. At first happy to receive Vancouver's cast-off heavy industry, they are recognizing the attractions of new sources of employment and regional product and are becoming more inclined to pick and choose.

The future

In these developments lies the great challenge of Vancouver's future. Can 'livability' be maintained – and at a price the average Canadian can afford in the face of continuing rapid growth? Clearly the public expects it to be and is pressing heavily on public and private leaders to ensure that it is. Fears of increasing crime, loss of farmland, destruction of neighbourhoods, and deterioration of the physical environment (the last particularly in view of the potential for air pollution from vehicle exhausts in the enclosed Fraser Valley) have all generated concerns.

This concern led, in part, to a 1990 GVRD planning review, which produced fifty-three recommendations to guide the region for the next two decades. The recommendations express the public will to enhance the livability of the region environmentally, socially, and culturally. But their implementation would bring even greater restrictions on urban growth and expansion. Urbanization has already spilled beyond the borders of the district into the Fraser Valley. There, many residents are calling for the creation of a 'green line' as the edge of urban development. They see the estuaries, the rural areas used for agriculture and wildlife, and the mountain views as irreplaceable resources.

Urbanization, they say, should flow in other directions. (Such views, incidentally, contrast sharply with those predominant only twenty years ago, when maps described tidal flats, bogs, and first-class agricultural land as 'potentially industrial.') The consequences of such policies are self-evident, particularly for land-extensive uses and for the price of housing, and many oppose them. The debate has been joined even as far away from Vancouver as Abbotsford-Matsqui, and outlying municipalities are weighing the virtues of becoming outer suburbs of a sprawling Vancouver, limiting development, or shaping a future as contained secondary attractor cities for the Lower Mainland.

For its part, the GVRD is seeking provincial government assistance in spreading economic development to towns and cities beyond the region, particularly to the outports of the Strait of Georgia. In a sense, this strategy follows the lead of the forest products industry, because much of the new plant built by that sector in the past generation has been in places such as Crofton, Nanaimo, Chemainus, and Campbell River on Vancouver Island, and Squamish, Port Mellon, and Powell River on the mainland. Farther afield, the cities of the Okanagan Valley have expanded and, with Kamloops, have become increasingly important secondary centres in the urban hierarchy of the southern prov-ince. This trend has escalated, particularly since the opening of the Coquihalla toll road through the Cascade Mountains brought these communities within a four-hour drive of Vancouver. Even farther afield, Prince Rupert has received a substantial economic boost from the acquisition of a major coal-loading terminal and the fastest operating grain terminal in the province, which, earlier, might have gone to Vancouver.

In the 1990s it is clear that the GVRD is too small an area for adequate planning. Among other initiatives, there have been suggestions that the Lower Mainland Regional Planning Board be resurrected with functions equivalent to those of the regional districts. Some have advocated a return to the idea of a 'Georgia Strait Region' as an appropriate planning region for the highly urbanized population of southwestern British Columbia. In the past, the waterway between Vancouver and Vancouver Island was generally seen as a barrier. From an industrial point of view, however, it provides efficient, low-cost transportation, particularly for logs, wood chips, and other forest industry commodities. The ferry system is already one of the most effective in the world, carrying thousands of automobiles and passengers each day on 300- to 400-car roll-on roll-off ferries, and fast passenger-only commuter ferries are under consideration. The Sunshine Coast and Bowen Island are within fifty minutes of downtown Vancouver, and a potential terminal on Gabriola Island would be within one hour; these times are competitive with rush hour driving time from the outer suburbs.

The implications of the Canada-United States Free Trade Agreement point to other possibilities. Bellingham already considers itself part of the Vancouver commercial region, and the mayor of Seattle has broached the idea of an integrated region stretching from Vancouver to Portland. Indeed, the possibility of a synergistic relationship among British Columbia, the Pacific Northwest, and California is a topic of intense interest in business and professional circles and is reflected in such journals as the *New Pacific*. While the Free Trade Agreement clearly threatens some sectors of Vancouver's economy – those based on fishing and agriculture, for example – it may improve the prospects for others.

The challenge bears repeating. In the face of rapid immigration, a limited land base, and an enormous potential for further growth and change, can the livability of Vancouver be maintained? Or will the geography of the region, at once a major source of its attractiveness and of its limitations, present obstacles with which the planners and politicians cannot cope technically, economically, or politically?

8

Time to grow up? From urban village to world city, 1966–91

David Ley
Daniel Hiebert
Geraldine Pratt

'**W**e have to grow up.' Arthur Erickson could scarcely have uttered a phrase closer to the contested public identity of Vancouver than he did in a public forum on waterfront development in October 1989. By 'growing up' he was referring not to just the height and density of urban development but also to the attitudes Vancouverites have about the kind of city in which they would like to live. Provocatively citing a population projection of ten million for the Vancouver region, he challenged his audience that 'you do not live in a suburb but in a major urban centre.' As a native son, and the city's most celebrated architect, Erickson spoke with authority, and if he was playing to a public arena, he was certainly successful. Two days later, the lead editorial in the *Vancouver Sun* ruminated on his pronouncement: 'Mr. Erickson has brought into sharp focus an issue that goes to the heart of the matter of development in the city: what kind of density can it sustain and still remain a good place to live?' Moreover, the newspaper knew its constituency as well as did Arthur Erickson: 'Ask any resident of Greater Vancouver if he or she would like to see the region's population even double its present level, and the answer in most cases would be that it's already big enough. But how can future growth be prevented? Mr. Erickson is right when he says we can't put gates around the city to keep people out.' The discussion did not pass out of public scrutiny, because, in many ways, it had been part of a perennial planning and political dilemma for over twenty years. Several months later, Design Vancouver, an umbrella group promoting the design professions, continued the discussion by organizing a public debate on the question: 'Should Vancouver plan for a future population of ten million?' Erickson and two land-use consultants spoke to a capacity crowd in support of the motion. On the other side stood Ray Spaxman (the city's director of planning from 1974 to 1989), Walter Hardwick (a liberal alderman between 1968 and 1974), and Carole Taylor of the 1990 council. The debate was light-hearted, but it moved across fundamentally contested terrain and exposed starkly contradictory visions of the city. One side accepted, even welcomed, rapid growth and an urban environment shaped by the dictates of economic development. The other side envisaged a city in which economic impulses would be tempered by social and environmental controls.

Neither vision was entirely new, but the former had held sway in the city for most of its first century and had been embodied in a never-realized design for Vancouver's West End, sketched by the young Erickson years before, when, following Le Corbusier, he imagined the entire neighbourhood compressed into a single high density mega-building. The alternative vision also had roots in the city's relatively distant past, and its proponents had occasional successes in implementing key elements of their program, which included land-use zoning and park designation. This viewpoint did not prevail, however, until the late 1960s, when civic leaders such as Hardwick, Art Phillips, Michael Harcourt, and, soon after, their appointee, Spaxman, brought social and environmental concerns to the fore. During the last half of the 1970s, these ideas were challenged but never fully displaced by a resurgent enthusiasm which, like Erickson's, was assembled around such tenets of modernity as universalism, cosmopolitanism, and a forward-looking belief in riding the wave of historical forces. Reflecting on these tensions in 1977, columnist Allan Fotheringham identified two sparring visions: the functional (the city of the bottom line) and, against it, the aesthetic (the city beautiful). In Vancouver, 'a city so gifted in its setting, so profligate with its advantages,' he found it amazing that there was a 'continual struggle...between the engineering mind and those who realize we live through our eyes and what they transmit.' A decade later, when a controversial world's fair, Expo '86, exposed the city to a global gaze, Vancouver's planning department likewise concluded that citizens were of 'two minds about the way their city should change.' In part they wanted 'Vancouver to be a thriving urban centre, a world city, with more jobs and business opportunities and with a greater variety of things to see and do.' But they also wished 'to preserve Vancouver's present small-city character: relatively low-density housing, peaceful neighbourhoods, and quiet streets.' They clung, as Fotheringham would later express it, to the belief that Vancouver was 'a village on the edge of the rain forest.'

For the past twenty-five years, then, these conflicting views have shaped conceptions of development strategies for Vancouver and its region. No discussion of social and political geography during this period can ignore them, and they provide a framework for the three principal topics addressed in this chapter: the changing status of households and the family in tandem with a fast-evolving labour market; the growing cosmopolitanism of the region's population and landscapes; and the political and planning debates about an appropriate vision for urban life.

'Village on the edge of the rain forest' or world city?

Although Fotheringham's 'the village on the edge of the rain forest' was a pithy sketch which included an alleged preference for leisure activities by its residents, the metaphor also contained an element of environmental stewardship, appropriate for the city in which both Greenpeace and Canada's first Green party were founded. Perhaps the metaphor may be extended further. The village economy is one which is rooted in local resources and raw materials, and the relatively small downtown of the mid-1960s served local needs and accommodated companies with interests in the city's resource hinterland; today, in the much larger downtown, a segment of the resilient service sector is beginning to uncouple itself from the remainder of the province. If the village mentality conveys a sense of localism and resistance to change, the planning and politics of the past quarter century have had much to do with a defence of the local and the familiar, of agricultural land, of the environment, of neighbourhoods, of heritage, and of the legacy of the past against both unwanted land uses and unwelcome newcomers. The idea of village life invokes also a sense of community, of social responsibility, of fair play, and of social justice manifest in the city's publicly inspired urban development and imaginative social housing policies of the early 1970s – the period of the high water mark of the Canadian welfare state.

Until the mid-1970s, the essential characteristics of Vancouver's families and households also bore a resemblance to those prevailing earlier in the century. Particularly in the relatively homogeneous postwar suburbs, most mothers remained at home. In Canada as a whole in 1975, only 35 per cent of mothers with preschool-age children and employed husbands were themselves employed outside the home, a far cry from the 63 per cent in the labour force by 1988. Tradition and homogeneity ruled also in terms of the simpler ethnic constitution of the population, with, in 1971, some 60 per cent of Greater Vancouver residents identifying a family ancestry in the British Isles. New arrivals did little to complicate this pattern, for in 1967 Britain maintained its long-established position as the leading source of immigrants. Huge changes were, of course, to follow, and, by 1989, the first four, and eight of the top ten, sources of immigrants to British Columbia were nations in Asia.

Indeed, by the early 1990s, traditional patterns had been challenged in many domains in line with the rhetoric of a world city. The internationalization of the regional economy has various expressions, of which the 1989 Free Trade Agreement is only the most recent. The province's resource sector is increasingly controlled by distant absentee landlords, and its real estate, including office space, apartment buildings, and resort complexes, has become part of a more diverse international investment portfolio. The economic boosterism of regional politicians, including overseas trade missions and investment offices, coincides with the erosion of an earlier social welfare system. The growing wealth of parts of the inner city jostles with an equally unfamiliar network of food banks. Affluence and homelessness exist cheek by jowl. New levels of street crime among gangs of marginalized adolescents are part of a rising wave of serious crime. Family life has become more varied, with a growing pluralism of household types. As more women enter the labour force, a growing personal service sector, ranging from child care to fast food, offers services once produced within the home. The stereotypical family suburbs of the 1950s and 1960s have become far more differentiated and include numbers of childless households as well as poorer households displaced by redevelopment pressures at the core. And both the city and the suburbs house more diverse populations, with more residents living at the extremes.

All of these trends have a palpable geography, expressed in the landscape of each neighbourhood (Figures 1 and 2). In Kitsilano, both Greenpeace and SPEC (the Society Promoting Environmental Conservation) had offices on the counter-cultural strip of Fourth Avenue in the early 1970s, together with the Divine Light Mission, health food stores, and coffee shops like the Soft Rock Café, where conversations on ecology and mysticism urged a limit to economic growth and its attendant evils. These messages were hawked up and down the street and beyond by the *Georgia Straight*, founded in 1967 by students and artists, and claiming sales of 60,000 within four months of its first issue. But the 'hippy haze' of the years around 1970, captured also in John Gray's Kitsilano novel, *Dazzled,* and by a 'tribe' of Kitsilano poets, has been replaced by the condo fever of Sherman Snukal's play, *Talking Dirty,* set in one of the 1,500 condominium apartments built in the neighbourhood between 1971 and 1976. The world of young professionals depicted by Snukal, with its designer icons of clothing, food, furnishings, and leisure, led to a transformation of Fourth Avenue and, in turn, other inner-city districts. By the mid-1970s, sales of the *Georgia Straight* were below 10,000 and did not recover until the 1980s when, like the neighbourhoods it serves, it turned from 'hippy' to 'yuppy' and became a free entertainment and leisure guide.

Or consider the landscape changes in the Downtown Eastside, adjacent to downtown, and the poorest district in the metropolitan area. In 1973, the Downtown Eastside Residents' Association (DERA) was formed to contend with the evils of skid row, and it has provided a remarkable model of community development for an area with a predominantly elderly male population, which includes other marginalized groups, such as discharged mental patients, teenage runaways, and growing numbers of Natives. Through a sophisticated political strategy of both protest (Figure 3) and negotiation, DERA succeeded in upgrading services, ensuring by-law enforcement, and securing the construction of hundreds of social housing units. By the mid-1980s however, this careful community-building was threatened, first, by its neighbour, Expo '86, and since then by mega-project development closing in on it from three sides – from the west the expanding downtown, from the east

Figure 1

Kitsilano: From hippy...

Figure 2

...to yuppy

Figure 3

DERA demonstration to support the establishment of a neighbourhood park, Downtown Eastside, mid-1980s

Figure 4

The two shores of False Creek. In the foreground, townhouses, including subsidized units on city-owned land. In the background, the BC Place covered stadium and the site of Expo '86, presently owned by the Hong Kong developer, Li Ka-shing

the redevelopment of the Port of Vancouver, and from the south, the Concord Pacific redevelopment on the Expo site.

Such patterns of change are not limited to the inner city. In the suburbs, the arrival of the 'festival market' is associated with the redevelopment for middle-class occupancy of old industrial waterfronts in North Vancouver, Richmond, and New Westminster (where Renaissance Venice is the unlikely design model). But it is on the two facing shores of False Creek that a former industrial landscape most eloquently tells a story of two eras (Figure 4). On the south shore, redevelopment of an obsolete industrial district was begun in the early 1970s on city-owned land. The city's plural objectives are apparent: low- to medium-density development, the ecological care of a design with nature, a social policy of unobtrusive and small-scale social housing, promotion of bicycle and pedestrian movement and discouragement of the automobile, and, above all, conscious social planning to build community – all of which express the mentality of the urban village. By contrast, the north shore was initially developed by the provincial government, in the spirit of the headline-grabbing mega-structure. First came the covered stadium, BC Place, which set the tone for Expo '86. Though resisted initially by the city, Expo '86 was propelled into existence as a loss leader, an invitation to international investment and tourism. Construction went ahead during an economic recession while there were severe cutbacks on social programs, resulting in a period of dissension which included increased labour militancy and the formation of the Solidarity Coalition (an alliance of labour and community groups), which brought the province within a few hours of a general strike in 1983.

Expo '86 exemplified central themes of the 1980s: privatization, polarization, and internationalization. The Expo Board awarded building contracts to non-union companies, a provocative act in light of the high level of union membership in the building trades. Polarization was particularly evident in off-site impacts on residential hotels in the surrounding districts. In the Downtown Eastside, between 500 and 850 long-term residents were evicted as hotels prepared themselves for a flood of tourists. Internationalization was the explicit intent of the fair, with marketing slogans such as: 'An invitation to the world' and 'What your world is coming to in 1986.' Following the fair, rather than developing the land itself as had been promised, it was privatized by the provincial government and sold to the highest bidder; only strenuous intervention from the city secured sites for social housing, but as yet there are no funds to build it; and the sale to the Hong Kong developer, Li Ka-shing, continues the process of internationalizing Vancouver's land market.

The reworking of gender relations

In the last thirty-five years, there have been radical transformations in the types of work that individuals do, the number of workers per household, the arrangement of work within households, family stability, the very idea of what constitutes a family, and in gender and sexual identities. These changes have been shaped by economic, social, and political forces as they are lived and shaped by individual women and men. In the process, Vancouver has been recreated as a place. Critical to geographers is the notion that these changes have occurred in different ways in different parts of the metropolitan area – that the city supports very different ways of life. Places are more than neutral backdrops; they can influence how individuals live their lives and how they manage work and family arrangements.

In 1966, there was a pervasive traditionalism in Vancouver, which was reflected, for example, in a 'traditional' arrangement of social groups in urban space, as typified in urban theory developed in the 1920s. In this simple conceptual framework, central city areas were containers of marginalized groups, poverty, the elderly, ethnic minorities, renters, individuals living alone, as well as cosmopolitan and subcultural lifestyles. In contrast with this central city diversity, suburbs were generally seen as relatively homogeneous environments of middle-class, home-owning families. We catch a glimpse of this broadly dichotomous city in Figure 5, which shows the percentage of married women among those over the age of 14 in each census subdistrict of Vancouver in 1966. Nowhere in the suburbs was this figure below 60 per cent. In some areas almost 9 in every 10 women over age 14 were married. In most central city subdistricts, on the other hand, fewer than 60 percent of women were married. Another type of traditionalism is revealed by figures that show that fewer suburban than central city mothers worked outside the home in the mid-1960s (Table 1). Suburban families were also stable (or at least together) and children were most often raised by two parents. This is indicated by the low divorce rates; only 15 adults (defined as age 15 or older) out of 1,000 were currently divorced in the metropolitan area in 1966 as compared to 59 of every 1,000 just twenty years later. In 1966, the City of Vancouver housed half of the adult population, but almost two-thirds of divorced individuals living in the metropolitan area. Finally, another aspect of Vancouver in the 1960s was that the criminalization of all homosexual acts until 1969 meant that traditional sexual identities of masculinity, femininity, and heterosexuality were maintained – in public appearance, at the very least.

The effects of changing gender relations since 1966 can be traced throughout the metropolitan area. Distinctive new neighbourhoods have emerged and the differences between city and suburb have blurred. Newly acknowledged gender identities and new household arrangements have accentuated the diversity of the central city and disturbed the homogeneity and stability formerly associated with suburbia. Perhaps the most obvious example of this is the transformation of Vancouver's West End into a 'gay space.' Although the West End houses a diversity of social groups, an institutional and residential concentration of gay men in one part of this area imparts a distinctive character to the neighbourhood. A location of choice among gay men in the city since the 1960s, but, despite that, hardly visible until the gay liberation movement of the 1970s, the area provided an almost ideal territorial niche for gay men opting to 'come out.' It offered a large stock of relatively inexpensive, one-bedroom rental accommodation in high-rise buildings close to downtown amenities, which allowed for relative anonymity and eased fears of discrimination on the basis of lifestyle by landlords or apartment managers. Before the end of the decade, an article in the *Vancouver Sun* proclaimed that West End gay men were 'Out of the Ghetto' and announced the 'coming of age' of Vancouver's gay community. During the 1980s gay men laid increasing claim to this neighbourhood. In 1985, 19 of 37 gay businesses, identified in a study

Figure 5

Percentage of women, 15 years of age or older, married or separated, in metropolitan Vancouver, 1966

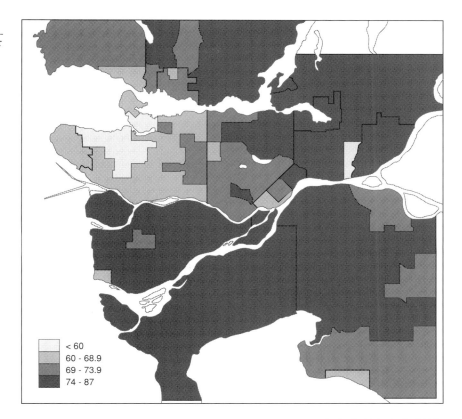

	< 60
	60 - 68.9
	69 - 73.9
	74 - 87

Table 1

Female participation in the labour force by municipality, 1961-86

in female	Participation rate**		Chang participation (%)
	1961	1986	
Vancouver, CMA	31.7	59.2	27.5
Vancouver	36.7	60.1	23.4
North Vancouver	28.5	64.1	35.6
West Vancouver	25.8	54.7	28.9
Richmond	27.4	63.0	35.6
Delta	20.8	59.0	38.2
Surrey	20.0	55.7	35.7
White Rock	16.8	41.1	24.3
New Westminster	34.1	54.3	20.2
Burnaby	28.1	59.6	31.5
Port Moody	22.7	65.8	43.1
Coquitlam	21.2	63.7	42.5
Port Coquitlam	23.4	61.3	37.9
Langley	n.a.	55.0	n.a.

Source: Census of Canada, Cat. 95-537, 95-758, 94-800-808, 95-978, E580, 95-168

Note: *Determined as the number of women working expressed as a percentage of the total female population 15 years of age and over, excluding institutional residents in an area, group, or category.

CMA = Census Metropolitan Area

of Vancouver's gay community, were located in the West End; the remainder were in adjacent downtown areas. In all, 70 per cent of the community services and voluntary organizations catering to one of the largest male homosexual populations in Canada are located in the West End. 'Neighbours,' a relatively long established gay bar encapsulates the wider pattern of landscape change that has transformed the West End as gay men have grown more comfortable with this place. Through several renovations, its architecture has become less defensive, and gay identity has been displayed more publicly (Figure 6).

Figure 6

The changing micro-geography of 'Neighbours,' a gay bar in Vancouver's West End

Phase One

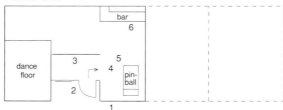

1. Small, poorly lit signage.
2. Dark entrance, one door locked.
3. Notices of gay events.
4. Dead end at pinball machine necessitates sharp turn upon entry.
5. Lighting level changes from bright to dim.
6. Bartender scrutinizes patrons as they enter, and may politely question their intent.

Phase Two

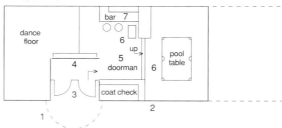

1. Large awning increases imageability.
2. Small but bright neon sign.
3. Both doors open.
4. Sign: "This is a gay bar."
5. Doorman scrutinizes patrons as they enter, but his primary function is to restrict the number of patrons to the bar's licensed limit.
6. Lighting levels remain constant; greater visibility of bar's interior from door.
7. Bartender no longer scrutinizes patrons entering.

Phase Three

1. Exterior, though still of dark wood, has been remodelled to increase imageability.
2. Large graphic posters, often of homoerotic content.
3. Doorman, primarily for the control of the numbers of patrons, usually sits out of sight of the entrance at position 4.
5. Bar has been moved to back wall, distant from door.
6. Large conversational area now open to visual penetration from doors.
7. Glass door allows vision of interior of the bar from the street, and may be used as an alternate, unrestricted entrance in early evening hours.

Figure 7

Rooming house for sale on the Fairview Slopes in the mid-1980s. The advertisement read: 'to be renovated or knocked down as building lot for your custom townhouses'

Figure 8

'Life on the Upslope': townhouses on the Fairview Slopes

On the Fairview Slopes, immediately south of False Creek, another distinctive landscape has coincided with changes in gender relations and the sociodemographic character of the city's population within the last fifteen years. In the 1960s, this area was a place of deteriorating single-family houses, many of which had been converted to rooming houses or communal homes, described by political scientist Donald Gutstein as a 'hidden village of mellow ambience' (Figure 7). Now it is a high-density area with expensive townhouses and condominium apartments offering high-amenity 'city living' (Figure 8). A disproportionate number of its residents are single, and over half the units in the newer, larger (and less expensive) developments (largely completed since 1985) have been purchased by single women. Many of the couples living in the area have postponed, or decided not to have, children. Two-thirds of all households in the district were childless in 1986, and many of those with children are headed by single parents; in 1981, single-parent families were twice as common in Fairview Slopes (where they accounted for almost one in every four families) as in the metropolitan area. In line with the theme of transformed gender relations, some 30 per cent of these families were headed by men (compared to 16 per cent in the City of Vancouver). Rather more than two-thirds of Fairview's husband-wife families had both partners in the labour force, although the comparable proportions for the city and the metropolitan area were approximately 50 per cent. It should be noted, however, that some 'traditional' characteristics of occupational segregation remain: a far smaller proportion of Fairview Slopes' women (12 per cent) than men (21 per cent) were employed in managerial occupations, and, in 1981, women residents of the area earned 20 per cent less than the $13,437 reported by the average male.

Many of those who live in Fairview Slopes regard it as an almost ideal environment. One resident interviewed (by Caroline Mills) in the 1980s claimed that Fairview Slopes would 'be a *very* desirable place when it firms up and people realize how convenient it is, how workable it is.' Another suggested that the distinctive environment would encourage new ways of urban living: 'I think that [the fact that] these homes are here is going to encourage people to change their lifestyle, rather than [that] these places are being built because people *have already* changed their lifestyle.' Among the opportunities offered by this setting are the availability of high-amenity rental accommodation, low housing maintenance demands, and the possibility of blending dual careers with consumption-oriented leisure activities due to the proximity of employment (downtown and in the hospital district) and of cultural and recreational facilities. The adult orientation of the neighbourhood is also valued by many. Again, the observations of residents point to the special character of the place. As one noted, 'the by-laws in this development are restrictive, and it's not very equitable: children are not permitted – only for rental people, legally you can't discriminate for them – but the by-laws say very strongly that this is an adult-oriented complex...In a way, that's a selfish lifestyle. But it's typical of the Slopes; the majority of the developments in the Slopes are designed with an adult orientation.' The new gender relations enacted on the Fairview Slopes are very much tied to an adult lifestyle, which mirrors both the increase in reproductive choice and the continuing difficulties that women and men have combining waged work and childcare.

These difficulties are experienced throughout the metropolitan area but are perhaps felt most acutely in suburban areas, where the traditional 'familial' ethic must now be reconciled with the reality of two-wage-earner families. Although women throughout the metropolitan area are more likely to be in the labour force today than they were in 1961, the most dramatic changes in household work arrangements have occurred in suburban areas such as Coquitlam (Table 1). In a recent sample of residents of Coquitlam (interviewed by Isabel Dyck), waged work was, for most women, episodic; all had spent some time, ranging from several months to several years, primarily engaged in mothering work. The networks they developed in the community during this time were vital to carrying out mothering work (through the exchange of information, support, and assistance) and later became essential resources for managing paid employment. Through the neighbourhood networks, women returning to the labour force often heard of child care arrangements. Sometimes neighbours took on this responsibility, in other cases they helped the woman in waged work augment arranged child care by driving her children to enrichment or special activity classes. Thus, the domestic workplace remains central to the experience of suburban mothers who work outside the home and is an essential means for accommodating child care and paid employment.

Yet networks of helping sometimes impose strains on individuals, households, and communities. In suburban Surrey in 1983, one woman, who rented a house in an area in which most were owned, complained of the other mothers' absence during the day. She remarked on the 'hostility' of those neighbourhood children with both parents in waged work and noted that these children were continually hanging around her home because they knew that she was often baking. A study of this area revealed that there were indeed far more full-time mothers in households occupying rental accommodation. Many renters referred to their desire to provide full-time mothering for their children when justifying their decision not to buy. One said: 'It's not worth the anguish to own a house. That's why some families don't stay together. It's chaos when you're raising children and both working. Both my parents worked and I never got over it. There was no family life...my parents just came home to sleep. Our family is a priority and we're not willing to sacrifice everything to own a house.' Such tensions are forcing a redefinition of landscape and lifestyle in suburbia. People are questioning the desirability of homeownership (previously a foundation of the 'suburban dream') at any cost. Partners find themselves engaged in the unfamiliar task of negotiating the allocation of domestic work among household members. Full-time mothers harbour doubts and resentments about informally absorbing the mothering work of neighbours who are balancing paid and domestic work.

The growing diversity of household types in suburban communities is also redefining the nature of suburban life. Many of the social boundaries that formerly divided city and suburbs have been blurred by the dispersion of various household types through the metropolitan area. Figure 9, which shows the percentage of married adult women in each census tract in 1986, reveals that the proportion rarely exceeded 74 per cent in suburban areas, and that there was a great deal of variability (compare Figure 5). This is partly attributable to the increasing number of elderly people living in suburban areas; between 1966 and 1986, the proportion of people over 65 years in suburban Vancouver rose from 8.2 to 10.8 per cent. But households throughout the metropolitan area are altogether more complicated than they were in the past. In 1987, 37 per cent of all marriages in Coquitlam, Port Coquitlam, and Port Moody involved a divorced person; roughly one in six households blended families from previous marriages. The number of single-parent families in all parts of Greater Vancouver has also climbed steeply. In suburban White Rock, Langley City, and North Vancouver City there are proportionately more single-parent families with children at home than in the City of Vancouver. In some areas, they account for almost one in three households, and the majority are headed by women.

These social changes are reflected in the suburban landscape. The single-family house is no longer the only dwelling type in suburban areas, as it was for the most part in the 1960s. There has been a remarkable growth of multifamily housing in every Vancouver municipality. But housing affordability and poverty are issues in all parts of the region (Table 2). The poverty rate in suburban areas has also risen sharply, partly as a consequence of the increase in elderly persons and single-parent families. In several suburban municipalities, including the City of Langley and New Westminster, the proportion of low-income families now exceeds that in the City of Vancouver (Table 3). Poverty rates among unattached individuals are astonishing. In the City of Langley fully 52 per cent fell below the low-income cut-off used in the 1986 census, and in West Vancouver, the wealthiest municipality in Canada, this was true for almost one in three. And food banks are as necessary in Surrey as they are in the central city.

One simple measure of the growing polarization of income in most districts of Vancouver is the spread between mean and median household incomes over time. As income disparities widen, so the gap between mean (average) and median (mid-point of the distribution) income figures increases. Between 1971 and 1986, this spread widened in all areas

Figure 9

Percentage of women, 15 years of age or older, married or separated, in metropolitan Vancouver, 1986

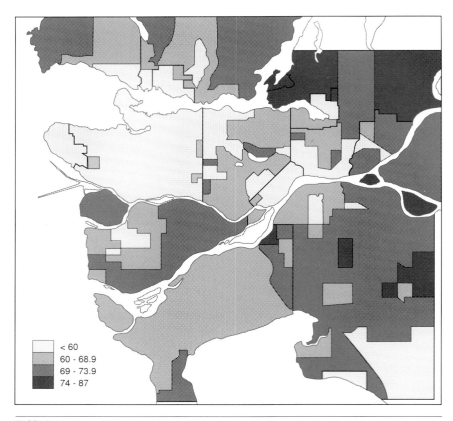

< 60
60 - 68.9
69 - 73.9
74 - 87

Table 2

Percentage of renters and homeowners paying housing costs exceeding 30% of household income by municipality, 1986

1986	> 30% of household income	
	Gross rent	Owner's major payments
Vancouver, CMA	39.06	15.05
Vancouver	38.53	13.86
Surrey	46.76	19.17
Burnaby	33.57	12.26
Coquitlam	39.80	13.19
Delta	40.38	14.06
Langley, City	54.43	20.67
Langley, District	42.71	18.63
New Westminster	38.91	12.08
North Vancouver, City	37.10	17.57
North Vancouver, District	35.04	15.00
Port Coquitlam	47.53	16.58
Port Moody	38.36	14.23
Richmond	34.67	13.82
West Vancouver	33.00	13.04
White Rock	45.50	12.36

Source: Census of Canada, Cat. 95-168, E580

Note: Estimates are based on one-family households without additional persons.

except Coquitlam, Port Coquitlam, and the City and District of Langley. Across the metropolitan areas it climbed by almost 50 per cent. In Burnaby, New Westminster, and Richmond the rate of increase was approximately 100 per cent. In Port Moody it approached 200 per cent, and, in the City of North Vancouver, it was 238 per cent.

In this context, it is hardly surprising that crimes against property are on the rise (Table 4). In 1977 only Vancouver, Surrey, the City of Langley, and New Westminster recorded more than 20 break-and-entry robberies per 1,000 people. Eleven years later, 7 districts crossed this threshold and 6 exceeded the rate of increase recorded by the City of Vancouver from 1977. In March 1989, a columnist in the *Vancouver Sun* reported on efforts to organize residents of Aldergrove to deal with an alarming increase in juvenile drug use and related property crime. For Vancouverites who might have been 'toying with the nifty idea that if you move out of downtown Vancouver and into the "burbs," you'll find cheaper housing and escape big town crime and drugs,' his message was 'good luck on the former and forget the latter.' Nor is physical safety any longer strictly a 'big city' problem. Reported sexual assault rates are well above the rate of one per 1,000 in most parts of the metropolitan area, and they are higher in Langley, Port Coquitlam, Coquitlam, Port Moody, and New Westminster than in the City of Vancouver itself (Table 5). Many more go unreported. In her attempts to

Table 3

Incidence of low income by municipality, 1981, 1986

	% of all families low income		% of all unattached individuals low income	
	1981	1986	1981	1986
Vancouver, CMA	11.0	14.8	34.9	40.7
Vancouver	14.3	19.4	35.9	43.6
Surrey	12.6	16.5	40.3	43.9
Burnaby	10.6	13.5	34.2	36.8
Coquitlam	8.2	12.9	30.2	37.0
Delta	7.5	9.7	34.5	38.5
Langley, City	14.8	22.5	46.7	52.3
Langley, District	7.2	10.4	34.0	36.8
New Westminster	14.0	21.1	36.9	46.2
North Vancouver, City	12.4	17.6	28.5	34.8
North Vancouver, District	6.3	7.5	26.6	30.5
Port Coquitlam	9.5	13.5	36.3	41.9
Port Moody	11.8	15.6	37.5	43.9
Richmond	7.9	11.1	28.5	30.0
West Vancouver	5.0	6.0	30.0	28.6
White Rock	8.3	7.9	37.8	36.6

Source: Census of Canada, Cat E580, 95-168

Note: Low-income cut-offs determined separately for families of different sizes and living in different degrees of urbanization. See Census for definitions.

Table 4

Rates for break-and-entry robberies by municipality, 1977-88

	Rate (per 1,000) 1977	% change 1977-83	% change 1983-8	Rate (per 1,000) 1988	% change 1977-88
New Westminster	26	35	-6	33	27
City of Langley	29	0	3	30	3
Burnaby	14	85	8	28	100
Vancouver	23	39	-12	28	22
City of North Vancouver	18	33	0	24	33
Surrey	23	15	-11	24	4
Abbotsford	14	57	-9	20	43
District of North Vancouver	12	-16	80	18	50
Coquitlam	14	71	-29	17	21
District of Langley	13	38	-10	17	31
Port Coquitlam	18	11	-32	15	-17
Port Moody	14	43	-25	15	7
White Rock	12	50	-33	14	17
Richmond	17	-23	7	14	-18
West Vancouver	9	44	0	13	44
Delta	10	40	-14	13	30

Source: Summary Statistics, Police Services Branch, Ministry of Solicitor General, Province of British Columbia, September 1989

Table 5

Incidence of sexual assaults in Metropolitan Vancouver, 1989

	Aggravated	With weapon	Without weapon	Other sexual offences	Total	Rate (per 1,000)*
Langley	n.a.	n.a.	n.a.	n.a.	113	1.61
Port Moody	0	2	22	0	24	1.52
Coquitlam and Port Coquitlam	0	3	117	13	133	1.35
New Westminster	0	7	47	0	54	1.35
Vancouver	3	64	464	38	569	1.32
Richmond	2	5	127	0	134	1.24
North Vancouver	0	5	74	0	79	.76
Delta	0	0	35	0	35	.44
West Vancouver	1	0	15	0	16	.44

Source: Statistics obtained from municipal police. The statistics for some municipalities are missing because relevant authorities refused to release them.

Note: *1986 population figures used. Figures are based on reported sexual assaults. Aggravated sexual assaults are violent crimes, sexual assaults include assaults causing bodily harm, and other sexual offences include touching, grabbing, incest, etc.

understand the murder of a young woman in an office building in downtown Vancouver in the mid-1980s, local poet Helen Potrebenko wrote, in 'Witness to a Murder':

> We should move to a nicer part of town,
> someone kept saying. Sure ...
> It was a domestic dispute, the men said,
> apparently comforted by the rolling resonance.
> Domestic dispute ...
> Some dispute. This was a dispute like there is a nicer part of town.

Finally, the growing cosmopolitanism of Vancouver offers new challenges to communities throughout the metropolitan area, including the following: the Lotus, the Red Eagles, the Gum Wah, the Vet Ching, the Dai Hun Jai, Los Diablos, Mara Latinos, the Russians, and the East Van Saints – all major juvenile gangs operating in Vancouver in 1990, at least four of which have emerged in the city in the last decade. Although gang activity is most concentrated in East Vancouver, it has spread in recent years to suburban areas such as Surrey, Richmond, and North Vancouver. City police have assembled dossiers on over 500 known gang members as well as on 500 believed to be in the process of joining gangs. Most are said to be of Asian origin – many the children of first-generation immigrants. The growth of these gangs, and their increasingly violent behaviour, is generally attributed to three factors: the illegal entry of a small number of professional criminals as immigrants or bogus refugees; the lack of community organization among some groups, such as the Vietnamese, who are relative newcomers to Vancouver; and the paucity of social services for non-English-speaking immigrants. The waiting list for English-language classes at Vancouver Community College, for example, can be up to two years. Similarly, resources for ESL (English as a Second Language) students are criticized as woefully inadequate: there is only one school guidance counsellor in Vancouver who speaks fluent Chinese, one who speaks Punjabi, and one who speaks Spanish; meanwhile, there is one counsellor for every 270 English-speaking students. Alluding to both the lack of social services and discrimination against those of Asian origin, a police corporal comments: 'Sure in the end we'll lock these kids [in gangs] up, but something should be done at the beginning to give their families the same opportunity as everyone else.'

On the edge of the Rim: The internationalization of Vancouver's population

If the 'village-on-the-edge-of-the-rain forest' metaphor was predicated on a traditional family-household structure, so, too, did it have selective ethnocultural connotations: in the 1960s metropolitan Vancouver was far less cosmopolitan than it is today, with an overwhelmingly 'white' cast. Nearly two-thirds of its population were of British or French background, and another 28 per cent claimed European ancestry. Less than 6 per cent were described by the 1961 census as non-Europeans – a group that included Aboriginal people as well as people of Asian, African, and Latin American descent. At a generalized scale, there was a distinct social geography of ethnocultural communities in Vancouver. Non-European-origin groups were largely concentrated in the inner city, where Little Tokyo and Chinatown were especially visible. There were a few pockets of Japanese-Canadian settlement outside the city, notably in Steveston and Port Moody, where Japanese workers were employed in resource-processing industries. To a degree, this clustering reflected the desire of Asian-Canadians to maintain their cultural and social networks, but it was also a response to prejudiced attitudes and discriminatory practices in the wider society.

The remainder of the City of Vancouver was essentially split in half, the east side populated by a mixture of British and other European working-class groups, the west side the preserve of an Anglo-Canadian middle class and élite. In the early 1960s Vancouver was surrounded by large tracts of sparsely settled suburbs; these too were almost exclusively 'white' in their ethnocultural composition. These broad social patterns remained largely intact through the 1960s, although the number of people of non-European origin doubled during the decade. In 1971, when approximately one in six Vancouver residents lived in the urban fringe, people of Chinese background accounted for only one in every forty suburban dwellers.

The size, economic impact, and political significance of Asian-origin groups in Vancouver began to change decisively during the 1970s, primarily as a result of new Canadian immigration regulations. Until 1967, Canadian immigration policy favoured those of European background. Then, following the influential 1966 White Paper on Immigration, the distinction between prospective immigrants from 'preferred' and 'non-preferred' nations was replaced by a points system designed to treat applicants equally, regardless of their nationality or ethnic background. Significantly larger numbers of non-European immigrants have entered Canada under the new policy. One simple, but revealing, measure of this is that in 1966, just prior to the change in policy, Britain, Italy, the United States, West Germany, and Portugal provided the largest numbers of Canadian immigrants. By 1989, the five most significant source countries for immigrants to Canada were: Hong Kong, Poland, the Philippines, India, and Vietnam.

This internationalization of immigrant flows to Canada coincided with a period of rapid growth in the economies of British Columbia and, especially, Vancouver. As a result, Vancouver has attracted growing numbers of non-European immigrants since 1971: the metropolitan population of Asian, African, and Latin American groups together exceeded 150,000 in 1981 and was nearly 250,000 by 1986. Thus, there was a 422 per cent increase in Metropolitan Vancouver's non-European

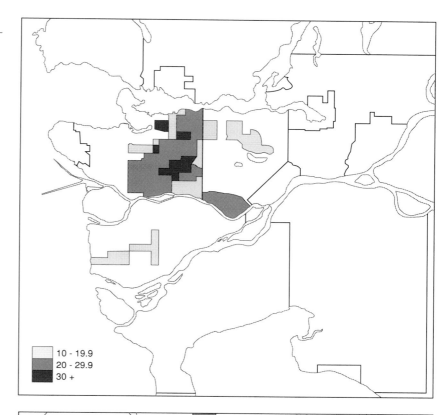

Figure 10

Percentage of population of Chinese
ethnic origin in metropolitan
Vancouver, 1986

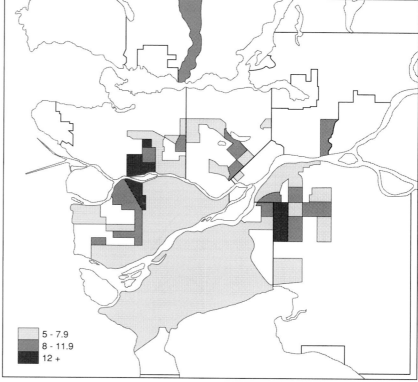

Figure 11

Percentage of population of South Asian
ethnic origin, metropolitan Vancouver,
1986

population between 1971 and 1986 (as opposed to a 28 per cent overall growth rate). Put another way, non-European groups accounted for over half of the net population growth in Vancouver during this fifteen-year period.

The transformation of Vancouver's social geography has been equally dramatic. The traditional distinction between deprived inner-city and affluent suburban neighbourhoods has been broken down, and the geography of ethnic groups in the region has become more complex. This is best explored via a brief examination of the Chinese and South Asian communities. In 1986, 109,000 residents of metropolitan Vancouver identified themselves as 'Chinese.' Note that this census category does not represent a homogeneous group and includes immigrants from Taiwan, Hong Kong, and various regions of China as well as the descendants of immigrants who arrived in Canada during the past 130 years. Taken as a whole, however, this Chinese population is clustered in certain parts of the city; according to one index of segregation, in which complete dispersal is represented by a value of 0 and absolute ghettoization by a value of 100, the Chinese population of Vancouver recorded a relatively high value (given comparative statistics calculated for various groups in different Canadian cities) of 52.5. Many, especially recent immigrants, live in Chinatown, where Chinese-Canadians account for over 60 per cent of the population. The cost and quality of housing in this neighbourhood are well below average for the metropolitan area, and the incidence of poverty is high. Other Chinese individuals and families live in a variety of settings, ranging from modest housing throughout the east side of Vancouver; through relatively new suburban areas in Richmond and Burnaby; to much more exclusive areas, such as Kerrisdale, Shaughnessy Heights, and the British Properties (Figure 10).

South Asians reside in a somewhat more dispersed pattern; in 1986 their index of segregation was 41.5 (Figure 11). Few live in the inner city, while many are located in the south-eastern quadrant of the City of Vancouver. There, the Main Street Market (Figure 12) and the Sikh Temple on Ross Street (Figure 13) provide visible evidence of a large Indo-Canadian presence. Since its completion in 1969 the temple has been a focal point for the Sikh community and has drawn Sikhs from other parts of Vancouver into the area. As immigration from the Punjab accelerated during the late 1970s and early 1980s, however, Sikhs also began to move to more peripheral areas of Greater Vancouver. By 1990 approximately 20,000 Sikhs, the largest concentration in the Lower Mainland, lived in eastern Delta and western Surrey. Scott Road, the boundary between these two municipalities, contains over seventy South Asian shops. Merchants on the street have formed the Punjabi Bazaar, a business association established to promote Punjabi and Sikh retail interests. And the nearby Bear Creek Community Centre, operated by the Sher-a-Punjab company, is the largest facility of its kind in

Figure 12

Indo-Canadian shopping district on Main Steet

Figure 13

Sikh Temple, Ross Street, south Vancouver

Vancouver, with its spacious banquet hall seating 600 people. The centre is patronized by members of Vancouver's Hindu and Sikh communities, two groups that have traditionally maintained separate cultural institutions. The community's Guru Nanak Sikh Temple, also in the Scott Road area, was extensively renovated in 1991 and can now accommodate up to several thousand for a single service.

The social geography of South Asian- and Chinese-origin groups partly reflects the operation of Vancouver's housing market. Land prices in inner-city neighbourhoods are prohibitively expensive for those with limited financial resources and/or those who value space. According to a South-Asian land developer: 'A lot of these people [living near Scott Road] are new immigrants. They like to live in big houses with big yards so they can have vegetable gardens. The place to get a house like that is Surrey, not Vancouver.' However, the residential distribution of these groups is also linked to their selective participation in Vancouver's labour market (Table 6). Chinese men tend to be clustered in three quite distinct types of work: medicine and the sciences; sales and service; and, to a lesser extent, manufacturing. Chinese women exhibit an

Table 6

Index of occupational over- or under-representation in the Vancouver CMA for Chinese and South-Asian origin groups, 1986

Occupational category	Chinese		South Asians	
	Males	Females	Males	Females
Managerial	0.79	0.81	0.63	0.30
Natural Sciences	1.48	1.23	0.53	0.71
Social Sciences	0.61	0.26	0.36	0.62
Teaching	0.81	0.74	0.26	0.95
Medicine	2.12	0.88	0.31	1.13
Arts/Lit	0.22	0.40	0.15	0.24
Clerical	1.11	0.81	0.87	0.73
Sales	1.02	1.06	0.89	0.50
Service	2.14	1.16	0.01	1.61
Farm/Hort	0.36	1.00	4.53	10.00
Other primary	0.35	0.00	0.95	0.00
Processing	1.06	2.02	2.91	1.19
Machine op	0.72	4.49	1.05	0.74
Construction	0.50	0.48	0.97	0.00
Transport	0.58	0.43	0.80	1.00
Other	0.66	1.44	0.94	1.11
Number	606	558	294	239

Source: 1986 Census of Canada Sample Tape

Note: Numbers in the table refer to the rate of over- or under-representation for the group and occupational sector. For example, Chinese-origin men were more than twice as likely as all other men in Vancouver to work in the service sector in 1986 but only half as likely as other men to work in the construction sector.

even more segmented pattern and are 4.5 times more likely than is the average woman to be hired as a machine operator (particularly in the garment trades). By contrast, male and female South Asians are extraordinarily clustered in agricultural work, although they, too, are frequently hired in factories. Much more research is needed, however, to determine the extent to which this labour-market segmentation is the result of ethnocultural discrimination. In any event, the agricultural orientation of South Asians helps explain their propensity to live in suburban areas, since large numbers of these men and women are hired in berry farms and truck-gardens on the fringe of the urban area. A growing number of South Asians are purchasing agricultural land and establishing related businesses; in the process, a South Asian ethnic enclave economy is emerging, complete with its own entrepreneurial class and labour force.

Occupational patterns also underpin the distribution of Chinese-Canadians in Vancouver, but because members of this group are found in a more disparate set of jobs than are those of South-Asian origin, their social geography is at once more clustered and more fragmented than that is of the South-Asian population. Overall, people of Chinese origin are heavily concentrated in East Vancouver and central Richmond, but there are distinct enclaves within the group. Most Chinese female garment workers interviewed (by Charles Mather) in a study of clothing production in Vancouver live in or near Strathcona, near their place of work. By contrast, a rapidly growing number of Chinese professionals and entrepreneurs have recently settled in Vancouver's élite neighbourhoods; in 1986, at least 10 per cent of Shaughnessy residents were of Chinese descent, and a quarter of the houses in upper middle-class Oakridge were occupied by Chinese-Canadian families (Figure 10).

Since 1986, a rapidly growing number of affluent Chinese families – many of whom are recent immigrants from Hong Kong – have also purchased homes in Kerrisdale and the British Properties. This represents a profound shift in the traditional experience of immigration. Until recently, most immigrants arrived in Canada with little money and few specific skills; typically, they accepted employment in blue-collar occupations and sought the cheapest housing available. Since 1978, when the Canadian government introduced a new program allowing those willing to invest at least $250,000 in a Canadian business venture to enter the country as 'entrepreneurial immigrants,' a sizable and growing proportion of Asian-origin immigrants landing in Canada have been both highly educated and affluent. This trend has been further encouraged by the addition of another category, 'investor immigrant,' in 1986. By 1988, the top ten source countries of 'investor immigrants' were: Hong Kong, Taiwan, West Germany, Macao, Lebanon, England, Iran, Kuwait, the Philippines, and Malaysia. Entrepreneurs and investors, together with their dependents, now comprise approximately 10 per cent of the entire flow of immigrants to Canada.

In addition to these federal programs, both the provincial government and Vancouver's city council have sought to attract investment and immigrants from Pacific Rim countries. The two levels of government regularly send representatives to Hong Kong, for example, in hopes of benefitting from the exodus of capital and population from that country as 1997 approaches. One of the primary goals of Expo '86 (the 'loss leader' showcase funded, in large part, by the province) was to advertise the amenities and economic opportunities of Vancouver and British Columbia to an international audience. Business interests have become equally aggressive in promoting Vancouver: consulting firms hoping to profit from trade, investment, and immigration are emerging on both sides of the Pacific; the number of trans-Pacific real estate transactions is increasing rapidly; and specialized educational institutions – targeting Asian students who want to improve their English literacy – have begun to appear in Vancouver.

These various initiatives are achieving considerable success. Certainly the ethnocultural composition of Vancouver's population is changing rapidly. Depending on how ethnic groups are classified, between 10.3 and 21.4 per cent of the metropolitan population was of Asian background in 1986. This proportion continues to rise, making the 'global city' metaphor appear more and more appropriate for Vancouver. Aside from the influx of new investment, this type of internationalization brings many benefits. Cultural institutions and the arts in Vancouver have both become more cosmopolitan and, many would argue, revitalized through cultural diversification. Even mundane habits are affected as the number and variety of ethnic shops and restaurants expands. On this level at least, multiculturalism is almost universally welcomed.

Yet immigration often precipitates conflict too, especially when groups entering a society are culturally, economically, or visibly distinct. There is a long history of prejudice against non-European immigrants in Vancouver, of day-to-day discrimination punctuated by episodes of xenophobic violence. The Chinese- and Japanese-origin communities in early Vancouver responded by turning inward, by creating their own economic opportunities, and by residing in densely populated ghettos. Fortunately, overt violence against visible minorities in Vancouver has all but vanished over time, but prejudiced attitudes and behaviour have dissipated much more slowly. For example, students in certain secondary schools complain of deeply felt divisions between 'white' and non-European students, and community forums on multicultural issues often draw a small contingent of vocal racists. Amidst these worrying symptoms of racism, several municipalities have made efforts to educate the public on multicultural issues and to provide special services for Asian immigrants.

In the past, ethnic conflict in Vancouver was closely enmeshed with the competition for working-class jobs. The most vociferous demands for the restriction of Chinese immigration in the late nineteenth century came from those who claimed that Chinese men took work from 'whites' by accepting below-average wages and working conditions. In contemporary Vancouver, ethnic tensions continue to reflect economic competition but now appear to be focused on real estate transactions and issues related to the housing market. The public disenchantment that followed the sale of the Expo '86 site, at a bargain price, to Li Ka-shing exemplified this concern. So, too, did criticisms of the offshore sale, in Hong Kong, of all units in 'The Regatta,' a new luxury condominium project on the south side of False Creek, by Victor Li (Li Ka-shing's son) and Terry Hui. Apparently, few approve of this aspect of Vancouver's drive to 'world city' status, because the sale of property in the global marketplace reduces local opportunities for the purchase and enjoyment of certain portions of the city.

The interweaving of economic and ethnic competition is nowhere more controversial than in Vancouver's established élite neighbourhoods, where immigrants from Pacific Rim countries (together with upwardly mobile Asian-Canadians) have begun to redevelop housing in non-traditional styles. Rising demand in these areas has had a variety of consequences. Dramatic increases in property values in Oakridge, Kerrisdale, and other select neighbourhoods have allowed long-term residents to reap huge financial rewards by selling their property and buying elsewhere. For example, a Kerrisdale house worth $200,000 in 1986 would have sold for well over $500,000 in 1989, at the peak of the market. Yet other residents of these areas have been adversely affected by house-price inflation. The assessed value of west-side houses has risen, sometimes by as much as 300 per cent. Senior citizens on fixed incomes in these areas have become 'house rich, cash poor'; some have even been forced to sell their homes for want of money to pay municipal taxes. High prices have also effectively closed the market on Vancouver's west side to all but a tiny fraction of new home buyers.

The degree to which immigrants contribute to the inflation of property markets is the subject of much debate in Vancouver. The weight of public opinion is that Asian immigrants, particularly those entering Canada as entrepreneurs and investors, are responsible for escalating values on the west side of the city. Further, some commentators suggest that the effects of offshore and immigrant investment on Vancouver's upmarket housing sector have rippled through the market, raising demand and prices throughout the urban area. These accusations have been eagerly taken up by the media, both in Vancouver and in the rest of the world; Hong Kong newspapers regularly report on the reception of Asian-based investment in Vancouver, and articles on this issue have even appeared in the European business press. Real estate interests in Vancouver worry that these perceptions will lead to a racist backlash against Asian immigration and will dampen interest in the housing market. In response, they have established the Laurier Institute, a privately funded research centre 'committed to fostering Canada's understanding of the benefits of cultural diversity.' Predictably, the investigation of Vancouver's housing market was its first priority. Two reports have been released; the first asserts that immigrants have had no impact whatsoever on house prices; the second advances a similar, though slightly more qualified, conclusion.

Beyond these debatable economic consequences, the changing ethno-cultural composition of élite neighbourhoods is also associated with profound changes in the character of these residential settings. The Kerrisdale landscape, for example, has been virtually transformed during the past decade. Initially established as a British, upper middle-class suburb of Vancouver, Kerrisdale came to be characterized by a relative uniformity of architectural style (in the picturesque tradition, incorporating motifs from rural England and the Tudor and Victorian periods) and landscaping (also in an English picturesque style; see Figure 14). The brisk pace of development in Vancouver between the Second World War and 1980 left the landscape and population mix of Kerrisdale remarkably unchanged; even in 1981, 63.3 per cent of the neighbourhood were of British origin, compared with 40.5 per cent in the City of Vancouver. This relative stability ended in the 1980s. As Vancouver's housing market rebounded after the recession of the early 1980s, long-term residents of Kerrisdale began to notice the transformation of their setting. Houses were being sold, demolished, and replaced by very large dwellings. Initially, only a few lots were affected by this practice, but redevelopment became pervasive. On a few streets, over half the houses were replaced by the end of the decade. The term 'monster houses' began to appear in the late 1980s; it satirized both the scale and the aesthetic qualities of the new buildings. 'Monster houses' are typically palatial (often containing over 5,000 square feet of living space) and virtually antithetic to the picturesque tradition. Most are clad in brightly coloured brick, and their design emphasizes large window areas and ostentatious entrance ways (Figure 15). The contrast between the English garden style of 'old' Kerrisdale and the landscaping used for 'monster houses' could hardly be greater. Trees, shrubs, hedges, and flowerbeds are replaced by stark cement, gravel, and grass yards, which are, in turn, surrounded by imposing fences and elaborate gateways.

Many long-term residents of Kerrisdale (i.e., those who have not sold their property for a huge profit) complain bitterly of these recent changes. Letters of protest sent to Vancouver City Council decry the loss of what their authors interpret as landscapes exemplifying harmony between people and nature. In the words of one petitioner: 'We moved to Churchill Street in December of 1982 because we valued the sense of neighbourhood that clearly existed there. Over the years, we have watched hundreds of joggers and strollers and dog walkers pass along our street enjoying its pastoral setting. The trees on our property

Figure 14

English picturesque: the traditional Kerrisdale landscape

Figure 15

The 'monster house': the new upper middle-class landscape of Kerrisdale

must exceed 20 m in height and are the home to many nesting birds and foraging squirrels ... help us keep this area a spot where all Vancouverites can leave the hustle of Granville Street and walk beneath the trees and perhaps smell a few roses.' This criticism meshed easily with a growing societal concern over environmental degradation and reached a peak in 1989, when a new Kerrisdale resident of Chinese origin hired a company to fell two 80-metre Giant Sequoia trees in his front yard. Petitions were circulated, protests were held, and one individual even offered to buy the property in question for $200,000 more than the owner had paid for it. The 'two trees' controversy reverberated throughout Vancouver and even made front-page news in Hong Kong. But to no avail; after a few days the trees were cut down.

City Hall has received thousands of letters addressing the issue of neighbourhood change in Kerrisdale. Almost all lament this change and urge controls against the removal of trees or the building of 'monster houses.' While most authors are careful to avoid criticizing any specific ethnic group, a few were forthright in attributing responsibility for these undesirable changes to wealthy immigrants from Hong Kong. The propensity for Chinese-Canadians to buy or build monster houses is also frequently noted in media reports, although little hard evidence is cited to corroborate this assertion. Yet there does appear to be a strong connection between ethnic and landscape change in Kerrisdale and other élite residential areas. In a compilation of field observation and street directory listings (by Niall Majury), it was shown that most 'monster houses' in Kerrisdale are indeed occupied by families with Chinese surnames (though this does not indicate that these property owners are recent immigrants). Moreover, monster houses incorporate aspects of *feng shui* in their design: for example, street addresses are often changed to more favourable numbers; emphasis is placed on the relationship between light and shadow (hence the removal of giant trees which block the sun); and special care is taken in arranging internal elements, so that front and back doors are not directly aligned, for example.

The 'monster house' issue is but one manifestation of ethnic/economic stress in Vancouver. Other examples include the aforementioned growth of adolescent gangs and associated violence; the difficulty of integrating immigrant children into the school system; and the need to suppress racist behaviour whenever it appears. In the end, however, the conflict surrounding the redevelopment of élite landscapes provides perhaps the most telling illustration of the basic paradox of cosmopolitanism in Vancouver. Anglo-Canadians see the picturesque tradition as the embodiment of a 'natural' order, while Chinese-Canadians see the geometric imperatives of feng shui as 'naturally' beautiful. In a truly multicultural society, these viewpoints should co-exist with one not excluding the other. But this is not easy to achieve; in practice, the reality of internationalization is far less simple, just as the notion of multiculturalism in Canada is inherently nebulous. While most agree that a mosaic of cultural traditions is appropriate, even desirable, few are willing to answer the essential question: How will the inevitable clashes between fundamentally different traditions be reconciled?

The politics and planning of urban development

In 1966 Vancouver was a city of 410,000 in a metropolitan area with a population of 933,000. During successive five-year intervals the metropolitan population has grown by 16 per cent (1966-71), 8 per cent (1971-6), 9 per cent (1976-81), and 9 per cent (1981-6). Inevitably, attitudes toward growth have been an important cornerstone of urban politics and planning over the past twenty-five years. In 1966, however, there was a simpler politics of urban development. In the city, a pro-development council continued in the tradition established in 1936 when a 'non-partisan,' at-large, system of municipal government had replaced the earlier ward system. As elsewhere in western Canada, this progressive 'reform' model was endorsed by business interests. The 'Non-Partisan Association' (NPA), founded in 1937, was a party dominated by small businessmen who favoured growth and regarded the city as a business corporation. Through the next thirty years, 90 per cent of candidates who ran for office under the NPA banner were elected, and the party was never out of power. Few embodied its growth ethic more thoroughly than Tom Campbell, mayor from 1966 to 1972, and a developer who held a limitless enthusiasm for growth: 'Vancouver is the San Francisco of Canada ... the New York. I can see it someday becoming the largest city in Canada. Montreal and Toronto have had it.' In addition to Campbell, four of the other six NPA members of the eleven-person council in 1971-2 held occupations related to real estate and land development.

But significant change lay ahead. During the 1966-71 period, rapid population growth was accompanied by the first postwar office boom, with a quadrupling of the rate of construction of new downtown office space over the level of the previous decade. With considerable prescience, Harland Bartholomew, author of the business-inspired master plan of 1929, had declared that his plan would require revision once Greater Vancouver's population reached a million – a figure that was attained in the late 1960s. Major planning issues did indeed face council: obsolete and polluted industrial land lay on the waterfront and around False Creek; growth-induced redevelopment confronted older neighbourhoods; and there was a need for an efficient transportation system in an ever-dispersing metropolis. Little wonder that a survey in advance of the 1972 civic election, when the NPA majority was toppled, found that development and the role of the real estate industry were the leading campaign issues in the minds of the electorate.

Already in 1967, however, there were new voices in the air. The maturing of the postwar baby boom brought a new population to the inner-city rooming houses and basement suites. In March 1967 merchants on Fourth Avenue bewailed a hippy invasion; two months later estimates were made that 7,500 'hippies' were living in Kitsilano. With others elsewhere in the city, they articulated an anti-growth and often adversarial ideology that embraced environmentalism, civil rights, Eastern religions, and the drug culture. Simultaneously, the city housed a rising middle class of educated professionals with some sympathy for the social movements of the day. But the major event of 1967 which launched the assault on growth boosterism (and the style in which it was conducted) was the 'great freeway debate.' Its resolution proved to be a significant and lasting success for urban conservation and an inspiring example for anti-growth, pro-neighbourhood coalitions.

Behind the freeway debate was a twenty-year downtown plan, launched by the city in the mid-1950s, which called for the redevelopment of a vast tract of 2,500 acres of land and a city-wide freeway network. After receiving a number of technical reports, city council announced in 1967, without public discussion, that a 120-foot-wide freeway would be built through the heart of Chinatown. This declaration raised a huge public furor, and, in the months that followed, a growing alliance challenged the freeway alignment; the broader downtown plan; and, also, the insensitive, centralized administration that was such an abiding feature of technocratic civic decisionmaking. With remarkable insensitivity that also revealed the cloistered decisionmaking process, Mayor Campbell denounced the demand for public meetings as 'a public disgrace and a tempest in a Chinese teapot,' and fumed 'do we have to hire a playhouse to put on a puppet show for objectors?'

Such political arrogance was increasingly out of phase with the political culture of a rising middle class of liberal, well-educated professionals. During 1967, and in reaction to the NPA council, three separate groups, connected with the University of British Columbia, the Community Arts Council, and downtown business-people, began regular informal meetings to discuss citizen participation and urban planning. Their deliberations were given urgency by the NPA council's announcement in the summer of 1968 of the third phase of urban renewal in Chinatown-Strathcona, which would raze a further fifteen blocks and displace 3,000 people. Encouraged by recent success in the freeway struggle and with the support of articulate professionals in planning and community development who brought the issue to the city at large, the Strathcona Property Owners and Tenants Association (SPOTA) was eventually successful in defending its neighbourhood when federal urban renewal funding was withdrawn by the new Trudeau government. Around this time, there was a third rebuff to a downtown megaproject. The Canadian Pacific Railway had revealed designs for Project 200, a vast redevelopment of Gastown, the old Vancouver townsite, which would have demolished the oldest buildings in the city. The Community Arts Council organized a well-attended walking tour of the district, and the strength of public support for preservation of Gastown seriously undermined support for Project 200. By summer 1969, a program of heritage restoration and an arts and crafts presence with a counter-cultural ambience (which included the office of the *Georgia Straight*) was approved. The preservation and protection of communities seemed to be in the ascendancy.

So this was a time when new voices were being heard, and new social and cultural agendas were being forced onto an increasingly beleaguered civic administration. Racial minorities began to effectively challenge their exclusion from a centralized political process. The 'new' middle class was mobilizing an alternative vision of community planning. And members of the counter-culture clamoured at the margins with their own environmental and cultural interests. All resisted any land-use change that was propelled by rapid growth and orchestrated by inaccessible politicians and their technical staff. It was also at this time that Greenpeace and other groups urged protection of the environment and the city's neighbourhood and heritage; and, a few years later, British Columbia's first provincial government led by the New Democratic Party would seek to protect high-grade farmland by establishing an agricultural land reserve as a green belt around the expanding metropolis.

It remained for political parties to keep pace with this volatile political culture. Nationally, the 1968 election of the Trudeau administration, with its promises of an open society, represented such a shift. The same year, in Vancouver, two new civic parties were founded. The Electors' Action Movement (TEAM) was a fusion of the three liberal middle-class groups which had been meeting informally to discuss planning issues. And the Committee of Progressive Electors (COPE), with a socialist program to serve working-class and tenant interests, was established. Though markedly different in terms of their constituencies and their emphases, both parties wanted local government to be more involved in growth management and social welfare; they were also committed to an open and democratic decisionmaking process. Each played a significant role in setting the political agenda in the period after 1972. TEAM held a majority on city council between 1972 and 1978, while COPE and its NDP ally held a slim majority between 1982 and 1986.

Earlier in this chapter, one of the features identified as expressing the village ideology was the defence of neighbourhoods. In response to the insensitivity of city-wide planning and vigorous grassroots reaction to it, Vancouver was divided into twenty-two planning districts in the early 1970s, and the TEAM council introduced a program of local area planning – a collaborative venture between neighbourhood committees and city planners to develop a local zoning map and priorities for physical planning. Extensive exercises in citizen consultation and participation registered some gains for communities, notably in down zoning and environmental improvement, although the empowerment of local committees fell far short of the community control model sought by some residents' groups. In Strathcona, Kitsilano, and other districts, new federal programs for housing rehabilitation and neighbourhood improvement, which replaced discredited urban renewal policies, provided significant funds for the preservation and enhancement of neighbourhood services and facilities. The high point of the neighbourhood movement was the period from 1972 to 1975, when all three levels of government endorsed neighbourhood initiatives. In locational conflicts brought before council during 1973-5, the neighbourhood voice was successful in two cases out of three – a success rate much higher than that for pro-development and other business groups. The NDP provincial government established Community Resources Boards, decentralized teams of professionals, and an elected committee to organize the local delivery of public services.

An important figure in bolstering neighbourhood defences was Ray Spaxman, the city's director of planning from 1974 to 1989. Appointed by the reform council of the early 1970s to develop a more participatory planning process, he promoted not only the neighbourhood but also 'neighbourliness' as a major planning principle: 'My contention is that the more we identify and reward neighbourly behaviour the more successful we (as a city) will be.' Social housing was required to follow, as far as possible, the massing and design texture of the neighbourhood in which it was built. Neighbourhood character was promoted through design guidelines intended, for example, to consolidate the ethnic atmosphere of Chinatown or the heritage landscapes of Shaughnessy and southern Mount Pleasant. Down-zonings to check high-rise redevelopment were a third tool to slow the scale of change in a number of neighbourhoods. But the 'neighbourliness' of Spaxman's small-town vision was increasingly threatened by rapid social and land-use change.

By the late 1980s neighbourhood defences had been overrun at a number of points. Economic expansion and the courting of international investment and wealthy immigrants by municipal and provincial governments brought significant development pressure to bear on many neighbourhoods. In upper middle-class Kerrisdale, the replacement of rental apartments by high-cost condominiums has disrupted a long stable neighbourhood; in the first eight months of 1989, 359 rental units faced demolition, to be replaced by 222 condominiums. In the Downtown Eastside, the poorest district in the city, considerable consolidation occurred in the early 1980s, when the city acquired ten sites for non-profit housing societies, creating 1,244 affordable units. But since the return of an NPA council in 1986, redevelopment pressures have impinged on the neighbourhood from all sides. According to Jim Green, community organiser for DERA, the last reserve of low-cost housing in Vancouver faced an invasion of 'major high-rise developments coming in from all three sides of us in the next few years, an invasion of well-off people, and nobody is giving a thought to the effect it will have on this neighbourhood. Expo was bad, the worst year of my life. But this will be worse because the evictions will be more massive, and more long-term – in fact, permanent.' In the east-side neighbourhood of Grandview-Woodlands, graffiti declaring 'Keep yuppies out of East Van' indicate a resolute challenge to a high-rise project at a rapid transit station that would bring high densities and wealthier households into a working-class district of family housing. Faced with the prospect of continuous in-migration, the council is preaching the virtues of residential intensification. New development on former industrial sites on the northern and eastern shores of False Creek and the Coal Harbour waterfront downtown include higher densities than those with which Vancouverites have been familiar.

Intensification of development is also occurring in the suburbs. In part, this was anticipated and, indeed, planned. In 1976, the Greater Vancouver Regional District (GVRD) released the Livable Region Plan. The document shared much with the planning ideology then promoted in the city: it was the outcome of a lengthy program of community consultation, and, in treating resident concerns seriously, it went far beyond an economic development document. If anything, growth was seen as a problem rather than as a solution: in the mind of the GVRD planning director: 'Planning can be against growth, if that is society's goal, rather than for it.' The plan sought to minimize work-residence separation and to enhance service and amenity levels in the suburbs by 'taking more jobs and services out to where people live.' This objective would be accomplished by decentralizing job growth from downtown Vancouver into four suburban town centres in Burnaby, New Westminster, Surrey, and Coquitlam. By 1986 this strategy was to have created relatively self-contained communities with abundant open space; Greater Vancouver would become 'a region of complete communities – livable cities in a sea of green.'

But there was no means of enforcing this vision of growth management, and development has paid scant attention to it. Of the four proposed town centres, only Burnaby (Metrotown) on the rapid transit Sky Train route has assumed significant status, although New Westminster is experiencing recent revitalization as a result of substantial public investment, including the opening of Sky Train in 1986. Considerable town centre growth and consolidation have also occurred in Richmond, which was not foreseen by the Livable Region Plan. Nor has there been any real diminution of the primacy of downtown Vancouver. The city's office space has more than doubled between 1978 and 1989, even if its share of the regional total has slipped a little from 73 to 64 per cent. More substantial has been the continuing decentralization of industrial and warehouse space, only a third of which was in the central city by 1989.

The Livable Region Plan has been quietly by-passed by the provincial government. In recent years, the planning function of the GVRD was removed by the province and centralized as a provincial responsibility. Permission has been granted for the Agricultural Land Reserve to be nibbled away by both urban and quasi-urban land uses; currently some forty new golf courses have been proposed for the Fraser Valley and Lower Mainland. Because population growth has been particularly rapid in the suburbs, which now account for some 70 per cent of the regional population, growth control is now as much a suburban as a city issue, particularly in Richmond, Delta, and Surrey. Anxieties over the pace of change have begun to have an impact on suburban politics. In Delta, determined citizens have persuaded the council to reject development of a large land parcel, the Spetifore lands, approved for withdrawal from the Agricultural Land Reserve by the provincial government under questionable circumstances. In Surrey, the fastest growing municipality, substantial controls were imposed by a new council elected in 1988. In Richmond, a bitter and protracted conflict has been waged between the Save Richmond Farmland Society and the municipal council over development of the Terra Nova farmlands. The Society pressed the case to an eventually unsuccessful hearing before the Supreme Court of Canada. While it was still unresolved in the courts, the developer began construction; indicative of the tensions involved, the first house on the site was destroyed by arson.

By 1990 there were signs that local government was once again prepared to take a broader view of its responsibilities. After public consultations, the GVRD issued a draft version of a new Livable Region Plan. Like its predecessor, this document promoted social rather than private rights, and it attacked pollution, discouraged use of the private car, backed the preservation of agricultural land, and reasserted the principle of directing employment growth to regional town centres. It also called for the restoration of its powers in land use, transportation, and social planning. The scope of the report is little short of global, aspiring to make Greater Vancouver 'the first urban region in the world to combine in one place the things to which humanity aspires on a global basis: a place where human activities enhance rather than degrade the natural environment, where the quality of the built environment approaches that of the natural setting, where the diversity of origins and religions is a source of social strength rather than strife, where people control the density of their community and where the basics of food, clothing, shelter, security and useful activity are accessible to all.' For the cynic, this is merely window dressing – an unattainable wish list. However, even as rhetoric, the breadth of the vision of the document gives pause for thought, particularly in a province where the message of the 1980s has been social restraint, privatization, and the often one-dimensional pursuit of economic development. Some of the public issues rated highly in the early 1970s, notably those related to the environment and the housing question, have returned to prominence in the public mind. There are signs that the 1990s may hold promise for new attempts to mould a more rounded vision of urban life in Greater Vancouver, although, given the much greater diversity of society and its interests now as opposed to twenty years ago, such an endeavour will undoubtedly tax the ingenuity of local democracy.

Conclusion

Many of the trends identified in this chapter are politically contentious, and this was certainly reflected in the municipal elections in 1990. In the city, the issues and candidates could not have been more clearly drawn. Jim Green, community organizer in the Downtown Eastside for a decade, was the mayoralty candidate for the COPE/NDP unity slate. An American draft-dodger, union shop steward, and representative of the poor (a group increasingly threatened by the polarizing tendencies of a world city), Green focused the sympathies of those critical of unregulated growth, gaining much support despite COPE's reputation as a party to the left of the NDP. His election slogan, 'the Neighbour-hood Green,' cleverly alluded to the twin issues which had led to elec-toral success for reform groups in the early 1970s: protection for neighbourhoods and for the environment. His opponent was the NPA incumbent, Mayor Gordon Campbell, a 1970s liberal who had become a 1980s conservative, and who had approved unprecedented develop-ment in his 1988-90 term of office. A developer and businessman, young, personable, well-educated, and well-travelled, he epitomized the free market internationalization of the world city. The election was fought essentially around the issue of development and housing afford-ability, and some novel alliances among groups newly marginalized and politicized were created. Elderly tenants in middle-class Kerrisdale sought advice from Green's Downtown Eastside Residents' Association. In Strathcona, poor Chinese residents were attentive to the pro-growth entreaties of the new Chinese business class, which supported the NPA. Aided by an advertising budget its opposition could not match, the NPA barely won an election that the party had expected to lose. Reflecting the deeply ambivalent attitude of Vancouverites, both the NPA and COPE elected five candidates to council – the single new NPA council member was a Hong Kong-born banker. The controlling vote in the 1990-3 council is held by Mayor Campbell, who was returned with a small majority – 54 per cent of popular support.

In several suburban municipalities, the same tensions generated an even more marked electoral response. Middle-class Richmond, which consistently elects conservative politicians in provincial and federal elections, returned seven out of eight Civic New Democrats to council as well as an NDP mayor. Elected on a platform to preserve farmland and control growth, the new council promised an immediate review of the Terra Nova development and issued a moratorium on other green belt incursions (including the rezoning of then-Premier Vander Zalm's theme park, Fantasy Gardens). In Delta, another incumbent and pro-development mayor was defeated by almost a 2-1 margin, and slow growth candidates constituted a majority in the newly elected council, where women were also in the majority.

The politics of alternative visions, whether urban village or world city, are related to traditional left-right divisions but cannot be reduced to them. In both the city and the suburbs, the 1990 election produced some novel alliances and unexpected crossovers. This was partly the result of ambivalent and even contradictory tensions within each vision. The unregulated growth implicit in the world city vision is a form of creative destruction: it holds the promise of job creation and (with political will) the funding of social programs, but it leads to rapid social and land-use change, and forsakes other values. The slow growth implicit in the vision of the urban village respects existing social and environmental values but at the cost of a certain provincialism which might divert growth to neighbours. Clearly, the management of change and the integration of divergent values is no easier to resolve in Vancouver than it is anywhere else.

9

The biophysical environment today

D.G. Steyn

M. Bovis

M. North

O. Slaymaker

*T*he scenic surroundings of Vancouver present a backdrop unparalleled in North American cities. High, snow-capped peaks in the Coast Mountains descend through steep, thickly forested slopes to deep fjord inlets. To the southeast, the imposing volcanic edifice of Mt. Baker dominates the broad expanse of the Fraser Lowland. To the west, the brilliant spectacle is completed by the Gulf of Georgia, studded with numerous islands and backed by the serrated backbone of the Vancouver Island Mountains. Everywhere the vegetation is lush, for winters are seldom harsh, moisture is abundant, and spring comes early in this land of equable climate. The splendour of its surroundings prompted local journalist Chuck Davis to compare his city with the world's great scenic cities: Cape Town, Hong Kong, Rio de Janiero, and Sydney, and, ultimately, to declare Vancouver the most scenic.

To the untrained eye, Vancouver is one of those rare cities which seems to have achieved a delicate balance between urban development and scenic preservation – a balance more characteristic of small settlements than of major metropolitan areas. The water supplied to Vancouver requires minimal treatment and is considered cleaner than that in any major urban area on the continent. Above all, the supply is abundant. Beaches within 100 metres of downtown Vancouver are safe for swimming; few maritime cities in the world can claim as much. A prodigious number of aquatic birds are nourished by extensive estuarine wetlands. The fish resources are legendary. Black bears, coyotes, and raccoons are frequent visitors to residential areas from nearby wild lands. And eagles circle overhead.

Yet, there are many problems – some serious, some emerging, some dimly perceived – which are associated with the always uneasy co-existence of nature and urban development. Paradoxically, these problems are closely linked to the very factors which lend the region its scenic appeal. The first set of problems is caused by human disruption of the chemical characteristics of the environment, giving rise to air and water pollution. While the levels of pollution are not as severe as are those in heavily industrialized regions, the original pristine state has been measurably degraded. This chapter will examine how the particular combination of topography, coastline, and position of the City of Vancouver conspires to produce episodes of air pollution far worse than might be expected. Similarly, the topography of the area, combined with the proximity of the river drainage system to agricultural and urban areas, has led to locally severe pollution of ground and surface waters. Then, there are problems stemming from human ignorance about, or disregard for, natural geophysical phenomena, such as landslides, earthquakes, and floods, which are the principal natural hazards of the Vancouver region. These problems exist because glacially steepened slopes, blanketed by unconsolidated material and subject to seasonally high runoff, are located in a tectonically active environment. A third class of problems – the loss of biological diversity caused by urbanization and mechanized agriculture – poses no immediate chemical or physical threat to human well-being but mirrors global concerns about the destruction of natural vegetation and the loss of habitat for native fauna. These issues are addressed in the concluding section of this chapter in the context of the Green Zone Plan recently proposed for the Greater Vancouver region.

Air and water pollution in the Lower Fraser Valley

Air Pollution

Vancouver and the surrounding Lower Fraser Valley are afflicted by the ozone pollution found in many large, mid-latitude coastal cities, most notably Los Angeles, Athens, and Tokyo. Vancouver is considerably less industrialized and has a significantly smaller population than those cities, but the particular combination of meteorology, topography, and emissions in the Lower Fraser Valley considerably exacerbates the air pollution problem (although is is not as severe as it is in the world's most polluted cities).

High concentrations of ozone in the lower atmosphere should not be confused with the ozone-rich air which occurs naturally in the stratosphere roughly 25 kilometres above the Earth's surface and in which the much-publicized 'ozone hole' occurs. Stratospheric ozone is generated by a dissociation of atmospheric oxygen through the action of solar ultraviolet radiation. Ozone in the lower atmosphere is produced by a complex sequence of chemical reactions that involve a wide range of volatile organic compounds and oxides of nitrogen. The chemical reactions are driven by ultraviolet radiation from the sun and consume the volatile organic compounds and oxides of nitrogen to produce ozone, a highly corrosive gas that affects human and plant health and corrodes rubber, plastics, and many building materials. The chemical reactions are strongly temperature-dependent and are faster at higher temperatures. Like all pollutants, ozone generated by these photochemical reactions will only accumulate to appreciable levels if atmospheric conditions are not conducive to rapid dilution of the gas by mixing. Therefore, a combination of emissions and appropriate meteorological conditions are necessary for ozone pollution to reach high concentrations in the atmosphere of the Lower Fraser Valley.

In order to provide protection for human and plant health and materials, the federal government has established a sequence of generalized ranges for the management of air pollution. The 'desirable' range is the long-term goal for air quality and provides a basis for an antidegradation policy for unpolluted parts of the country and for the continued development of control technology. For hourly averages of ozone, this range is 0 to 50 parts per billion (ppb). The 'acceptable' range is intended to provide adequate protection against effects on soil, water, vegetation, materials, animals, and visibility as well as personal comfort and well-being. For hourly averages of ozone, this range is 50 to 80 ppb. The 'tolerable' range denotes a concentration of an air contaminant that requires abatement without delay to avoid further deterioration to an air quality that endangers the prevailing Canadian lifestyle or, ultimately, to an air quality that poses a substantial risk to public health. For hourly averages of ozone, this range is 80 to 150 ppb.

The Air Quality and Source Control Department of the Greater Vancouver Regional District (GVRD) is responsible for monitoring air quality in its jurisdiction. It operates a network of twenty-one ambient air monitoring stations and receives data from four similar stations operated by the provincial Ministry of the Environment in the eastern portions of the Lower Fraser Valley, which is outside its jurisdiction. Figure 1 indicates the density and distribution of the monitoring network in 1990. Data from these stations reveal that the maximum tolerable hourly

Figure 1

GVRD air quality monitoring network showing stations at which air pollutants are monitored on a regular basis. The numbers adjacent to some of the stations and the isopleths show the concentrations of ozone (in ppb) at 1600 PDT on 1988.09.03 during a particularly severe multi-day episode.

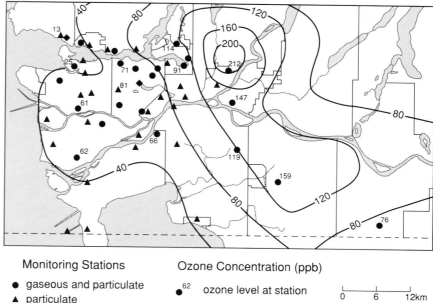

average concentration (150 ppb) was exceeded an average of four times per year in the 1980s, and the maximum acceptable hourly average concentration (80 ppb) was exceeded an average of 160 times a year. These figures have no absolute significance because they are a function of the density of the monitoring network, but they do indicate that the air we breathe is significantly polluted.

Emissions of volatile organic compounds from industrial sources, such as petroleum refining, paint, and other solvent manufacturing and dry cleaning, are less important in the Lower Fraser Valley than are those from automobile fuels, which account for over two-thirds of the total. These derive from partial combustion of fuel during normal operation of automobiles as well as during fuel-handling (e.g., refuelling). Most of these emissions occur in the densely populated western portions of the Lower Fraser Valley. In addition to anthropogenic emissions, many volatile organic compounds are emitted into the atmosphere in signifi-cant quantities by vegetation in hot weather, precisely the conditions under which ozone pollution occurs.

In summer, southwest British Columbia is dominated by an eastward extension of the Pacific High Pressure System. This anticyclonic weath-er is characterized by light winds and a slowly subsiding motion in the lower atmosphere. This subsidence results in clear skies because it produces warming, which evaporates clouds and also produces a layer of higher temperatures a few hundred metres above the surface – a phenomenon known as a temperature inversion.

Under these conditions (light winds and strong solar heating of the Earth's surface), a sea breeze blows from the cool sea to warmer land at roughly three metres per second. This is constrained by the topography of the Lower Fraser Valley to blow in a roughly south-southwest direc-tion for much of the day (Figure 2), and it carries the oxides of nitrogen and volatile organic compound emissions from the western Lower Fraser Valley eastward as photochemical reactions generate ozone. As the Lower Fraser Valley funnel narrows, the cloud of pollutants converges horizontally. The temperature inversion (which is usually 100 metres above the surface at the coast and rises to 800 metres in the centre of the Lower Fraser Valley, well below the crests of the surround-ing coast ranges) acts as a lid and ensures that pollutants cannot escape vertically.

Sometimes these circumstances produce very severe episodes of ozone pollution. Characteristically, ozone concentrations are quite low at the coast and increase to a maximum in the northern parts of the Lower Fraser Valley, roughly 45 kilometres from the coast. Under the most extreme conditions (very light winds, high temperatures, and a very low temperature inversion), the concentration of ozone can exceed 200 ppb – as it was at 1600 PDT (Pacific daylight time) on 3 September 1988

Figure 2

The average hodograph at Vancouver International Airport during all sea-breeze days in the period January 1973 to December 1982. Wind at the times given is shown by the vectors from the plotted points to the centre of the figure.

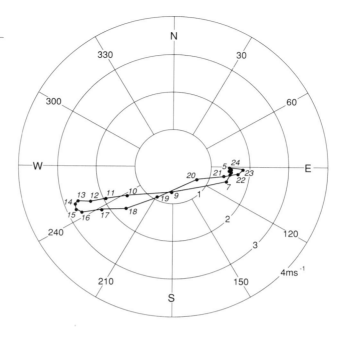

Figure 3

Seven-day moving averages of daily maximum ozone at Rocky Point Park in Port Moody (top) and Abbotsford Airport (bottom) for the years 1978 to 1990. Note the alternation of high summer and low winter values and the generally higher values at Port Moody.

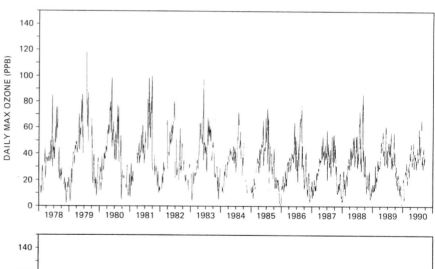

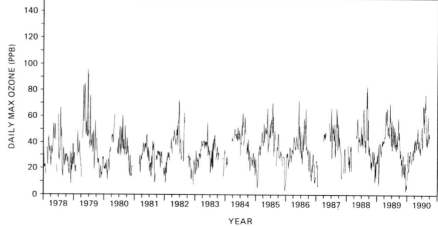

(Figure 1). Since ozone episodes depend on the occurrence of high temperatures and light winds, their severity is greatest in summer (Figure 3).

In order to control ozone pollution, the major noxious atmospheric pollutant in the region, the GVRD has undertaken preliminary studies of emissions of compounds classed as 'ozone precursors' and initiated a vehicle emissions testing program, subsequently taken over by the provincial Ministry of the Environment under the name 'Air Care.'

To be effective, the control of emissions must be carried out within a region that contains the major contributing sources and that experiences the major effects of the resultant air pollution. The present boundary that divides the Fraser Valley into two jurisdictional units (as far as air quality is concerned) is clearly in conflict with the physically unified nature of the region. The creation of a single air quality management body with monitoring, regulatory, and enforcement powers is the key to the amelioration of the problem of deteriorating air quality in the Lower Fraser Valley.

Water Pollution

Although the quality of water from mountain watersheds used for municipal supply in Vancouver is high, there has been significant degradation of water in the Lower Fraser River and its tributaries. Still, levels of pollution in the Fraser are well below those found in most rivers draining through large urban areas on this continent. A report on the Lower Fraser Valley in the early 1970s, entitled 'The Uncertain Future of the Lower Fraser,' warned that many pollutants were already at detectable levels in the main-stem water, and that, although serious ecological impacts were not yet obvious in most areas, future increases in pollutant concentrations would certainly produce adverse impacts on the estuarine ecology.

In discussions of water quality, the Lower Fraser is normally divided into three distinct segments or reaches. The main stem is the channel upstream of New Westminster, below which point it bifurcates into the North Arm and Main Arm. These are the major distributaries which traverse the Fraser Delta. The bulk of the annual flow is conveyed by the Main Arm, which is the main navigable waterway.

The flow, and therefore ability of the Fraser River to dilute and convey wastes, is generally high, although appreciable seasonal variation occurs. Annual mean daily flow is almost 3,500 cubic metres per second, but during the spring freshet, mean daily flows have reached 15,180 cubic metres per second at the Hope gauging station, about 140 kilometres east of Vancouver. Winter low flows have dipped as low as 340 cubic metres per second. Flows are tidally influenced as far upstream as Mission, and in the lowermost reaches, tidally reversed flow occurs. This, and weak longshore currents in the Strait of Georgia, can slow the purging of toxic compounds from the water column.

Figure 4 (opposite)

Vertical air photograph of the plume of suspended sediment entering the Gulf of Georgia from the North Arm of the Fraser River, 20 June 1978. The photo encompasses the southern part of English Bay, Point Grey Peninsula, the North Arm of the Fraser River and Sea Island. These conditions are typical during the time of highest snowmelt runoff during late May, June, and July, when large amounts of sediment are delivered to the delta front. Note the influence of the river-training wall on the pattern of sediment delivery from the North Arm.

Nº 153   A37597

Fraser River water is naturally turbid in comparison with the mountain tributaries which join it in the Lower Fraser Valley. Turbidity results from suspended silt, fine sand, and clay eroded from glacial materials upstream. Sediment yield from human activities, such as logging and agriculture, apparently have little influence on the average suspended load. A clearly visible plume of suspended material usually extends several kilometres into the Strait of Georgia from May to August and is carried northward and westward by currents into English Bay (Figure 4). In a water quality context, fine material is important because it bonds with heavy metals and other pollutants. As fine sediments are deposited in the sloughs and tidal marshes of the Fraser estuary, pollutants are trapped in the bottom sediments and then enter the food chain via bottom-feeding organisms. In the main stem and Main Arm, flow velocity and scour are usually sufficient to maintain fine sediments in suspension, and the channel bed consists of relatively clean, coarse sand; the North Arm is significantly more polluted. Flow velocity and turbulence also influence dissolved oxygen content. In the main stem and Main Arm of the river, dissolved oxygen is usually close to or above 90 per cent of saturation. In some of the sloughs and back channels of the Fraser Delta, however, biochemical decomposition of organic wastes, derived mainly from log storage and transport, sometimes depresses dissolved oxygen levels below the threshold required for fish respiration. A similar problem exists in the sluggish, lower reaches of many smaller tributaries of the Lower Fraser.

There are three major pathways by which pollutants enter waterways. The first, and the least detrimental, is the deposition of material directly from the lower atmosphere. In the Vancouver area this takes the form of dry deposition of suspended particulates during the summer months, when the lowest air quality conditions prevail, and wet deposition by precipitation during the remainder of the year, when air quality is notably better. In comparison with eastern North America, however, the lakes and streams of the Vancouver area are relatively unaffected by acidic precipitation.

The second source of detriment to the area's waterways is general runoff from the urbanized region and nearby agricultural areas, usually referred to as 'non-point-source' pollution. Copious winter rainfall entrains a wide range of suspended and dissolved substances as it runs off residential and industrial areas and through storm sewers to the river. Although these flow volumes are very high during the winter period, concentrations of pollutants are generally low. The majority of the storm sewers discharge to the North Arm and the north shore of the main stem (Figure 5). In the agricultural areas of Richmond, Delta, Surrey, and Langley, runoff from fields and agricultural buildings is conveyed to drainage ditches, which then discharge either directly to the main stem or to the slow-flowing tributaries.

The third and most damaging source of water pollution in the Lower Fraser River is the effluent discharged from industrial plants and municipal sewage treatment facilities located along the river banks. Many of these so-called point sources exhibit much less seasonal variation in discharge than do non-point-source flows but contain pollutant concentrations thousands of times greater than those in winter storm-water flows. Although the largest single point discharges of waste occur either directly into the Strait of Georgia or into the main stem and Main Arm,

Figure 5

Principal sewer trunk lines and sewage treatment plants administered by the GVRD. Numerous stormwater outfalls also exist but are not shown on this map.

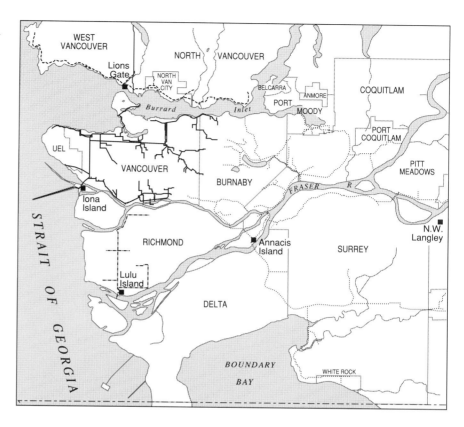

where waste dilution is usually very rapid and effective, many smaller industrial outfalls occur into the North Arm, and water quality there is noticeably poorer than in most other parts of the estuary.

There are two main types of pollutants in the Lower Fraser estuary: substances known, or thought to be, toxic to aquatic life; and organic wastes. The toxic substances can be divided into two sub-groups: heavy metals such as lead, zinc, chromium, nickel, and to a lesser extent copper and iron; and synthetic organic compounds such as petroleum distillates and others, collectively known as chlorinated hydrocarbons, which include pesticides and wood preservatives. Attributes common to all toxic substances are their ability to kill fish and other aquatic life forms if concentrations are locally high enough; their ability to significantly impair the health of organisms when introduced at non-lethal concentrations; and their tendency to persist in the aquatic realm, partly on account of their chemical affinity for fine sediments and partly through their progressive accumulation in plant and animal tissues. As a result, biomagnification of toxin concentrations tends to occur at progressively higher levels in the food chain.

Organic wastes, on the other hand, are principally derived from human sewage, agricultural runoff, log storage, and food processing industries such as fish and vegetable canning. Decomposition of these wastes by water-borne micro-organisms consumes dissolved oxygen and therefore exerts a biochemical oxygen demand on the receiving waters. If oxygen consumption exceeds the natural rate of re-aeration, progressive depletion of dissolved oxygen occurs. The principal organic waste discharges occur from three municipal sewage treatment plants: Annacis Island, Lulu Island, and Iona Island (Figure 5). With the completion of a deep-water outfall from the Iona plant in 1988, discharge of organic and other wastes to the Sturgeon Bank area of the Fraser Delta has been dramatically reduced by depositing wastes further out into the Strait of Georgia.

Two parameters – dissolved oxygen and fecal coliform count – are used as indicators of organic wastes. The latter are pathogenic organisms discharged in municipal sewage. Both parameters are measured close to the Annacis Island and Lulu Island treatment plants, and coliforms are monitored at a number of beaches during the summer period. In recent years, dissolved oxygen levels through most of the system have been above those stipulated by the provincial 'Water Quality Objectives,' although there have been temporary violations of these standards below the Annacis plant, at various points along the North Arm, and within Deas and Ladner sloughs.

Although several water quality measures showed improvement during the 1980s, there are wide temporal fluctuations in pollutant concentrations caused by variations in urban runoff as well as by the changing volume, composition, and concentration of municipal and industrial wastes. To provide estimates of long-term changes, alluvial and marine sediments have been sampled and scientists have undertaken bioassays of fish tissue known to be prone to the accumulation of toxic residues. Even this work has been hampered by the deliberate targeting of specific compounds in the sediment and tissue analyses. Ideally, a large number of substances would be monitored frequently, but the realities of provincial and federal funding dictate that these studies be relatively selective.

In the North Arm, where metal pollution of water and sediment has been the highest since records began, some metals, such as chromium, show a decrease over the past five years; others, such as manganese, show a notable increase within sediments. Similar results have been obtained from sediments in the Main Arm, where concentrations of many metals (notably nickel and zinc) have decreased over the past five years, although others have increased. A serious complicating factor in some of these comparisons is the variable particle size obtained in grab samples taken at different times. Concentrations of adsorbed metals will always be higher in the finer sediments, and much more extensive sampling will be required before firm conclusions about time-trends of pollutants in sediment can be drawn .

Bioassays integrate the effects of both water and sediment pollution over periods of months to years, so that spatial and temporal sampling variations are less crucial. Specific bottom-feeding organisms have been tested for contamination by synthetic organics, notably chlorophenols, as well as heavy metals. In general, metal concentrations appear to be relatively constant over time, with some evidence of a decrease if maximum values are compared over the past ten to fifteen years. Chlorophenols, however, show some notable increases in life forms as distinctively different as worms, bivalves, and fish, particularly along the North Arm and in the sloughs and other tidal inlets tributary to the Main Arm.

Several strategies for pollution abatement have been pursued on the Lower Fraser River since 1970. Some have sought to reduce pollution loadings. Among the most significant of these were construction of the Lulu Island and Annacis Island sewage treatment plants, opened in 1973 and 1974, respectively, and the discharge of wastes to deep water from the Iona plant in 1988. All plants provide advanced primary treatment of wastes, which removes floating and coarse suspended solids. Upgrading to secondary treatment would be extremely expensive and is not contemplated at present. However, it must be remembered that before the opening of the Iona plant in 1963 there was no sewage treatment at all in the Vancouver area, and that, in 1991, the provincial capital, Victoria, was still discharging raw sewage to Juan de Fuca Strait.

The setting of Water Quality Objectives by provincial legislation has established effluent and stream standards for both municipal and industrial discharges and has provided means for monitoring them. There has been a steady increase in the levels of effluent treatment required by waste-discharge permits. Enforcement of regulations and imposition of fines are increasing over time, as funds to support monitoring have increased. Yet more frequent monitoring and research designed to assess long-term change and degradation of aquatic ecosystems are urgently needed. With so many substances being discharged to the Fraser River system, there is legitimate concern that the toxic effects of compounds in combination may exceed the sum of their individual toxicities as measured in bioassays.

Landslides, earthquakes, floods, and other natural hazards

Instability of rock and soil

Debris flows have caused over one hundred deaths in the Howe Sound area since 1900. The most recent fatalities, in 1981 and 1983, prompted the redesign and replacement of virtually every rail and road bridge along the east shore of this fjord and the construction of mitigation works close to residential and commercial areas. The costs of such catastrophic events, in terms of human lives and suffering, and in finding engineering 'solutions' to minimize the impact of their recurrence, are enormous. Yet, to most people, the chances of inundation by flood, burial by rock avalanche or debris flow, or devastation by large storms or earthquakes are considered so improbable as to be beyond contemplation. The major obstacle to public perception of these hazards is the fact that the time interval between landslides, earthquakes, and floods may exceed a human lifespan.

Mountain basins prone to the sudden onset of debris flows are typically less than 5 square kilometres in area, have mean channel gradients of 20–30 degrees, and a total basin relief of 500–1,500 metres. Many basins along Howe Sound and within the North Shore Mountains have these physiographic attributes (Figure 6). Rock and debris slides involve the detachment of several hundred to several thousand cubic metres of material from steep slopes above debris flow channels. Rock slides are controlled by the inclination and orientation of natural joints and fractures within bedrock. Debris slides usually involve ablation till or colluvium, which slides across a much harder sub-stratum of basal till or bedrock. Both types of hillslope failure may trigger major debris flows, which then grow considerably in size as they scour loose material from the channel downstream of the initial point of failure. Tree stumps, logs, entire trees, and organic mulch are regularly delivered to the stream channel along with the rock material of a debris or rock slide. Large organic material serves to buttress coarse angular material in steep channels, promoting debris build-up in the short term. Typically, therefore, a debris flow is a water-saturated, heterogeneous mass of rock and organic debris which moves at speeds of 1–5 metres per second. Debris torrents are characterized by a high water content, low concentration of silt and clay, and high concentrations of both large and fine organic debris. Both tend to be large, frequently attaining depths of 10 metres and total volumes close to 20,000 cubic metres. Large boulders, 2–5 metres in diameter, are easily transported by these flows and add to their great destructive power.

Because major debris flows are correlated with the passage of Pacific frontal storms in the winter months, most flows occur between October and March. Orographically enhanced frontal rainfall recharges groundwater within the loose colluvium which lies above relatively impermeable glacial till. Recharge promotes instability of the colluvial layer by increasing both its weight and the water pressure within the soil voids. High stream discharge throughout most of the winter period ensures that material delivered to channels by debris and rock slides will be quickly saturated and transformed into a debris flow. Yet, prediction of debris flows is still extremely difficult. Part of the problem lies in forecasting mountain precipitation. Another consideration is antecedent moisture: if moisture levels are already high as a result of recharge from

Figure 6

Map of steep debris-flow channels along
the east shore of Howe Sound

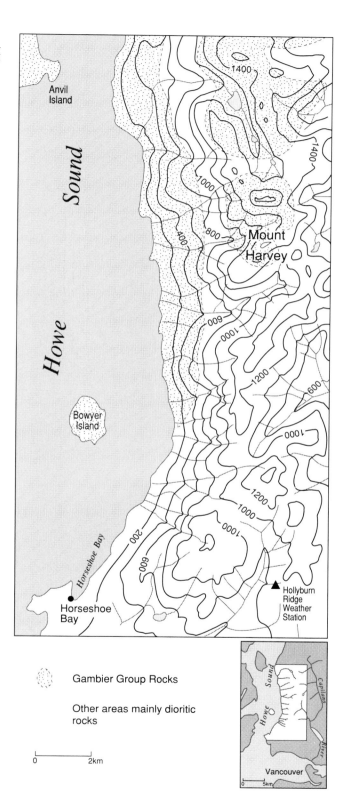

previous storms, then less rainfall is required to destabilize the soil layer. A further complication is the quantity and stability of debris in a steep channel system prior to a storm. This will control the magnitude of a debris flow if hill-slope failure occurs.

Conditions are somewhat different within ravines carved into the river bluffs and coastal cliffs of the Lower Fraser Valley. The ravines are much smaller features, carved exclusively in Pleistocene sediments (Figure 7). Both precipitation and runoff are notably lower in ravines than in the steep mountain streams, and sidewall stability tends to be much greater.

Most debris flows in these locations seem to have been triggered by slumping at the head of the ravine or by detachment of the relatively thin, weathered layer which mantles the ravine sidewalls. These slumps and slides then avalanche down the ravine, entraining debris, water, and organic material in a similar fashion to mountain debris flows.

Figure 7

Map of principal uplands underlain by Pleistocene sediments in the Vancouver area. Typical stratigraphic sequences are indicated in each area.

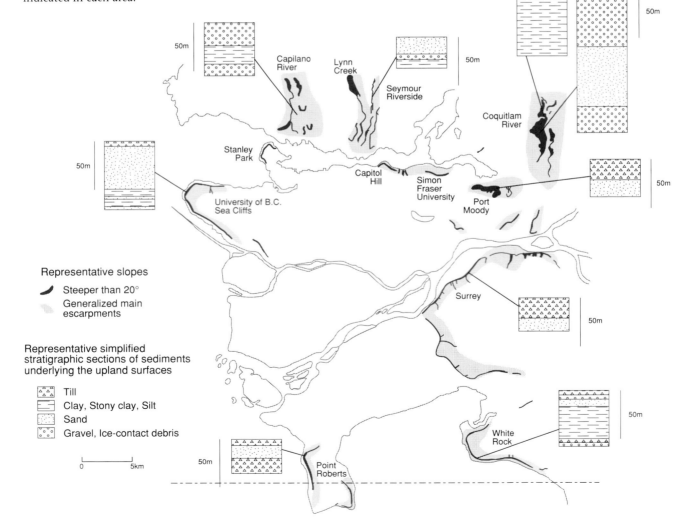

Figure 8

Photograph of Graham's Gully after the 'grand campus washout' of January 1935. Note the depth of the Gully as it crosses Marine Drive.

Since the channels in many ravines are less than 10 metres wide, they are prone to blockage by sidewall slumps. When the resulting landslide dam is breached, a considerable volume of water and sediment is flushed down the ravine.

Urban development has contributed to the degree of instability in many ravines. Notable in this respect are: the deforestation of ravine sidewalls, which lowers the stability of the weathered soil layer; the placement of unconsolidated fill at the head of ravines for road and residential construction; and inadvertent or intentional diversion of urban runoff down ravines, which causes slope undercutting as well as an increased capacity to transport debris. A dramatic case of human disturbance occurred in January 1935 in Campus Canyon, which cuts through the Point Grey cliffs at the northern end of the University of British Columbia campus (Figure 8). Here, the triggering event was the intentional release of a huge volume of water that had accumulated in the central part of the campus, east of the Main Library, as a result of prolonged rain falling on a thick snowpack. In a few days this vast increase in stream discharge carried away 76,000 cubic metres of unconsolidated sands and silts, carving a spectacular gully that extended 500 metres eastward from the cliff edge to the vicinity of the present Law Building.

The less spectacular, but ongoing, failure of the Point Grey cliffs is due to changes in groundwater seepage. When construction of the UBC campus on the western tip of Point Grey began in 1917, building stones for the first library and the Science Building were hauled up the cliff face from barges. Earlier, original forest cover had been selectively cut, and logs had been skidded down existing gullies to the sea. A series of fires had occurred in the early 1900s, and the campus land was progressively cleared and built over. All these events changed the nature of the vegetation of both the cliff top and face. The earliest photographs, taken in 1927, show the cliff face covered by a closed canopy of conifers and large maples that reached down to the beach. There was no indication of cliff failure caused by undercutting at the base by waves. The only failures of the cliffs were at the edges of the large mansion grounds, where stones had been hauled up from the beach or where the road came close to the cliff top.

In 1976, when the Museum of Anthropology (located between NW Marine Drive and the cliff top) opened, the university administration commissioned a study of the stability of the Point Grey cliffs. The study recommended the expenditure of $14 million to prevent further erosion. Very few of the suggestions were implemented. In particular, there was no attempt to control groundwater flow out of the cliff face, although this is the main, if not only, cause of continuing failure. As the water moves out of the face, it carries fine sediments, creating a niche, or overhang, which will fail because of the lack of internal cohesion in the unconsolidated sediments.

By all the evidence, cliff failures are a recent occurrence caused by changes in groundwater recharge. When forest covered the area on top of the cliffs, the forest canopy intercepted much of the rainfall, and the dense root system withdrew large amounts of water from the soil-water zone year round. Thus, relatively little water was able to infiltrate to any depth in the permeable sediments. What seepage there was on the cliff face did not disturb the stability of the existing trees. Removal of the forest and installation of hard surfaces, which initially shed water directly over the cliff top or into the ground adjacent to the cliffs, appears to have started the failures.

The solution to the problem seems obvious: replant the forest. Unfortunately, this would not meet with general approval. The drive around Point Grey is a favourite sight-seeing route; much of its attraction derives from ocean and mountain views that would be lost if trees were to grow into forests with closed canopies. Those living in residential areas on the north-facing slopes of the point also wish to maintain their views of the North Shore Mountains, which are now being lost as the successional alder woodlands are overtopped by taller conifers at the edge of the cliffs. The tree cutting that has been initiated to protect these views may further destabilize the cliffs.

Earthquakes

During earthquakes, relatively low-density, saturated soils may undergo rapid and almost total loss of strength in response to seismic movement. This process is known as liquefaction. The loss of soil strength is caused by a progressive rise in water pressure within the soil pores, accompanied by a concomitant drop in the contact pressure between soil grains. When water pressure becomes equal to the total overburden pressure, the soil loses frictional strength and the capacity to bear loads. It deforms in a fluid-like manner and structures begin to sink into the liquefied mass or are driven upward by the pressure of the escaping pore water.

Low-lying areas in coastal British Columbia, such as the Fraser River floodplain and delta, are vulnerable to liquefaction, because they are underlaid by recent fluvial and marine deposits of relatively low density and high water content. The water table is usually within about a metre of the surface, since the average elevation of the Fraser Delta is just over two metres above mean sea level. By contrast, the uplands, which are occupied by the most densely populated parts of the Vancouver metropolitan area, are relatively immune to soil liquefaction, since the underlying Pleistocene deposits were strongly compacted during the last glaciation.

Over large areas of the delta, liquefaction is most likely to occur in sandy material at depths ranging from 10 to 20 metres. In organic silts, having both lower strength and poorer drainage than sands, liquefaction might occur closer to the surface. In any event, widespread damage to structures and lines of communication could occur, together with localized ground subsidence of approximately half a metre. This would render some areas subject to tidal inundation. Three other factors could serve to magnify the risk of inundation. In the short term, seismic sea waves, or tsunamis, generated by a large earthquake could inundate parts of the delta. Second, permanent tectonic depression could occur as a result of a very large earthquake. Finally, the sea level rise, which is predicted to accompany general global warming in the next century and beyond, will only aggravate the effects of seismically induced inundation in the low-lying areas.

In an attempt to minimize structural damage, recent construction on the delta and on similar soils in the region now requires the placement of thick 'pre-loads' of sand. These serve to increase the density and decrease the water content of the underlying materials, both of which increase the resistance of soils to seismically induced liquefaction.

Floods

In the Vancouver region, two general sets of conditions produce flooding of the local rivers and creeks. The most potent with regard to flood magnitude and area of inundation is snowmelt. Peak flows usually occur in late May or early June, when abrupt warming causes rapid melting of winter snowpacks over huge areas of the province. The most extreme floods occur when melting of an abnormally thick snowpack is delayed by below-normal temperatures in April and May, resulting in large snowmelt rates when the seasonal warming occurs. This combination of conditions has caused all of the major historic floods along the Lower Fraser River. The largest of these occurred in 1894, when the Vancouver region was sparsely populated and extensive development of floodplain areas had not yet taken place. A somewhat smaller, but much more devastating, event took place in June 1948, when the monthly mean discharge of the Fraser River at Hope exceeded 10,000 cubic metres per second, with a peak discharge of over 15,000 cubic metres per second. Extensive areas of floodplain upstream of Vancouver were inundated when inadequate earthen dykes collapsed or were over-topped (Figure 9). All road and rail arteries to the east were severed. One-tenth of all agricultural land on the floodplain was inundated with

Figure 9

Map of the Lower Fraser Valley showing the major areas of inundation during the June 1948 snowmelt flood

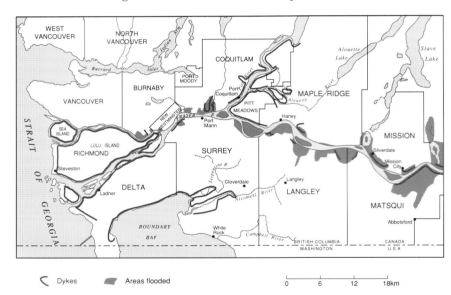

Figure 10

Flood hydrograph of Squamish River at Brackendale, near Squamish, October-November 1981. The high discharge peak was produced by heavy rainfall at lower elevations and rain on snow in the surrounding mountains. The rapid raise of the hydrograph is typical of many coastal basins subjected to prolonged, intense frontal precipitation.

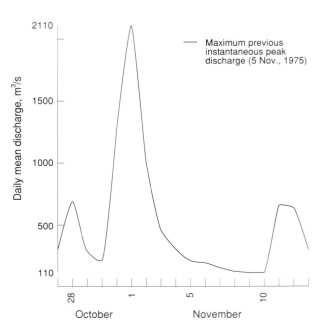

considerable losses of both stock and machinery. And more than 2,000 homes were damaged or destroyed. A comparable inundation today would cause losses totalling hundreds of millions of dollars.

The second flood-generating mechanism is winter rainfall, sometimes aggravated by rain on snow. This affects most of the rivers and creeks which drain the southern Coast Mountains, including some of the larger rivers, such as the Lillooet and Squamish. The combination of orographically enhanced frontal precipitation and steep topography produces rapid, voluminous runoff (Figure 10). Floods are usually generated by Pacific storms which entrain unusually warm, moist oceanic air. Precipitation is high by virtue of the higher moisture capacity of warm air. The potential for snowmelt augmentation of rainfall runoff is great, although unseasonably warm air is only likely to reach the coast between October and December, well prior to the peak of snow accumulation in the mountains.

The difference between snowmelt- and rainfall-generated floods has important implications for flood mitigation measures. Snowmelt floods involve the relatively slow build-up of runoff over periods of a week or more. It is therefore possible to provide adequate warning of high snowmelt water levels, although their magnitude far exceeds that of rainfall floods. Second, the cost implications of a major snowmelt flood on the Fraser River far outweigh those of rainfall floods on much smaller coastal rivers and creeks. Expensive structural solutions are therefore more economically viabile on the larger river systems, although the enormous value of the Fraser River salmon stocks has imposed severe constraints on many structural solutions proposed for the mitigation of Fraser River floods.

After the 1948 flood, dykes in the Lower Fraser Valley, and elsewhere in the province, were rehabilitated as quickly as possible. Since then, considerable expansion and reinforcement of the dyke network has taken place under a federal-provincial shared-cost arrangement. The existing network of dykes and emergency pumping stations would be adequate to convey the 1948 flood without serious incident. It might also prove capable of handling a flood of the 1894 magnitude (an event calculated to have a 140-year return period), although some inundation, as a result of seepage through dykes and pump failure, would be inevitable.

Between 1950 and 1976, many flood-control schemes involving upstream storage were proposed. The boldest of these proposals, unveiled in 1958, called for a large earth-fill dam on the Fraser River just north of Lillooet at Moran Canyon. This and other structures on the Thompson River would undoubtedly have solved the flood problem in the Lower Fraser but at the expense of the Fraser salmon runs. Public protest was loud, and the reaction was repeated five years later when a scaled-down storage proposal, labelled System E, was advanced by the federal-provincial Fraser River Board.

Interest in Fraser River projects waned in the late 1960s as attention (and money) were focused on the Columbia River Treaty dams and the Peace River power project. However, a recurrence of high water levels in 1972 led to a series of amendments to the provincial Land Registry Act. When proclaimed in 1974, they called for the delimitation of 200-year floodways along the major rivers of the province, thus controlling further encroachment onto floodplains. The regulations are administered under the Municipal Act and apply primarily to unincorporated areas not already committed to urban development. Since 1974, the municipalities of Richmond and Delta, both squarely within the 200-year floodway, have virtually doubled in population. Very large commercial and industrial districts have grown up, especially in Richmond. The same pattern has been repeated in other flood-prone communities elsewhere in the province, such as Chilliwack, Kamloops, and Prince George.

To date, the only significant structural works completed to mitigate flooding are systems of dykes. Apart from the Nechako River power development, owned and operated by Alcan, and the much smaller Bridge River system of dams, operated by BC Hydro, the Fraser River is still essentially uncontrolled. In this respect it is unique in western North America. Floods larger in magnitude than the 1894 event will eventually occur. Then, no doubt, the debate concerning the need for major dams upstream will be rejoined with vigour.

Land-use planning and parkland as responses to environmental challenges

In light of the transformation of an enormously varied biophysical environment into a landscape dominated by social artefacts, such as impermeable paved surfaces, underground pipe systems, and buildings, it would be useful to place the effects of these developments in a broad comparative frame. To this end, it is helpful to recognize, as some planners have done, four major land-use classes: urban-industrial; productive (e.g., forestry/agriculture); conserver (e.g., protected watershed areas); and aesthetic (e.g., parks and recreation). Of the GVRD's 3,468 square kilometres, about 1,000 square kilometres fall into the category of producer land and almost half (1,700 square kilometres) is considered urban-industrial land. The rest – approximately one-quarter – falls into the conserver and aesthetic categories. While, by the standards of metropolitan regions around the world, Vancouver has an exceptionally high proportion of conserver and aesthetic landscapes, it must be noted that these types of land use are unevenly distributed within the GVRD. Almost 60 per cent of the Coast Mountains are, in some form, of protected status, whereas only 2.45 per cent of the Fraser Lowland is in conserver or aesthetic land uses.

The human impact on the natural flora and fauna of the Fraser Delta and associated floodplains of the Fraser, Serpentine, and Nicomekl rivers has been both rapid and irreversible. By 1976 about 70 per cent of the original saltmarsh and 30 per cent of the original tidal marsh had been dyked and essentially destroyed, the dyking preventing the daily tidal inundation essential for the maintenance of these plant communities, upon which the food chain of the estuary depends. Fortunately for the maintenance of the abundant life forms of the delta-estuary, more islands were created in the mouth of the South Arm of the Fraser, and the tidal flats advanced further into the Strait of Georgia, which provided fresh surfaces for colonization by tidal marsh plants.

Since 1976, any development that affects the remaining marshes has been carefully vetted, and mitigation, at least, is mandatory if loss of marshland cannot be avoided altogether. Yet, this essential habitat continues to be threatened. The pollution of Fraser sediments by heavy metals, human wastes, and chemicals derived from both agricultural and industrial sources has had a long-overlooked effect on marsh plants. These trap sediments and draw on, and hence re-cycle, the nutrients carried by the river. These plants are now producing tissue which is qualitatively different from the pre-1868 detritus. Recent investigation of toxic chemical accumulation in marsh plants in the delta indicates that the sedge *Carex lyngbyei* concentrates PCB's and cadmium to levels above those found in the sediments supporting them. So, even if the quantities of marsh vegetation are maintained, the quality of the detritus yielded must now be a matter of increasing concern. As early as the 1960s, the oyster fishery of Boundary Bay was closed because of coliform contamination. The crab fishery off Iona Island was closed soon after. Yet these shellfish are still being consumed by organisms higher in the food chain that may eventually be consumed by people.

The most complete habitat loss has occurred in the grasslands of the Fraser Lowland, an area which first attracted agricultural settlement. Before dyking converted almost all of the area, first to farmland and more recently to residential, commercial, and industrial uses, these areas were flooded each year. It is already too late to recover these plant habitats. Further threats to the remaining grasslands, now agricultural fields at the front of the Fraser Delta, have recently been the cause of significant public opposition. In Richmond, attention is focused on the Terra Nova lands, and in Delta Municipality it is focused on Boundary Bay and the conversion of farmland to golf courses. In both municipalities the issue of preserving habitat for wildlife has been used to create political platforms for individuals intent on gaining aldermanic seats in municipal elections.

In 1976 the GVRD commissioned a survey of the remaining natural vegetation within its jurisdiction. Alder woodland was the most common vegetation type on the formerly forested lower slopes. The report noted extensive and rapid development, and associated loss of natural vegetation, of both upland and lowland areas not formally designated as parks or reserve land. Comparison of the map that accompanied this report with recent aerial photographs of the GVRD shows a 24 per cent reduction in the area of natural vegetation over a sixteen-year period. This rate of loss has not gone unnoticed by the municipal planners. Both Richmond and Surrey have prepared inventories of the remaining natural areas with a view to developing long-term plans for managing their existing natural resources, and the GVRD is developing its own 'Green Zone' concept as part of a region-wide planning process to update the Livable Region Plan.

In the Fraser Lowland, where competition for scarce land is extreme, the emphasis has been on creating a large number of parks (39), averaging 1.4 square kilometres in area. The lowland parks are easily accessible to the public and include one of the world's loveliest urban parks, Stanley Park, and the area's largest, Pacific Spirit Park (7.8 square kilometres), immediately adjacent to the UBC campus. Originally set aside as University Endowment Lands but never conveyed to the university, the area was the focus of a Musqueam land claim before being transferred by the provincial government to the GVRD to operate as a park in 1989. It contains early successional alder, mixed deciduous and coniferous forest, second-growth conifers (approximately 90 years old), isolated old-growth Douglas fir, and a fascinating bog ecosystem at Camosun Bog.

The park is intensively used by students, local hikers, young families, older people, cyclists, and horse riders. Conflicts between user groups are increasing and park managers will have to steer a careful course to preserve the natural systems and to allow use of the park for educational and recreational purposes.

The twenty-nine protected areas on the North Shore Mountains are far larger, averaging 25.5 square kilometres in area. Access to the Coast Mountain parks and reserves is limited to a few major trails, and water supply areas (of which there are 470 square kilometres within the GVRD) are off-limits to the public. Thus, less than half of that part of the region dedicated to conserver and aesthetic landscapes is accessible to the public.

Among the many gems of parkland accessible to the public is Lighthouse Park at Point Atkinson. This land has never been logged and provides an example of the wilderness environment of 1886. Although it covers only a square kilometre or so, it sits astride two of the most important biogeoclimatic zones in the Fraser Lowland – the Coastal Douglas Fir and Coastal Western Hemlock – as well as the lower slopes of the Coast Mountains. It epitomizes the close interaction between the sea, glaciation, and the mountains. Everywhere there is evidence of glacial erosion and wave action, moulding the granitic bedrock. Rocky headlands and outcrops contrast with valleys and draws, ephemeral streams, and shaded and humid rock cliffs – all differentiated by the ways in which runoff, seepage, and exposure to wind and sun produce differences in the availability of moisture. Trees, ferns, rushes, sedges, grasses, lichens, mushrooms, birds, mammals, amphibians, reptiles, insects, and marine life exist in essentially the same ecosystem relationships as they did in the nineteenth century.

Ecological reserves are also part of the conserver and aesthetic land uses in the Vancouver area. Bowen Island (4 square kilometres) contains bedrock associations of Douglas fir and red cedar – areas representative of the dry subzone of the Western Hemlock bioclimatic zone. The Endowment Lands reserve (0.9 square kilometres) provides a representative Puget Sound lowlands forest, and Pitt Polder (0.9 square kilometres) is an undisturbed peat ridge relict bog. These reserves come under the Ecological Reserve Act of 1971 and are used as benchmark areas against which to measure environmental changes stemming from human activities.

The largest protected area in the region is the Vancouver Water Supply Area, comprising the Capilano, Seymour, and Coquitlam watersheds. These watersheds have undeniable aesthetic attributes, such as magnificent stands of red cedar, western hemlock, amabilis fir, and mountain hemlock; glacial cirques, arêtes, and peaks; and small areas of alpine tundra. Although public access is denied, sustained-yield logging is permitted in these watersheds. Conversely, in the forested lowland parks, where preservation of the natural state is the management goal, the pressure of public access will inevitably lead to degradation of the very systems the parks were intended to preserve. This concern is heightened by the fact that the lowland parks are relatively small, isolated from one another, and affected by urban air and water pollution.

In response to such concerns, the GVRD is developing a Green Zone concept to 'conserve the scarce land resource in the region, and to preserve the sense of harmony with nature and proximity to the wilderness that the region's residents cherish.' According to a draft statement:

> The Green Zone... is more than open space to meet the needs of people living in the region. [It] is intended to limit the geographical extent of urban development... to acceptable areas, and to protect and maintain open spaces that serve ecological, social or economic functions in the region ... These include maintenance of ecologically important areas, provision of outdoor recreation opportunities, protection of scenic views and landscapes, protection of public health and safety, maintenance of culturally or economically important activities, and prevention of the financial and social costs of urban sprawl.

These are encouraging words, but there is a real danger that the accessible conserver/aesthetic lands in the GVRD will be seriously degraded by the sheer pressure of human appreciation.

Epilogue

Derek Gregory

*I*n her rather grudging essay on Vancouver in the 1980s, travel writer Jan Morris finds the defining characteristic of the city to be 'niceness.' Far from sharing the travel industry's all-too-familiar enthusiasm for the spectacular scenery of the city and its region, Morris describes Vancouver as a 'victim' of its glorious setting. In her view, places such as Vancouver and San Francisco suffer from their much-vaunted surroundings because visitors expect too much of them. More than this, she suggests, there is less to the landscape of Vancouver than meets the eye.

This view of the landscape, which reduces it to a placid 'setting' or static 'backdrop,' fails to recognize both the extraordinary dynamism of the biophysical world and the intricacies of human interaction with it. As *Vancouver and Its Region* shows, it is quite wrong to conjure up a pristine and unchanging pre-settlement environment: snow and ice, wind and running water, fire and falling rock scoured this land long before people occupied it. The first humans to live in its valleys and coastal plains established intimate relations with the natural world, seeing themselves as part *of* its rhythms and moving with its currents and seasons. Those who came after typically set themselves apart *from* 'Nature.' It was something to be mastered, exploited, and commodified. Yet even now, in the last decade of the twentieth century, we cannot escape our fundamental, yet precarious, dependence on the biophysical environment.

The concept of landscape offers one of our most powerful ways of thinking about this relationship. It helps us to understand and give form to our implication in what we all too glibly call 'Nature.' The literary critic Raymond Williams once described Nature as the most complex word in the English language, and so it is. But it is made all the more unsettling today by two realizations. First, that some changes in the regional environment originate far beyond its borders – like the ever-so-gradual movement of vast crustal plates that will some day trigger a devastating earthquake. Second, and perhaps even more disturbing, that some of these environmental changes result from our own unknowing or uncaring encroachments on the global ecosystem – as, for example, the use of aerosol sprays has damaged the stratospheric ozone layer and increased the incidence of ultra-violet radiation, or the destruction of tropical rain forests may alter climates far removed from the equator. To negotiate our way, intellectually and politically, through these complex chains of cause and effect and their implications for human life on earth demands an understanding not only of the physicality of the landscape, but also of that intersection of the human and natural economies where Nature is lived and worked with that we might call, following Alexander Wilson's splendid book on the North American landscape, the culture of Nature.

Art historians and historical geographers have shown that the modern concept of landscape was born in the city – in the cities of the European Renaissance to be exact – and that it refers to a specific way of looking at the world, a way of situating oneself simultaneously inside and outside the frame, of arranging a composition before oneself as a viewer and surrendering oneself to the pleasures of the gaze. In Vancouver, the temptation to do just that is everywhere. There is scarcely a morning on which the view across Howe Sound from Point Grey fails to set my spine tingling. On some, the mountains soar into the sky and range themselves in serried mauves and purples and pinks into the far distance. On others, scudding clouds, cat's-paws of wind on water, and the ever-shifting grey light of winter create entirely different, yet nonetheless entrancing, tableaux.   But when I think of the irony of memorializing this – 'Beautiful British Columbia' on the license plates of thousands of automobiles – I realize that the concept of landscape also carries implications of power, domination, and representation. The ribbons of summer haze riding against the mountains of the North Shore are wraiths of pollution; the lights twinkling in the mountain dark above the city are the illuminated signs of industrial skiing; and as Margaret Atwood has it in *Cat's Eye*, 'out of sight of the picture windows ... you come to the land of stumps.'

Whenever I pick my way past the mobile dressing rooms parked nose to tail along the sidewalk, as I do more and more often these days, my eyes invariably run along the heavy cables snaking toward a cluster of high, radiant lights and the clutter of folding chairs and reflecting umbrellas at their base. A voice calls 'Quiet!', a sound boom is lowered, camera platforms rise into the air; and I am reminded again that I live in the third major centre of movie and TV production in North America. Its streets and buildings, its mountains and forests are filtered through the soft Pacific air and made to stand in for New York or New Guinea; the landscape is framed, cut up, and spliced into a placeless montage to be projected onto video screens around the world.

This is probably not how most residents of Vancouver and its region think of the landscape, but these unfamiliar and unsettling perspectives are forceful reminders that landscapes are major cultural products of our time. Wandering through the crowded gangways of Granville

Island market or underneath the glassy bubbles of light in Vancouver's Gastown, it is all too easy to overlook how our 'ways of seeing' – for that is what landscape is – have been structured to encourage the selective consumption of space and time. And this is critical. Because once we reduce landscape to scenic backdrop or theme park we become distracted from the multiple, sometimes conflicting, but sometimes mutually enriching, cultural meanings that are embedded in other ways of looking at place. We become accustomed to looking at landscapes through windows of opportunity, as so many sites for spectacular display or speculative investment. To flick through the pages of *Vancouver* magazine, or to follow the signs along the scenic Marine Drives of Vancouver and West Vancouver is to place a transparent screen between oneself and the landscape. Expertly crafted by commentators and critics, this screen encourages consumption of the landscape as a single, particularly powerful – but ultimately partial and thus profoundly distorted – symbol of a highly specific 'West Coast' lifestyle whose assumptions and priorities are by no means shared, let alone endorsed by everyone who lives within the region. For these reasons, it is important to think critically about the landscapes of Vancouver and its region, as this book encourages us to do.

It is also important not to lose sight of Vancouver's ethnoscape, 'the landscape of persons who make up the shifting world in which we live.' The concept was developed by an anthropologist, Arjun Appadurai, to capture something of the flickering, fleeting, and almost phantasmagoric quality of place in the modern world. It reminds us of the multiple and compound ways in which the stabilities of place are shot through with mobilities – of people, of capital and commodities, of information and images – that traverse global space and continually remake places and the identities of their occupants. This does not mean that the particularities of place are erased; social life still 'takes place' in the most physical of senses. It does imply that places are local distillations of vibrant global processes that travel through them and whose effects are inscribed within them. Put most simply, it is no longer possible (if it ever was) to make sense of what happens in a place without raising one's eyes beyond the local horizon. And of this, too, this book reminds us, just as it demonstrates that however intensely a 'sense of place' is evoked, felt, or cultivated, it is something of a holding operation: a strategic closure, a sort of false summit beyond which lies a series of higher elevations that impact on the apparent immediacies and intimacies of the present moment.

If all of this seems unduly abstract, let me try to make it more concrete. I hesitate to describe my move to Vancouver from Cambridge, England, in 1988 as a flight from the medieval to the modern (or even the postmodern), but the transition from one city to the other has prompted me to think about the connections between people and the places in which they live in new and more personal ways. How, I wonder, will my identity and the identities of members of my family be re-shaped by our new geographies? At first, I thought of the Cambridge I had left as a cloistered and cobwebbed world, a city of 100,000 people set among the colleges of a university founded by medieval monks on the preter-naturally flat Fenland; and I saw Vancouver, in more or less complete contrast, as emblematic of a new cosmopolitanism, a city-region of more than one million people, rimmed by mountains and reaching across the Pacific Rim through the global circuits of the information economy. But to draw the contrast in such terms now seems both facile and misleading, because in our late twentieth-century world the famil-iar, taken-for-granted distinctions between 'here' and 'there' have become blurred.

This blurring is not simply the product of registering, at some deeply sedimented level, an enduring continuity between Britain and British Columbia. Although that colonial legacy is not as deeply buried as I would wish, it is the differences between these two places that seem sig-nificant to me. Nor is the blurring entirely attributable to the troubled but still truculent hegemony of the United States. Coke is 'it' – whatev-er 'it' is – on both sides of the Atlantic and on both sides of the border. Safeway 'brings it all together' in Britain much as it does in America. McDonald's golden arches glitter virtually everywhere. And yet I know here, as I knew there, that I am not living in the fifty-first state.

The difficulties of disentangling 'here' from 'there,' of mapping the world in the old ways, owe much to the fact that we are caught up in a dizzying process of global de-territorialization and re-territorialization. In the late twentieth century, we build our castles and shape our identi-ties on perpetually shifting sands. All of us, whatever our origins and wherever we live, are constantly in-between, forever displaced and unsettled; our cultures are made and re-made; they touch and interpen-etrate. We share with countless others a transnational public culture and a global marketplace; we ride on the same economic rollercoaster (though in conspicuously different cars) and our daily lives, in all their concrete specificity, are affected by and reach into those of nameless others, millions of miles away. And this is as true of life in Cambridge as it is of life in Vancouver. One of the central dilemmas of modernity (as James Clifford suggests in *The Predicament of Culture*) is the challenge of coming to terms with a world in which 'difference is encountered in the adjoining neighbourhood, the familiar turns up at the ends of the earth.' And yet I feel much more forcefully aware of that predicament, of its imperatives and responsibilities, in Vancouver than I ever did in Cambridge. I have been a geographer all my working life, but it is per-haps only now, in this place, that I have begun to understand what that might mean and what immense horizons are opened up by a geographical imagination.

When I was in Cambridge, the history of geography was viewed, not altogether surprisingly, as a chronicle of European intellectual endeavour. In this view, geography was a European science. Although its roots reached back into classical antiquity, the founding moment of the modern discipline was often taken to be Captain James Cook's entry into the South Pacific in 1769. Cook sailed with a company of illustrators, collectors, and scientists whose work was distinguished by a concern for realism in description, for systematic classification in collection, and for the comparative method in explanation. In his lively commentary *On Geography and Its History*, David Stoddart argues that it was the extension of these scientific methods of observation, classification, and comparison to peoples and societies that made possible the development of geography as a distinctively modern and avowedly 'objective' science.

But this was so in a very particular sense. Stoddart attributes the extraordinary power of the reports from Cook's voyages and their successor expeditions to their basis in direct observation rather than fancy, and sees their authority as being further enhanced by the practice of collecting, which attested in the most insistently material of ways that these people had indeed visited remote places and observed novel phenomena. Yet the observations of these explorers were direct only in the sense that they were produced by people who had been to the lands they described; they continued to be informed by European conceptual categories and European ways of seeing. This was thus a distinctly Eurocentric science. More important, the process of observation and representation on which it relied was shot through with assumptions and implications of power that were every bit as material and intrusive as the more obvious sequences of occupation and dispossession that followed in their wake. However disinterested its practitioners considered themselves to be, this 'objective science' was an essential instrument of colonization. And this was so because the very act of capturing other people on the pages of a notebook, of writing their 'otherness' without their consent, effectively seized, silenced, and transformed them into actors in a drama conceived in Europe and presented before a European audience

This is an awkward legacy with which to come to terms. As part of the story of modern geography – and of the western intellectual tradition more generally – it is only now being challenged and displaced by other voices, both inside and outside the academy. These days most geographers are more attentive to other, less imperial traditions of inquiry, and are less content to parade the world as exhibition and entertainment. Moreover, and still more directly, these issues are also part of the story of Vancouver and its region.

Captain Cook sailed into Nootka Sound in 1778, accompanied by George Vancouver, and fourteen years later, Vancouver raked much of the northwest coast with English place names. The King's English (such as it was) echoed across the Strait of Georgia; one of Vancouver's lieutenants was rhetorically inscribed in Puget Sound, and another on

Mount Baker. The very act of naming was a way of bringing the landscape into textual presence, of bringing it within the compass of a European intelligibility and folding its valleys and inlets into the charts of a distinctive constellation of power and knowledge. It was a way of domesticating its order with the instruments of a European rationality, making it at once familiar to its explorers and alien to its native inhabitants.

Alienation and dispossession soon assumed more tangible forms, and by the middle decades of the nineteenth century a formidable apparatus of colonial, commercial, and Christian power had been installed in southwestern British Columbia. This constellation of power and knowledge became so deeply engrained that its assumptions and consequences came to seem not only rational but also 'natural,' at least to those who profited most from them. This process was not monolithic, though it may well have seemed so to those who were on the receiving end of it. The entanglements between white and Native cultures were complex, shifting, and diverse. They took place in a wide variety of settings and for many different purposes, and white discourses were themselves modulated by differences of class, occupation, and gender. Still, it is difficult to avoid the conclusion that a powerful and pervasive 'structure of feeling' was quickly established among the European population of the Lower Mainland. Thus, a sort of practical consciousness did not have to be formalized to provide those who shared it a mental landscape within which to make sense of what was happening around them. And when they needed to articulate their place in the world more directly, it also afforded a familiar and even reassuring vocabulary of progress and privilege that gave shape, substance, and direction to the development of what an older generation of geographers would have called a 'regional consciousness.'

This was scarcely all of a piece, and it was by no means inclusive. In general, however, the webs of meaning in which so many white settlers, their families, and their heirs were suspended provided a bridge of ideological stability and security that spanned the tumultuous developments of the late nineteenth and twentieth centuries. Through these often dramatic changes, many residents of Vancouver and its region no doubt mourned the passing of the old days; but their unease remained embedded in popular attitudes and everday practices that provided a deep sense of continuity between a particular past and what seemed to be a predictable future.

Unelaborated though it is, this sketch achieves two things. First, it catches the hard edge of the discourse from which Native peoples and Chinese and Japanese immigrants in particular were routinely excluded, and through which they were constructed as 'other,' given that most contradictory of colonial constructions, an anonymous identity, and allocated a place – 'their' place – on the periphery. Fortunately, the arrogance of these dreadful inscriptions has been challenged in recent years, from without as well as from within, and increasingly powerful traditions of critical scholarship have sought – as do contributors to this volume – to recover the intricate historical geographies of the first nations and to disclose the ways in which the 'othering' of the Chinese and the Japanese took place.

Second, this sketch suggests that the rhetoric surrounding Vancouver's recent emergence as a world-city is challenging the once dominant structure of feeling in new ways. The old practical consciousness no longer fits the new local-global realities. In his diagnosis of *The Condition of Postmodernity*, geographer David Harvey argues that one of the most challenging, bewildering, and even threatening dimensions of our turbulent present is an overwhelming sense of 'time-space compression.' In his view, objective relations in time and space have been transformed to such a degree that we are forced to alter how we represent the world to ourselves and, indeed, how we understand ourselves. Where international exchange rates were once reported in the press every six months or so, they are now monitored constantly and broadcast hourly; where once it took weeks and sometimes months to transfer capital or information from one country to another, they can now be flashed from one hemisphere to another in an instant. In so many ways, the world sometimes seems to collapse in on us and its new, multiple and compound geographies wreak havoc with our once supposedly reliable maps. For these reasons, it is vitally important to recognize that in thinking about the geography of Vancouver and its region we are also thinking about our place in the world and the place of the world in us. And in that sense, the project represented by this volume is a beginning, not an end.

Bibliographical notes

Introduction

Any review of recent issues of the *Vancouver Sun*, the *Vancouver Province*, or monthly magazines (such as *Western Living* and *Vancouver* magazine) published in the city will reveal examples of the concerns discussed in the early part of this introduction. Michael Kluckner's *Paving Paradise: Is British Columbia Losing Its Heritage?* was published by Whitecap Books of Vancouver in 1991. UBC landscape architect Moura Quayle identified the choice between 'Mickey Mouse' or Stanley Park at a 'green-zone' conference reported in the *Vancouver Sun* (2 December 1991), and UBC Community and Regional Planning faculty Alan Artibise and Michael Seelig were the principal authors of the Pacific Fraser Region study reported in the *Vancouver Sun* (10-17 November 1990), with the assistance of reporter Daphne Bramham and graphic artist David MacLean. The prospect that the Vancouver region might end up 'like every other unattractive place on earth' was raised by William Rees, a colleague of Artibise's and Seelig's, in an article in the same series. Cole Harris's observations about the writing of BC history are in *BC Studies* 49 (Spring 1981), 86. T. Roosevelt's *The Winning of the West* was published in New York and London by Putnam in 1905; the relevant material is in volume 4, pp. 234-7. The citation of Adna Ferrin Weber is from *The Growth of Cities in the Nineteenth Century: A Study in Statistics* (Ithaca: Cornell University Press 1963), 1. The 'sea of sterile mountains' phrase is taken from James Morton, *In the Sea of Sterile Mountains: The Chinese in British Columbia* (West Vancouver: J.J. Douglas 1974). Tony Hiss, *The Experience of Place* (New York: Random House 1990) is an argument that 'whatever we experience in a place is both a serious environmental issue and a deeply personal one.' For more on the idea of 'World-Systems' see Emmanuel Wallerstein's brief overview 'World Systems Analysis,' in A. Giddens and J.H. Turner (eds.), *Social Theory Today* (Stanford: Stanford University Press 1987), 309-24, and any of the several books by Wallerstein on this theme. Northrop Frye posed the great Canadian question, 'Where is here?' in the conclusion to C.F. Klinck (ed.), *Literary History of Canada* (Toronto: University of Toronto Press 1965).

Chapter 2: The primordial environment

Geology

The theory of plate tectonics envisages that the outermost layer of the Earth is subdivided into about a dozen rigid plates that move independently. Understanding local plate movements is the object of intensive research by geophysicists. The article by R.P. Riddihough, 'One Hundred Million Years of Plate Tectonics in Western Canada,' in *Geoscience Canada* 9 (1982):28-34, is a readable overview of the topic. A summary of what is known about vertical and horizontal movements of the local landmass is provided by J.J. Clague and W.H. Mathews, in 'Geomorphic Processes in the Pacific Coast and Mountain System of British Columbia,' a review chapter in W.L. Graf (ed.), *Geomorphic Systems of North America* (Boulder, CO: Geological Society of America 1987), 528-39. The most recent explosive volcanic activity in the Vancouver region occurred in the mountains in the headwaters of the Bridge River Valley about 2,400 years ago. Information on the spatial distribution of the volcanic ash associated with this explosion is provided in R.W. Mathewes and J.A. Westgate, 'Bridge River Tephra: Revised Distribution and Significance for Detecting Old Carbon Errors in Radiocarbon Dates of Limnic Sediments in Southern British Columbia,' in the *Canadian Journal of Earth Sciences* 17 (1980):1,454-61. G.C. Rogers, 'An Assessment of the Megathrust Earthquake Potential of the Cascadia Subduction Zone,' in the *Canadian Journal of Earth Sciences* 25 (1988):844-52 offers a sobering account of the possibilities of the 'Big One.'

Pleistocene glaciation

Over the past fifty years, W.H. (Bill) Mathews has carried out research into the history of British Columbia's natural landscape. His 'Development of Cordilleran Landscapes During the Quaternary,' a review in R.J. Fulton (ed.), *Quaternary Geology of Canada and Greenland* (Ottawa: Geological Survey of Canada 1989),32-4, is a definitive summary. For a more detailed discussion of the glacial impact on the North Shore Mountains the reader is referred to J.M. Ryder, 'Geomorphology of the Southern Part of the Coast Mountains of British Columbia,' in *Zeitschrift fur Geomorphologie* 37 (1981):120-47. The best description of British Columbia's fjords remains M.A. Peacock, 'Fjord-land of British Columbia,' in *Bulletin of the Geological Society of America* 46 (1935):633-96. The most extensive research on the region's pollen records has been carried out by R.W. Mathewes and is reported in part in 'Paleobotanical Evidence for Climate Change in Southern British Columbia During Late-Glacial and Holocene Time,' in *Syllogeus* 55 (1984):397-422.

Climate and hydrology

An overview of Vancouver's climate is available in J.E. Hay and T.R. Oke, *The Climate of Vancouver* (Vancouver: Tantalus Press 1973), 49 pp. Precipitation in the Vancouver region is discussed by J.B. Wright and C.H. Trenholm, *Greater Vancouver Precipitation* (Atmospheric Environment Service, Technical Memo No. 722, 1969), 36 pp. The original stream network of Vancouver is described by G. Harris, 'The Salmon and Trout Streams of Vancouver,' *Journal of the Vancouver Aquarium (Waters)* 3 (1978):4-31. The earliest discussion of local debris torrent hazards associated with these streams is found in O. Slaymaker and H.J. McPherson, *Mountain Geomorphology* (Vancouver: Tantalus Press 1972), 74 pp.

Vegetation

V.J. Krajina was the originator of the idea of biogeoclimatic zones. See V.J. Krajina, 'Biogeoclimatic Zones and Classification of British Columbia,' in *Ecology of Western North America* 1 (1965):1-17. Much of the mapping of British Columbia's biogeoclimatic zones has been effected by K. Klinka. See, for example, K. Klinka, F.C. Nuzdorfer, and L. Skoda, *Biogeoclimatic Units of Central and Southern Vancouver Island* (Victoria: Ministry of Forests 1979), 120 pp. M. North, D. Holdsworth, and J. Teversham's 'A Brief Guide to the Use of Land Surveyors' Notebooks in the Lower Fraser Valley 1859-1890,' in *BC Studies* 34 (1977):47, details the usefulness of these notebooks in determining early vegetation distribution. R.J. Hebda and W.G. Biggs,

'The vegetation of Burns Bog, Fraser Delta, BC,' in *Syesis* 14 (1981): 47-66 provides a detailed account of this area.

Acknowledgments: The authors acknowledge the professional assistance of Anne-Marie Landry, who typed several drafts of the chapter, and Paul Jance, who designed and drafted four of the figures.

Chapter 3: The Lower Mainland, 1820-81

The early 1820s

There are two main sources of data on the Lower Mainland in the 1820s: historical records and ethnographic accounts. Of the former, the most important is the Fort Langley Journal, 27 June 1827-30 July 1830, by James MacMillan and Archibald McDonald, in the British Columbia Archives. Although primarily about the fort and its trade, it is also a major source of information about Native life. There are three accounts of a Hudson's Bay Company expedition to the Lower Fraser in 1824: 'Extracts from Mr. Chief Trader MacMillan's Report of His Voyage and Survey from the Columbia to Frazer's River,' in George Simpson, *Fur Trade and Empire*, ed. and intro. Frederick Merk (Cambridge MA: Harvard University Press 1931), App. A, 248-50; T.C. Elliot (ed.), 'Journal of John Work, November and December, 1824,' *Washington Historical Quarterly* 3 (July 1912):198-228; and François-Noel Annance, Journal of Voyage from Fort George to Fraser River, 1824-25, Hudson's Bay Company Archives, B.76/a/1, fos. 2-10 (microfilm reel 1M55). The David Thompson map, referred to in the text, is in the Toronto Public Archives, and *The Letters and Journals of Simon Fraser, 1806-1808* are edited by W. Kaye Lamb (Toronto: Macmillan 1960).

The point of entry to modern studies of the Halkomelem-speaking peoples is Wayne Suttles, 'Central Coast Salish,' in Suttles (ed.), *Handbook of North American Indians*, Vol. 7, *Northwest Coast* (Washington: Smithsonian Institution 1990), 453-75. See also Suttles, *Coast Salish Essays* (Seattle, University of Washington Press 1987), and, more specifically, his *Katzie Ethnographic Notes* (Memoir 2, British Columbia Provincial Museum 1955), as well as, in the same volume, Diamond Jenness, 'The Faith of a Coast Salish Indian.' The most careful general ethnography is Homer G. Barnett, *The Coast Salish of British Columbia* (Eugene: University of

Oregon Studies in Anthropology, 4, 1955; reprinted by Greenwood Press, Westport, CT: 1975). Barnett's field notes, in Special Collections at the University of British Columbia Library, should also be consulted. Earlier and more idiosyncratic is Charles Hill-Tout, 'Ethnological Studies of the Mainland Halkome'lem, a Division of the Salish of British Columbia' (London: *Report of the British Association for the Advancement of Science,* 72, 1902), 355-449, and reprinted with an introduction by Ralph Maud (Vancouver: Talonbooks 1978). James A. Teit, a remarkable amateur ethnographer associated with Franz Boas, wrote little on the Coast Salish, but his field notes contain relevant material. Teit's discussion of coast-interior trade routes is in the Boas Collection in the American Philosophical Society in Philadelphia, 372, reel 4.

The maps of seasonal population distribution in the early 1820s give a highly schematic picture of a much more complex pattern of seasonal movement and resource use. They are largely based on the ethnographic and archival materials described above. The distributions indicated in Burrard Inlet are probably the most contentious. In the 1930s Chief Khatsahlano told Major J.S. Matthews, the city archivist, that the Squamish had always lived there. On the other hand, Homer Barnett was told by his Squamish informants that the Squamish settled there in winter only when the sawmills were built (Field Notes, box 1, folders 4-5). Gilbert Malcolm Sproat, who spent some time in Burrard Inlet in the late 1870s as Indian Reserve Commissioner, also reported this (e.g., Department of Indian Affairs, RG 10, reel C-10106, vol. 3611, file 3756-7; or reel C-10113, vol. 3645, file 7936). I therefore adopt this position.

The concept of lifeworld is taken from Jürgen Habermas, *The Theory of Communicative Action*, Vol. 2, *Lifeworld and System: A Critique of Functionalist Reason* (Boston: Beacon Press 1987). Others, however, have argued similarly: for example, Claude Lévi-Strauss, *The Savage Mind* (Chicago: University of Chicago Press 1966), ch. 1; or Robin Riddington, *Little Bit Know Something: Stories in a Language of Anthropology* (Vancouver/Toronto: Douglas and McIntyre 1990).

Fort Langley in Native territory, 1827-58

Only a small fraction of the Hudson's Bay Company records relating to Fort Langley has survived. The Fort Langley Journal, cited above, is invaluable, and is the basis of much of this section. Other important sources are the Correspondence Relating to Fort Langley, 1830-59 (A/B/20, L 3A) in the British Columbia Archives; and the Fort Langley Correspondence Book, 1830-71 (B113/b/1-4), Correspondence Inward, 1844-70 (B113/c/1), and Miscellaneous Items, 1830-78 (B113/z/1-2) in the Hudson's Bay Company Archives.

The secondary writing on Fort Langley includes: Mary K. Cullen, 'The History of Fort Langley, 1827-96,' *Canadian Historic Sites: Occasional Papers in Archaeology and History*, No. 20 (Ottawa: National Historic Parks and Sites Branch 1979), 5-122; and Jamie Morton, 'Fort Langley: An Overview of the Operations of a Diversified Fur Trade Post 1848 to 1858 and the Physical Context in 1858' (ms. prepared for Parks Canada, n.d.). More generally relevant are: Richard Mackie, 'Colonial Land, Indian Labour, and Company Capital: The Economy of Vancouver Island, 1849-1858' (MA thesis, University of Victoria, 1984); James R. Gibson, *Farming the Frontier: The Agricultural Opening of the Oregon Country, 1786-1846* (Vancouver: UBC Press 1985); and Barry M. Gough, *Gunboat Frontier: British Maritime Authority and Northwest Coast Indians, 1846-1890* (Vancouver: UBC Press 1984). The paragraph on the Royal Navy and the Cowichan is derived from Gough. The place of Fort Langley within the transcontinental fur trade is sketched in R. Cole Harris (ed.) and G. Matthews (cart.), *Historical Atlas of Canada*, Vol. 1, *From the Beginning to 1800* (Toronto: University of Toronto Press 1987).

Essentially, there is no writing on the response of Native peoples in the Lower Mainland to the establishment of Fort Langley. The account of Coyote is in James Teit, 'Mythology of the Thompson Indians,' *Memoirs of the American Museum of Natural History*, Vol. 8, Part 2 (New York: 1912), 416. The southward expansion of the Lekwiltok is discussed in Robert Galois, 'Kwagulth Settlement Sites, 1775-1920: A Geographical Survey and Gazetteer' (ms. under consideration by UBC Press).

Town, countryside, camps, 1858-81

The best general study of early New Westminster is Margaret McDonald, 'New Westminster, 1858-1871' (MA thesis, UBC, 1947). Moody's comments about the forest are quoted on p. 25. More specialized is Laura E. Scott, 'The Imposition of British Culture as Portrayed in the New Westminster Capital Plan of 1859 to 1862' (MA thesis, Simon Fraser University, 1983). The first opportunity for a close examination of New Westminster's occupational, ethnic, and racial structure is provided by the 'Nominal Return of the Living' in the Canadian Census of 1881 (Public Archives of Canada, C-13285). My remarks about racism follow Kay J. Anderson, *Vancouver's Chinatown: Racial Discourse in Canada, 1875-1990* (Montreal: McGill-Queen's University Press 1991). Oblate administration is discussed in Melanie Jones, 'The Ste-Marie Mission, 1860-1900' (MA thesis, UBC, 1992).

The basic works on land policy are Robert E. Cail, *Land, Man, and the Law: The Disposal of Crown Lands in British Columbia, 1871-1913* (Vancouver: UBC Press 1974) and Phyllis Mikkelsen, 'Land Settlement Policy on the Mainland of British Columbia, 1858-1874' (MA thesis, UBC, 1950). The sections on rural settlement in John E. Gibbard, 'Early History of the Fraser Valley, 1808-1885' (MA thesis, UBC, 1937) are helpful; as is Donna H. Cook, 'Early Settlement in the Chilliwack Valley' (MA research paper, UBC, 1979). A compilation by F.W. Laing, 'Agricultural Notes From Records in the Provincial Archives,' is in the Vancouver City Archives. The first somewhat systematic inventory of agriculture in the Fraser Valley is in the *Second Report of the Department of Agriculture of the Province of British Columbia* (Victoria: 1892) because, by some horrible blunder, the nominal censuses of agriculture in BC for 1881 and 1891 have been destroyed. The alienation of farmland in the Fraser Valley until 1871 can be analyzed using an inventory meticulously prepared by F.W. Laing, 'Colonial Farm Settlers on the Mainland of British Columbia, 1858-1871' in the BCARS, and the distribution of rural settlement in 1881 can be worked out from the nominal census returns cited above. The diary of Fitzgerald McCleery, the only such document from the Lower Mainland in the colonial period, is published in *Vancouver Historical Journal*, No. 5 (Vancouver City Archives, 1965).

There are two censuses of the Natives in the Lower Mainland in 1876-7 – one by James Lenihan, Indian Superintendent, taken in May and June 1877, DIA, RG 10, reel C-10114, vol. 3650, file 8302; and the other by George Blenkinsop, apparently taken at the beginning of December 1876, NAC, RG 88, vol. 494. The map of 1877 and details in the text are based on these censuses and accompanying notes. The basic discussions of the evolution of Native land policy and of the reserve system are in Robin Fisher,

Contact and Conflict: Indian-European Relations in British Columbia, 1774-1890 (Vancouver: UBC Press 1977) and Paul Tennant, *Aboriginal Peoples and Politics: The Indian Land Question in British Columbia, 1849-1989* (Vancouver: UBC Press 1990). Barbara Weightman, 'The Musqueam Reserve: A Case Study of the Indian Social Milieu in an Urban Environment' (PhD thesis, University of Washington, 1972) is useful; the 1913 statement of a Musqueam Chief before the Royal Commission on Indian Affairs for the Province of British Columbia (the McKenna McBride Commission) is taken from Weightman, 90. Sproat describes his meeting with the six chiefs in a letter to Forbes Vernon, 12 April 1878, Canada, Indian and Northern Affairs, Sproat's Letterbook No. 2, 10-12.

A general account of the first decade of salmon canning along the Fraser River is now relatively accessible. The basic secondary sources are: Cicely Lyons, *Salmon Our Heritage: The story of a Province and an Industry* (Vancouver: Mitchell Press 1969); Keith Ralston, 'Patterns of Trade and Investment on the Pacific Coast, 1867-1892: The Case of the British Columbia Salmon Canning Industry,' *BC Studies* 1 (Winter 1968-9):37-45; Edward N. Higginbottom, 'The Changing Geography of Salmon Canning in British Columbia, 1870-1931' (MA thesis, Simon Fraser University, 1988); and Duncan Stacey, 'Technological Change in the Fraser River Salmon Canning Industry, 1871-1912' (MA thesis, UBC, 1977). The details are much more elusive, as no company records survive for this period. The most precise annual information is in the 'Annual Report of the Ministers of Public Works' (or, after 1876, of the Department of Marine and Fisheries), *Sessional Papers*, Parliament of Canada. The quotation from A.C. Anderson, commissioner of Fisheries, is in the *Sessional Papers* of 1879. Information about the terms of work in the early fishery is best obtained from the hearing of the Fisheries Commission of 1892, printed verbatim in the *Sessional Papers* of 1893.

The only information I have found on Native housing at the canneries is in James Lenihan's census, cited above.

The literature on logging and sawmilling in Burrard Inlet includes: F.W. Howay, 'Early Shipping on Burrard Inlet, 1863-1870,' *British Columbia Historical Quarterly* 1 (1937):3-20; James E. Flynn, 'Early Lumbering on Burrard Inlet, 1862-1891' (B.Sc. thesis, UBC, 1942); and James Morton, *The Enterprising Mr. Moody, the Bumptious Captain Stamp* (North Vancouver: J.J. Douglas 1977). Stamp's cited correspondence is in the BCARS, Colonial Correspondence, B-1366, 1643 (letters of 21 Dec. 1859, 4 April 1861, and 17 May 1865); his surveyor's report is in B-1343, 969 (J.B. Launders). The timber lease for Hastings Mill is published in the *British Columbia Sessional Papers*, 1885, 387-9. References to forest fires in the Moody correspondence, also in the BCARS, Col. Corr., B-1347, 1159, are in letters of 5 Jan. 1870 and 12 May 1873. Dawson's description of the Moodyville mill is in Douglas Cole and Bradley Lockner (eds.), *The Journals of George M. Dawson, British Columbia, 1875-1878* (Vancouver: UBC Press 1989),115. For Sproat's views on Native access to timber lands see his letters of 7 Dec. 1876 (DIA, RG 10, reel C-10106, vol. 3611, file 3765-5), and of 24 Nov. 1877 (DIA, RG 10, reel C-10122, vol. 3699, file 16665). On changing individual space-time geographies associated with industrial work, see Allan Pred, *Making Histories and Constructing Human Geographies: The Local Transformation of Practice, Power Relations, and Consciousness* (Boulder, San Franciso, and Oxford: Westview Press 1990), especially ch. 3.

Interpretation

Bulwer Lytton's imperial exhortation to Colonel Moody (29 Oct. 1858) is in the Moody papers, BCARS, Col. Corr. B-1346, 1149b. The organization of much of the chapter and the points made in conclusion draw partly on ideas about this country and about European social change in new environments overseas that I have put elsewhere – eg., 'European Beginnings in the Northwest Atlantic: A Comparative View,' in David Hall and David Grayson Allen (eds.), *Seventeenth Century New England* (Boston: The Colonial Society of Massachusetts 1984) and 'The Pattern of Early Canada,' in Graeme Wynn (ed.), *People, Places, Patterns, Processes: Geographical Perspectives on the Canadian Past* (Toronto: Copp Clark Pitman 1990) – and partly on the literature on the nature and geographical relationships of social power as societies become modern. Stewart R. Clegg, *Frameworks of Power* (London: Sage 1989) is a helpful introduction to the latter. I am more intrigued by the work of Anthony Giddens, particularly *The Nation State and Violence* (Berkeley: University of California Press 1987) than are many of his recent critics. Robert A. Dodgshon, *The European Past: Social Evolution and Spatial Order* (London: Macmillan

1987) is also suggestive. My remarks about disciplinary power mark the beginning of an attempt to think about the tactics of such power in a type of setting that Michel Foucault never considered.

Acknowledgments: This chapter grows out of the Historical Geography of British Columbia Project, funded by the Social Sciences and Humanities Research Council of Canada. Although written by Cole Harris, it is to a considerable extent a collective effort that reflects the general advice of Robert Galois, Daniel Clayton, and Richard Mackie, and the particular investigations of Donna Cook (rural settlement), Robert Galois (population distribution, 1881) Edward Higginbottom (fisheries), Werner Kaschel (New Westminster), and Yasmeen Qureshi (logging and sawmilling). I would also like to thank Robin Fisher, Michael Kew, and Wayne Suttles, each of whom commented on a draft of this chapter.

Chapter 4: The rise of Vancouver

Much has been written on Vancouver between 1886 and 1951, but a great deal of it remains unpublished and there have been few syntheses. The best guide to the literature on all of the topics covered in this chapter, and one that has been invaluable in tracking down material on particular topics, is Linda Hale, *Vancouver Centennial Bibliography*, 4 vols. (Vancouver: Vancouver Historical Society 1986). Among studies that attempt to present an integrated treatment of the role of the city, Patricia E. Roy, *Vancouver: An Illustrated History* (Toronto: James Lorimer and Company 1980) is far and away the most comprehensive and accessible and has been referred to throughout the preparation of this chapter. Norbert Macdonald reports the results of many years of detailed research in *Distant Neighbours: A Comparative History of Seattle and Vancouver* (Lincoln: University of Nebraska Press 1987). Walter Hardwick, *Vancouver* (Don Mills, ON: Collier-Macmillan 1974) deals relatively briefly, but suggestively, with the development of the city. L.J. Evenden (ed.), *Vancouver, Western Metropolis* (Victoria: University of Victoria 1978) treats the past lightly. J. Lew Robinson wrote a brief historical geography of the city ('How Vancouver has Grown and Changed') for the *Canadian Geographical Journal* (1974):40-8. Two very different essays on other cities that influenced my approach to this chapter are S.B. Warner

'The New Freedom' [Los Angeles], in *The Urban Wilderness: A History of the American City* (New York: Harper and Row 1972), 113-49, and David Harvey, 'Paris 1850-1870,' in *Consciousness and the Urban Experience* (Baltimore: John Hopkins University Press 1985), 63-220.

Early studies by Alan Morley, *Vancouver: Milltown to Metropolis* (Vancouver: Mitchell Press 1961) and Eric Nicol, *Vancouver* (Toronto: Doubleday 1970) are anecdotal and say little about the central concerns of the chapter. Michael Kluckner's two books, *Vancouver: The Way It Was* (Vancouver: Whitecap Books 1984) and *Vanishing Vancouver* (Vancouver: Whitecap Books 1990) include many interesting facts and some splendid illustrations, but they do not provide a coherent historical treatment. A substantial selection of the enormous number of photographs of Vancouver is presented in Anne Kloppenborg et al., *Vancouver's First Century: A City Album, 1860-1960* (Vancouver: J.J. Douglas 1977; 2nd ed. Douglas & McIntyre 1991).

'More room, more homes ... more people'

The quotations that begin this section are from Frank Yeigh, *Through the Heart of Canada* (Toronto: Gundy 1913), 303, and from Ethel Wilson's novel *The Innocent Traveller* (Toronto: Macmillan 1944), 124, set in the city's West End. A wide-ranging treatment of the images that contemporaries attached to fast-growing New World cities in the nineteenth and early twentieth centuries is offered by David Hamer, *New Towns in the New World. Images and Perceptions of the Nineteenth Century Urban Frontier* (New York: Columbia University Press 1990).

More on the development of transportation networks in the Vancouver region can be found in Patricia E. Roy, 'The Changing Role of Railways in the Lower Fraser Valley, 1885-1965,' in Alfred H. Siemens (ed.), *Lower Fraser Valley: Evolution of a Cultural Landscape* (Vancouver: Tantalus Press 1968), 51-68, and in Ronald H. Meyer, 'The Evolution of Roads in the Lower Fraser Valley' (BA essay, Department of Geography, UBC, 1967); also useful is Meyer's 'The Evolution of Roads in the Lower Fraser Valley,' in Siemens (ed.), *Lower Fraser Valley*, 68-88.

Population growth is the subject of N. Macdonald's, 'Population Growth and Change in Seattle and Vancouver, 1880-1960,' *Pacific Historical Review* 39 (October 1970):297-321, reprinted in J. Friesen and H.K. Ralston (eds.), *Historical*

Essays on British Columbia (Toronto: McClelland and Stewart 1976), 201-27. Much information on population distribution is contained in a 1966 Geography MA thesis completed at UBC by Glynn I. Howell-Jones, 'A Century of Settlement Change: A Study of the Evolution of Settlement Patterns in British Columbia,' and the article derived from it, 'The Urbanization of the Fraser Valley,' in Siemens (ed.), *Lower Fraser Valley*, 139-61. The beginnings of urban sprawl in Surrey are treated in A.D. Crerar, 'Population Density and Municipal Development – The Vancouver, BC, Metropolitan Area,' *The Canadian Geographer* 9 (1957):1-6.

The discussions of Richmond and Maple Ridge are drawn in large part from five studies completed in the Department of Geography at UBC: R.M. Bone, 'A Land-Use Study of the Steveston District' (BA essay, 1955); Philip H. Connolly, 'A Geographical Analysis of Historical Events in the Maple Ridge District' (BA essay, 1953); Henry I. Ivanisko, 'Changing Patterns of Residential Land Use in the Municipality of Maple Ridge, 1930-1960' (MA thesis, 1964); John M. Read, 'The Pre-War Japanese Canadians of Maple Ridge: Landownership and the "Ken Tie"' (MA thesis, 1975); and Arno L. Ulmer, 'A Comparison of Land Use Changes in Richmond, British Columbia' (MA thesis, 1964). The early vegetation of the Richmond area is mapped in M.E.A. North, M.W. Dunn, and J.M. Teversham, *Vegetation of the Southwestern Fraser Lowland, 1858-1880* (Ottawa: Environment Canada 1979). Leslie J. Ross, *Child of the Fraser* (Richmond: Richmond Centennial Society 1979) and *Maple Ridge: A History of Settlement* by the Maple Ridge Branch of the Canadian Federation of University Women (1972) are useful local histories. The warning that Maple Ridge would soon be enveloped by Vancouver was sounded in *A Report to the Council of the District of Maple Ridge on Subdivision Planning Policy* by the Lower Mainland Regional Planning Board (New Westminster, 1959).

Vancouver's history was compared to 'that of Topsy' in the report of Harland Bartholomew and Associates. This work, *A Plan for the City of Vancouver* (Vancouver: Town Planning Commission 1929) is also the source of my estimate that some 90 per cent of the 1929 population lived near rail tracks. Many of the points made in this section, and that which follows it on the urban fabric, rest on work reported by undergraduate students who completed projects on the development of small parts of the city in my Geography of Urbanization class at UBC in the late 1970s and 1980s; among them, Grant Atkins, Kathryn Barr, Kevin Danks, Leslie Gilbert, Eva Lee, Chris Went, Richard H. White, Greg Young, and George Yu in particular, may recognize some of their work. Other details and the material on DL 301 are drawn from Donna McCririck's UBC MA thesis in Geography, 'Opportunity and the Workingman: A Study of Land Accessibility and the Growth of Blue Collar Suburbs in Early Vancouver,' completed in 1981 and reported in part in D. McCririck and Graeme Wynn, 'Building "Self-Respect and Hopefulness": The Development of Blue-Collar Suburbs in Early Vancouver,' in Graeme Wynn (ed.), *People, Places, Patterns, Processes: Geographical Perspectives on the Canadian Past* (Toronto: Copp Clark Pitman 1990), 267-84. George McSpadden's endeavours are recounted in *British Columbia Pictorial and Biographical* (Winnipeg: S.J. Clark 1914), Vol. 1, 397-9, which also includes a biography of James Z. Hall (169-70); Vol. 2 includes material on Charles W. Tait (263-6). Kluckner, *Vancouver: The Way It Was,* contains brief notes on J.Z. Hall and James L. Quiney (154-62). Sidney Wybourn appears in the pages of Edward M. Gibson, 'The Impact of Social Belief on Landscape Change. A Geographical Study of Vancouver' (PhD thesis, UBC, 1972). The Dunbar case study is drawn from a research project of my own; Kathleen Bednard and George Yu, research assistants in the summer of 1985, will recognize the importance of their labours to this discussion.

By far the best work on Vancouver housing, upon which I have drawn heavily in this text as well as in my own explorations of the city, has been done by Deryck W. Holdsworth. His UBC Geography theses – MA, 'Vernacular Form in an Urban Context' (1971); and PhD, 'House and Home in Vancouver: The Emergence of a West Coast Urban Landscape' (1981) – are rich fare indeed for anyone interested in this topic. Many of the quotations in this section are taken from Holdsworth, and page 126 of his PhD thesis is the source of the 'carpenter versions of Gothic revival' quote. Among several articles presenting parts of this work in more readily available form are Holdsworth's 'House and Home in Vancouver: Images of West Coast Urbanism, 1886-1929,' in G.A. Stelter and A.F.J. Artibise (eds.), *The Canadian City: Essays in Urban History* (Toronto: McClelland and Stewart

1977), 186-211; 'Cottages and Castles for Vancouver Home-Seekers,' *BC Studies* 69-70 (Spring-Summer 1986):11-32; and 'Regional Distinctiveness in an Industrial Age: Some California Influences in British Columbia Architecture,' *American Review of Canadian Studies* 12 (1982), 64-81. See also G.E. Mills and D.W. Holdsworth, 'The BC Mills Prefabricated System: The Emergence of Ready-Made Buildings in Western Canada,' *Occasional Papers in Archeology and History*, No. 14 (Ottawa: Canadian Historic Sites 1975), 127-69. Graeme Chalmers and Francis Moorcroft describe characteristic house types in *British Columbia Houses: Guide to the Styles of Domestic Architecture in British Columbia* (Vancouver: Wedge 1981). Harold Kalman, *Exploring Vancouver 2* (Vancouver: UBC Press 1978) is also useful for its comments on historical (and modern) buildings.

Prosperity by sea and land

The early competition between Vancouver and Victoria is discussed in L.D. McCann, 'Urban Growth in a Staple Economy: The Emergence of Vancouver as a Regional Metropolis, 1886-1914,' in Evenden (ed.), *Vancouver*, 17-41. R.A.J. McDonald, 'Victoria, Vancouver and the Economic Development of British Columbia, 1886-1914,' in A.F.J. Artibise (ed.), *Town and City: Aspects of Western Canadian Urban Development* (Regina: Canadian Plains Research Centre 1981), 31-55, reprinted in W.P. Ward and R.A.J. McDonald (eds.), *British Columbia. Historical Readings* (Vancouver: Douglas and McIntyre 1981), 369-95, is a most useful treatment of its topic. The effect of the Klondike gold rush is discussed in N. MacDonald, 'Seattle, Vancouver and the Klondike,' *Canadian Historical Review* 49 (September 1968):234-46. Cole Harris, 'Locating the University of British Columbia,' *BC Studies* 32 (1976-7): 106-25 gives a good deal of insight into provincial rivalries and attitudes in the early twentieth century.

My discussions of the Vancouver labour market and of wage and price levels depend, in the main, on Eleanor A. Bartlett, 'Real Wages and the Standard of Living in Vancouver, 1901-1929,' *BC Studies* 51 (Autumn 1981): 3-62 and McCririck, 'Opportunity and the Workingman.' R.A.J. McDonald's UBC PhD thesis, 'Business Leaders in Early Vancouver, 1886-1914,' (1977) is invaluable for any treatment of the urban economy (as well as several other topics) for the period before 1914. The best published study of working people in early Vancouver is R.A.J. McDonald, 'Working Class Vancouver, 1886-1914: Urbanism and Class in British Columbia,' *BC Studies* 69-70 (Spring-Summer 1986):33-69. P. Roy, 'Vancouver: "The Mecca of the Unemployed," 1907-1929,' in A.F.J. Artibise (ed.), *Town and City* (Region: Canadian Plains Research Centre, University of Regina 1981), 393-413, is useful. The collaborative efforts of The Working Lives Collective, *Working Lives 1886-1986* (Vancouver: New Star Books 1985) also contributed to this discussion. R.M. Galois 'Social Structure in Space: The Molding of Vancouver 1886-1901' (PhD thesis, Geography, Simon Fraser University, 1979) is meticulous but more difficult.

Leah Stevens, 'Rise of the Port of Vancouver, British Columbia,' *Economic Geography* 12 (January 1936):61-70 and Charles N. Forward 'The Functional Characteristics of the Geographic Port of Vancouver,' in Evenden, *Western Metropolis*, 57-77, trace the development of port facilities and traffic. L. Stevens 'The Grain Trade of the Port of Vancouver, British Columbia,' *Economic Geography* 12 (1936):185-96 was also useful, as was P.D. McGovern, 'Industrial Development in the Vancouver Area,' *Economic Geography* 37 (1961):189-206. My understanding of labour unrest in Vancouver rests on Paul Phillips, *No Power Greater: A Century of Labour in British Columbia* (Vancouver: BC Federation of Labour 1967); R.C. McCandless, 'Vancouver's "Red Menace" of 1935. The Waterfront Situation,' *BC Studies* 22 (Summer 1974):56-70; S. Rosenthal, 'Union Maids: Organized Women Workers in Vancouver 1900-1915,' *BC Studies* 41 (1979):36-55; and S. Jamieson, *Times of Trouble: Labour Unrest and Industrial Conflict in Canada, 1900-1966* (Ottawa: Task Force on Labour Relations 1968), especially p. 269.

The city: healthy, beautiful, efficient

The discussion of sewer and water provision in Vancouver is drawn for the most part from Louis P. Cain, 'Water and Sanitation Services in Vancouver: An Historical Perspective,' *BC Studies* 30 (Summer 1976):27-43. Thomas Mawson's ideas can be found in his 'Vancouver. A City of Optimists,' *Town Planning Review* 4 (1913):7-12 and especially as outlined in this chapter in his 'Civic Art and Vancouver's Opportunity,' *Canadian Club of Vancouver. Addresses and Proceedings* (1911-12), 39-46. Much has been written on the City Beautiful movement. M.A. Meek, 'History

of the City Beautiful Movement' (MA thesis, UBC 1979) sketches something of the Canadian context and treats Vancouver (as well as several eastern cities) briefly on the way to more detailed analysis of Prairie cities. The discussion of Thomas Adams is drawn from his 'Report on the Planning of Greater Vancouver,' *The Contract Record* 29, no. 5 (3 February 1915):122-3 and his 'Vancouver Civic Centre Competition,' *Town Planning Review* 6 (1915):31-3, as well as from small items in *The Contract Record* 28, no. 4 (1914):1,244 and 29, no.1 (1915):8.

Two useful, more general, studies of the development of town planning in Vancouver are John Weaver, 'The Property Industry and Land Use Controls: The Vancouver Experience, 1910-1945,' *Plan Canada* 19 (September-December 1979):211-25 and John Bottomley, 'Ideology, Planning and the Landscape: The Business Community, Urban Reform and the Establishment of Town Planning in Vancouver, British Columbia' (PhD thesis, Geography, UBC, 1977). Bartholomew and Associates, *Plan for the City of Vancouver* is central to understanding the emergence of the 'city efficient.' A broader context is provided by M. Christine Boyer, *Dreaming the Rational City: The Myth of American City Planning* (Cambridge, MA: MIT Press 1983) and W. Van Nus, 'Towards the City Efficient. The Theory and Practice of Zoning, 1919-1939,' in A.F.J. Artibise and G.A. Stelter (eds.), *The Usable Urban Past* (Toronto: Macmillan 1979), 226-46. I have also drawn upon material in the Vancouver City Archives, including items in the City Clerks' Correspondence, Inward, vols. 26, 50; the City Clerk's Records, Series 1, Operational Files: Correspondence Inwards; and the Town Planning Commission, Series III, Operational 1 Files 1925-52, Series A-1-Correspondence. Here, for example, are J.A. Walker's 'History of Town Planning in Vancouver' (vol. 22, file 9) and Alfred Buckley's lecture proposal (vol. 13, file 3). The ideas of F.E. Buck are expressed in two articles in the *Journal of the Town Planning Institute of Canada:* 'Advantages of Town Planning,' 3 (1924):8-11 and 'Presidential Address,' 7 (1928):116.

Insiders and outsiders

Harry Archibald's views on Vancouver are in W.C. Archibald, *Home-Making and Its Philosophy, as Recounted by a Nesting Branch of the Archibalds* (Boston: Archibald 1910), 499. My treatment of the élite society of the West End relies very heavily on Angus E. Robertson's outstanding MA thesis, 'The Pursuit of Power and Privacy: A Study of Vancouver's West End Elite, 1886-1914' (Department of Geography, UBC, 1977), although particular points have been taken from Kalman, *Exploring Vancouver* 2; Kluckner, *Vancouver: The Way It Was*; and the city directories. Ann McAfee, 'Evolving Inner City Residential Environments: The Case of Vancouver's West End,' *BC Geographical Series* 15 (1972):163-181 is a useful brief statement. A useful, more general, statement about life in the 'Nob Hills' of North American cities is provided by M. Klein and H.A. Kantor, *Prisoners of Progress: American Industrial Cities, 1850-1920* (New York: Macmillan 1976), 204-42. J. Barman, *Growing Up British in British Columbia* (Vancouver: UBC Press 1984) is an excellent assessment of the role of private schools in sustaining the values and community of the élite.

The Chinese are the most studied of Vancouver's ethnic communities. *From China to Canada* (Toronto: McClelland and Stewart 1982) is a general history edited by Edgar Wickberg et al. D.C. Lai includes some material on Vancouver in his *Chinatowns: Towns Within Cities in Canada* (Vancouver: UBC Press 1988). Paul R. Yee has illuminated the activities of Chinese business in Vancouver with his 'Sam Kee: A Chinese Business in Early Vancouver,' *BC Studies* 69-70 (Spring-Summer 1986):70-96; 'Business Devices from Two Worlds: The Chinese in Early Vancouver,' *BC Studies* 62 (Summer 1964):44-67; and 'Chinese Business in Vancouver 1886-1914' (MA thesis, UBC, 1983). For this discussion, however, I have drawn especially fully upon Kay J. Anderson, '"East" as "West": Place, State and the Institutionalization of Myth in Vancouver's Chinatown, 1880-1980' (PhD thesis, Geography, UBC, 1986) and two articles drawn from this work: 'The Idea of Chinatown: The Power of Place and Institutional Practice in the Making of a Racial Category,' *Annals, Association of American Geographers* 77, no. 4 (1987):580-98 and 'Cultural Hegemony and the Race Definition Process in Chinatown, Vancouver, 1880-1980,' *Society and Space* 6 (1988):127-49. Anderson's *Vancouver's Chinatown: Racial Discourse in Canada, 1875-1980* (Montreal: McGill-Queen's University Press 1991) makes much of this material more readily available. A similar shift in views of American chinatowns is traced by I. Light, 'From Vice District to Tourist Attraction: The Moral Career of American Chinatowns, 1880-1940,' *Pacific Historical Review* 43

(1974):367-94. G. Cho and R. Leigh deal with 'Patterns of Residence of the Chinese in Vancouver' in J. Minghi (ed.), *Peoples of the Living Land: Geography of Cultural Diversity in British Columbia* (Vancouver: Tantalus 1972), 67-84.

British Columbians' attitudes toward Asians have been discussed at length. See especially W.P. Ward, *White Canada Forever* (Montreal: McGill-Queen's University Press 1970); P.E. Roy, 'British Columbia's Fear of Asians,' *Histoire Sociale/Social History* 13 (May 1980):161-72, reprinted in Ward and McDonald (eds.), *British Columbia*, 657-70; and P.E. Roy, *A White Man's Province: British Columbia Politicians and Chinese and Japanese Immigrants, 1858-1914* (Vancouver: UBC Press 1989). Ward's 'Class and Race in the Social Structure of British Columbia, 1870-1939,' *BC Studies* 45 (Spring 1980):17-35, reprinted in Ward and McDonald (eds.), *British Columbia*, 581-99, is also pertinent here. Further information on the Jewish community in Vancouver is provided by Christine B. Wisenthal, 'Insiders and Outsiders: Two Waves of Jewish Settlement in British Columbia, 1858-1914' (MA thesis, Geography, UBC, 1987). There is a fairly long literature on the Japanese and a smaller one on the Sikhs in Vancouver, but very little of this writing is geographical. For the ideas of Henry Angus see his 'Underprivileged Canadians,' *Queens Quarterly* 38 (1931):445-60 and 'A contribution to International Ill-will,' *Dalhousie Review* 13 (1933):23-33. It was F. Pemberley, 'Vancouver, a City of Beautiful Homes,' *British Columbia Magazine* 7 (1911):1313-15, who saw homes as the index of the city. As the first histories of Vancouver, W.N. Sage's 'Vancouver: The Rise of a City,' *Dalhousie Review* 17 (April 1937):49-54 and 'Vancouver: 60 Years of Progress,' *Journal of Commerce Year Book* 13 (1946):97-115 are perhaps appropriate end points for this chapter.

Acknowledgments: Thanks are due to the following: research assistants Mike Ripmeester, Maria Cavezza, and Yasmeen Qureshi, who toiled independently on small but important facets of this chapter in the early, middle, and late phases of its production; Kathleen Bednard and George Yu for their work on the Dunbar project; my colleague, J. Lewis Robinson, who shared with me maps and files reflecting his longstanding interest in Vancouver; Tom Zuber of Mountain Animation, whose sketches capture the variety of Vancouver housing; helpful librarians and archivists at UBC, City of Vancouver Archives, Vancouver Public Library, City of

Richmond Archives, and the New Westminster Public Library; generations of graduate and undergraduate students in Geography at UBC whose theses and essays provided many of the building blocks for my interpretation of the city; and two hard-working secretaries, Kelly Littlewood and (more recently and especially) Jade Wong, who reduced the clamour of competing demands on scarce time in order that these pages might be written.

Chapter 5: Primordial to prim order

The concept of resistance to growth and examples of the impact of topographic controls on the form of other cities can be found in *Urbanization and Environment*, by geographers T.R. Detwyler and Melvin Marcus (Belmont, CA: Duxbury Press 1972). Conversion of land from rural to urban uses is a major concern in the region, especially since the Fraser Delta supports some of the richest agricultural land in Canada. Statistics on the rates of conversion are found in several government publications, including *Vancouver Urban-Centred Region, 1976-82* and *Land Use in 1982 and 1986* (Ottawa: Environment Canada, Canada Land Use Monitoring Program, 1989); an assessment of the land available for expansion in the GVRD is found in *Greater Vancouver Land Resource: Current Use and Land Supply* (Burnaby: Greater Vancouver Regional District 1982).

An appreciation of the extent of the transformation of the original drainage system of the region is provided in an excellent illustrated account of 'Vancouver's Old Streams,' originally published in *WATERS*, the *Journal of the Vancouver Aquarium* (1978), and subsequently republished in revised form by the Vancouver Public Aquarium Association (in 1989). The special concerns of the Fraser River are covered in the government reports: L. Hoos and G.A. Packman, *The Fraser River Estuary: Status of Environmental Knowledge to 1974* (Ottawa: Environment Canada 1974); *Fraser Delta Environmental Management Plan* (Surrey: BC Ministry of Environment 1985); and K. Kennett and M. McPhee, *The Fraser River: An Overview of Changing Conditions* (New Westminster: Fraser River Estuary Management Program 1988). Early information on the stability of the distributary channels of the Fraser Delta has been generated by W.A. Johnston, 'Sedimentation of the Fraser River Delta,' *Geological Survey of Canada*, Memoir 125, published

in 1921. This was subsequently updated by the work of J.E. Armstrong, W.H. Mathews, and F.P. Shepard. Ken Hall of the Westwater Research Centre has spearheaded work on the quality of water and sediment in the Brunette River Basin. The most accessible summary of his findings is published in Anthony Dorcey (ed.), *The Uncertain Future of the Lower Fraser* (Vancouver: UBC Press 1976).

The air temperature heat island of Vancouver has been the subject of several publications originating in the Department of Geography at UBC. Relatively non-technical accounts include Tim Oke, 'The Distinction Between Canopy and Urban Boundary-Layer Heat Islands,' *Atmosphere* 14 (1976):268-77 and a section of *The Climate of Vancouver*, by John Hay and Tim Oke, in the BC Geographical Series, No. 23 (Vancouver: Tantalus Press 1976). Rather striking patterns of surface temperature heat islands 'seen' in colour images from weather satellites are to be found in Matthias Roth, Tim Oke, and Bill Emery, 'Satellite-Derived Urban Heat Islands from Three Coastal Cities and the Utility of Such Data in Urban Climatology,' *International Journal of Remote Sensing* 10 (1989):1,699-1,720.

The interesting case of the fog climatology of Vancouver was first covered by Jack Wright and Norm Penny in 1970 in a report entitled 'The Incidence of Fog at Vancouver International Airport,' *Technical Memorandum*, TEC 748 (Toronto: Department of Transport, Meteorological Branch); and then by Don Faulkner in 1978: 'On the Incidence of Radiation and Advection Fog at Vancouver International Airport,' *Technical Memorandum*, TEC 862. The subject fascinated Kadzuo Fukaishi, a visitor to the Geography Department at UBC so he extended the analysis to cover synoptic influences on fog along the coast of BC in 'A Climatological Analysis of the Fog at Vancouver, BC, Canada,' *Journal Hokkaido University of Education*, Sect. IIB, 30 (1979):25-44.

A short history of air pollution monitoring and air quality trends in the GVRD can be found in *GVRD Air Management Plan – Stage 1: Assessment of Current and Future Air Quality*, prepared for the GVRD Pollution Control Division by Concord Scientific Corporation and B.H. Levelton & Associates, 1989.

Illustrated histories of the False Creek basin are available in R.K. Burkinshaw, *False Creek*, City of Vancouver Archives, Occasional Paper No. 2,

1984 and C. Gourley, *Island in the Creek* (Madeira Park, BC: Harbour Publishing 1988). The history of Stanley Park is only one chapter in the fascinating history of the Vancouver Board of Parks and Recreation; see Richard Steele, *The First Hundred Years: An Illustrated Celebration* (Vancouver: Vancouver Board of Parks and Recreation 1988). The state of the forest in Stanley Park is examined in D.R. Blakewell, *Forest Maintenance Program for Stanley Park* (Vancouver Board of Parks and Recreation 1988). The efforts made to preserve the Camosun Bog are based on the research and subsequent proposals of Audrey Pearson, originally as part of her MA thesis in Forestry at UBC, and later published in 1985 by the GVRD under the title, *Ecology of Camosun Bog and Recommendations for Restoration.*

The isobel map of Stanley Park shown in Figure 15 is from the Vancouver Soundscape Project, directed by composer R. Murray Schafer of the Communications Studies Department of Simon Fraser University. There are several reports and recordings associated with this novel and fascinating project.

Acknowledgments: Special thanks to Wendy Hales for her assistance in researching materials on several topics, especially False Creek and photographs in the Vancouver City Archives; to Julie Orban and Yasmeen Qureshi, who constructed composite maps; and to the UBC Geography 417 class of 1990-1 who contributed many ideas.

Chapter 6: Vancouver, the province, and the Pacific Rim

Discussions of Fordism and post-Fordism are numerous. The two terms were first systematically explored by M. Aglietta in *A Theory of Capitalist Regulation* (London: New Left Books 1979), but a more accessible account is found in A. Lipietz, *Mirages and Miracles: The Crisis of Global Fordism* (London: Verso 1986). The idea of a new international division of labour was proposed by F. Froebel, J. Heinrichs, and O. Kreye in their book *The New International Division of Labour* (Cambridge: Cambridge University Press 1980), while B. Bluestone and B. Harrison popularized the notion of deindustrialization in their monograph *The Deindustrialization of America* (New York: Basic Books 1982). There are two very useful edited collections that apply all these ideas to the recent Canadian

context: W. Clement and G. Williams (eds.), *The New Canadian Political Economy* (Kingston: McGill-Queen's University Press 1989) and D. Drache and M. Gertler (eds.), *A New Era of Global Competition* (Kingston: McGill-Queen's University Press 1991).

The general framework for the discussion of Fordism in British Columbia comes from David Harvey's essay, 'The Geographical and Geopolitical Consequences of the Transition from Fordist to Flexible Accumulation,' in G. Sternlieb and J. Hughes (eds.), *America's New Market Geography* (Rutgers, NJ: Center for Urban Policy Research 1988), 101-34. The discussion of the forest products sector is based on a number of sources. The most important is Pat Marchak's book, *Green Gold: The Forest Industry in British Columbia* (Vancouver: UBC Press 1983), which is usefully updated in two other articles she has subsequently written: 'The Rise and Fall of the Peripheral State: The Case of British Columbia,' in R. J. Brym (ed.), *Regionalism in Canada* (Richmond Hill, ON: Irwin Publishing 1986), 123-59 and 'Public Policy, Capital, and Labour in the Forest Industry,' in R. Warburton and D. Coburn (eds.), *Workers, Capital, and the State in British Columbia* (Vancouver: UBC Press 1988), 177-200. The recent history of the mining sector is dealt with in R.W. Payne's article, 'Corporate Power, Interest Groups and the Development of Mining Policy in British Columbia, 1972-77,' *BC Studies* 54 (1982):3-37. His account is usefully supplemented by T.I. Gunton's *Resources, Regional Development and Provincial Policy: A Case Study of British Columbia* (Ottawa: Canadian Centre for Policy Alternatives 1982); and J. Bradbury's 'British Columbia: Metropolis and Hinterland in Microcosm,' in L.D. McCann (ed.), *Heartland and Hinterland: A Geography of Canada* (Scarborough: Prentice Hall 1987), 400-40.

The effects of the early 1980s recession in BC are discussed in R.C. Allen and G. Rosenbluth (eds.), *Restraining the Economy: Social Credit Policies for BC in the Eighties* (Vancouver: New Star Books 1986). A perhaps crisper analysis is offered by P. Resnick in two trenchant essays on the province's international economic repositioning and the rise of political neoconservatism in British Columbia: 'BC Capitalism and the Empire of the Pacific,' *BC Studies* 67 (1985):29-46 and 'Neoconservatism on the Periphery: The Lessons for BC,' *BC Studies* 75 (1987):3-23. The case study of Chemainus is drawn, in part, from Michelle Stanton's MA thesis, 'Social and Economic Restructuring in the Forest Products Sector: A Case Study of Chemainus, British Columbia' (Department of Geography, UBC, 1989). Finally, the rise and nature of the service sector in Vancouver is set out well by D. Ley and T. Hutton in 'Vancouver's Corporate Complex and Producer Services Sector: Linkages and Divergence Within a Provincial Staple Economy,' *Regional Studies* 21 (1987):413-24.

Several recent general works deal with the emergence of a Pacific Rim economic system or, at least, the Asian quadrant of it. Among them are David Aikman, *Pacific Rim: Area of Change, Area of Opportunity* (Boston: Little, Brown and Co. 1986); Staffan Burenstam Linder, *The Pacific Century* (Stanford: Stanford University Press 1986); Joel Kotkin and Yoriko Kishomoto, *The Third Century* (New York: Crown 1988); Brian Kelly and Martin London, *The Four Little Dragons* (New York: Simon and Schuster 1989); Robert Elegant, *Pacific Destiny* (New York: Avon Books 1990); Gerald Segal, *Rethinking the Pacific* (Oxford: Clarendon Press 1990); and Simon Winchester, *The Pacific* (London: Hutchinson 1991). Peter Nemetz (ed.), *The Pacific Rim*, 2nd ed. (Vancouver: UBC Press 1990) is more specialized but addresses Canada's role in the Pacific Rim.

Vancouver's links with the Asia-Pacific are covered in two monographs by Thomas A. Hutton and H. Craig Davis: *Vancouver as an Emerging Centre of the Pacific Rim Urban System* and *Immigration and Ethnic Conflict in Metropolitan Vancouver: A Challenge and Response* (Vancouver: School of Community and Regional Planning, UBC, 1989). See also David W. Edgington and Michael A. Goldberg, *Vancouver and the Emerging Network of Pacific Rim Global Cities* (Vancouver: Canadian Real Estate Research Bureau, Faculty of Commerce and Business Administration, UBC, 1990). Detailed statistics on trade can be obtained from the *Annual Reports* of the Vancouver Port Corporation. Statistics Canada was the source of data on overseas visitors to British Columbia, and Canada Immigration was the source for immigration numbers. Airline links, as well as passenger and cargo statistics, were obtained from Transport Canada and the International Civil Aviation Organization (based in Montreal).

The dynamics of Chinese investment across the Pacific Rim are discussed in Michael A. Goldberg, *The Chinese Connection: Getting Plugged in to Pacific Rim Real Estate, Trade, and Capital Markets* (Vancouver: UBC Press 1985); John De Mont and Thomas Fennall, *Hong Kong Money: How Chinese Families and Fortunes are Changing Canada* (Toronto: Key Porter 1989); and Donald Gutstein, *The New Landlord: Asian Investment in Canadian Real Estate* (Victoria: Porcépic 1990). Japan's trading prowess in the resources sector and the importance of the *sōgō shōsha* are covered in Ray Vernon, *Two Hungry Giants* (Cambridge: Harvard University Press 1983) and Kiyoshi Kojima and Tetsuro Ozawa, *Japan's General Trading Companies: Merchants of Economic Development* (Paris: OECD 1984). Recent Japan-Canada trade and investment relations are set out in Charles J. McMillan, *Bridge Across the Pacific: Canada and Japan in the 1990s* (Ottawa: Canada Japan Trade Council 1988). The Tumbler Ridge coal development has been documented in the *Report on Business* magazine, *Globe and Mail* (November 1988), 86-95.

The most comprehensive work addressing Vancouver's future economic development is by the Vancouver Economic Advisory Commission, 'Vancouver: A Strategy for Vancouver in the 1990s,' mimeo (Vancouver: Vancouver Economic Advisory Commission 1989). The Greater Vancouver Regional District, *Creating Our Future: Steps to a More Livable Region, Technical Report* (Vancouver: Greater Vancouver Regional District 1990) deals with a variety of long-term planning issues. Commentary on the city's current growth, and of likely prospects, is provided in the monograph by Alan F.J. Artibise and Michael Seelig, *From Desolation to Hope: The Pacific Fraser Region in 2010* (Vancouver: School of Community and Regional Planning, UBC, 1991).

Acknowledgments: The authors would like to thank Scott MacLeod, Michelle Stanton, Yaolin Wang, and Midori Yamamoto for their good work as research assistants in gathering data and information for this chapter.

Chapter 7: Vancouver since the Second World War

The sources used for this chapter fall into three broad groups. The first consists of books and academic papers obtainable in most libraries. The second comprises reports and working papers produced by the Greater Vancouver Regional District, departments of the various municipalities (especially Vancouver), and consultants. Most can be found in major libraries such as the Vancouver Public Library and the University of British Columbia Library. The third includes statistics produced by the GVRD, the municipalities, the three regional port authorities, and other government bodies.

In the first group, two books, Walter G. Hardwick's *Vancouver* (Don Mills: Collier-Macmillan 1974) and Donald Gutstein's *Vancouver Limited* (Toronto: James Lorimer & Company 1975) analyze growth, change, and sources of power from the perspective of the early 1970s, when political alignments were shifting and TEAM was changing planning agendas. V. Setty Pendakur, *Cities, Citizens and Freeways* (Vancouver: V.S. Pendakur 1972) addresses a major debate of the era, the outcome of which has had a lasting impact. L.J. Evenden (ed.), *Vancouver: Western Metropolis,* Western Geographical Series, Vol. 16 (Victoria: Department of Geography, University of Victoria, 1978) also contains chapters on the changes in the early 1970s. In addition, K.G. Denike examines the role of the city as a financial centre for western Canada, C.N. Forward analyzes its port function, S.W. Hamilton explains the operation of the land market in metropolitan Vancouver, and R.W. Collier and L.J. Evenden describe the evolution of the downtown and suburbs, respectively. The last two chapters can be compared with an earlier publication, John R. Wolforth's *Residential Location and the Place of Work* (Vancouver: Tantalus 1965), which has a set of maps on the relationships between work and residence locations. Finally, among publications of the 1970s, Edward M. Gibson, *The Urbanization of the Strait of Georgia Region*, Geographical Paper No. 57 (Ottawa: Lands Directorate, Environment Canada 1976) sets the growth of Vancouver in its historical and regional context. It includes a valuable sheet of maps by Louis Skoda.

Two publications of the 1980s deal, from different viewpoints, with the de-industrialization of central Vancouver and the concomitant growth of industry along the Main Arm of the Fraser River. These are Catherine Gourley, *Island in the Creek: The Granville Island Story* (Madeira Park, BC: Harbour Publishing 1988) and James S. Randall, *Metropolitan Vancouver. Industrial Decentralization and Port Activity on the Lower Fraser River,* Working Paper No. 12 (University of Toronto/ York University Joint Program in Transportation, 1983). They are worth comparing with an article on a much earlier stage in the process: Guy P.F. Steed, 'Intrametropolitan Manufactur-

ing: Spatial Distribution and Locational Dynamics in Greater Vancouver,' *The Canadian Geographer* 17, no. 3 (Fall 1973):235-58, which examines the decade from the mid-1950s to the mid-1960s.

There is a multitude of municipal and regional government and consultants' reports on industry from the 1970s and 1980s. Examples are Phil Paulson, *Industrial Development: Views and Issues of the Industrial Community* (Vancouver: GVRD, May 1976); Len Tennant, *Industry in Vancouver. A Preliminary Report for Discussion* (Vancouver: Planning Department, July 1977); and *The Demand for Industrial Land in the Lower Mainland* (Vancouver: Currie, Coopers and Lybrand, January 1983). Another body of reports deals with the 'Livable City' and 'Livable Region' concepts. They include Ruth Rogers, *Creating a Livable Inner City Community: Vancouver's Experience* (Ottawa: Ministry of State for Urban Affairs 1976); *The Livable Region 1976/1986* (Vancouver: GVRD 1976); *Greater Vancouver Economic Strategy. A Vision and Action Plan for the Livable Region. A Draft for Discussion* (Vancouver: GVRD Development Services Department, March 1989); and *Creating Our Future: Steps to a More Livable Region* (Vancouver: GVRD, September 1990). A final group of reports, which has become more prominent in the last few years, focuses on Vancouver's service sector and the city's role in the wider world. Two examples are Thomas A Hutton, 'A Profile of Vancouver's Service Sector,' presentation to the 'Megalopolis 90' Workshop on the role of tertiary industries in metropolitan developments, Plymouth, England, June 1989 (Vancouver: Economic Development Office) and 'Vancouver as an Emerging Centre of the Pacific Rim Urban System,' working paper presented to the inaugural Pacific Rim Urban Development Council Conference, Los Angeles, California, 6-8 August 1989 (Thomas A. Hutton and H. Craig Davis).

Most statistics were obtained directly from such bodies as the Fraser River Harbour Commission, the North Fraser Harbour Commission, the Vancouver Port Corporation, and the GVRD, or were found in their published annual reports. Supplementary materials came from Census of Canada and Statistics Canada publications and from the annual reports of federal agencies such as the National Harbours Board.

Acknowledgments: The authors wish to thank Maria Cavezza and Yasmeen Qureshi for their excellent work as research assistants for this chapter.

Chapter 8: From urban village to world city

Arthur Erickson's speech was reported in the *Vancouver Sun* by Daphne Bramham ('Erickson Tells City to Plan for 10 Million,' 11 October 1989); the *Sun* editorial, 'Planned Density Better All Round,' followed on 12 October 1989. The Design Vancouver debate was covered by Elizabeth Godley in the *Sun:* 'City Urged to Plan Growth' (12 February 1990). Sean Rossiter, joining the issue, alerted readers to Erickson's longstanding fascination with mega-projects, including his concept of the West End contained within a single building, in his article 'Pop.: 10,000,000' in *Vancouver* magazine (February 1990). Allan Fotheringham's contrast between the engineering and the aesthetic mind appeared in his column in the *Sun* on 21 December 1977. His representation of Vancouver as the 'village on the edge of the rain forest' was developed in his book *Malice in Blunderland* (Toronto: Key Porter 1982).

Planners and politicians in the city have frequently found themselves hoisted on the petard of growth versus livability. See, for example, the City of Vancouver Planning Department, *The Vancouver Plan: The City's Strategy for Managing Change* (1986) and City of Vancouver Planning Commission, *Downtown Vancouver: Planning Strategies for a Changing World* (1986), where the contradictions of growth and livability are recognized but not resolved.

The changing landscape of Kitsilano is discussed by David Ley in 'Inner City Revitalization in Canada: A Vancouver Case Study,' in the *Canadian Geographer* 25 (1981):124-48. Among literary works set in Kitsilano are John Gray, *Dazzled* (Toronto: Irwin 1984) and Sherman Snukal's successful play, *Talking Dirty* (Toronto:Playwrites Canada 1982). For the story of DERA see the chapter entitled 'The Downtown Eastside: One Hundred Years of Struggle,' in Shlomo Hasson and David Ley, *Neighbourhood Organizations and the Welfare State* (forthcoming). The contrasting shores of False Creek are interpreted in Ley, 'Styles of the Times: Liberal and Neoconservative Landscapes in Inner Vancouver, 1968-1986,' in the *Journal of Historical Geography* 13 (1987):40-56. A more detailed treatment of Expo '86 appears in Ley and Kris Olds, 'Landscape as Spectacle: World's Fairs and the Culture of Heroic Consumption,' in *Society and Space* 6 (1988):191-212. The subsequent history of the Expo lands, still undeveloped in 1991, has been steeped in controversy. From amidst a large stock of newspaper stories, see Gordon Hamilton, 'Breathtaking: Bouquet for $2 Billion Pacific Place,' *Vancouver Sun* (28 April 1988); 'Auditor General Says Expo Lands Sold at a Loss,' by Keith Baldrey in the *Sun* (2 May 1990); and Jack Moore's story 'Expo Land Profit Drop in Li Kashing's Bucket,' in the *Vancouver Courier* (17 March 1991). The sale of public land to Li Kashing has attracted attention further afield, including an editorial in the *Globe and Mail*, 'The Hasty History of the Expo Lands' (23 June 1989) and coverage in the *New York Times*, 'Prosperity from Asia has West in Conflict' (8 May 1989).

Our discussion of changing gender relations in the West End draws upon T. Fairclough's unpublished MA essay, 'The Gay Community of Vancouver's West End' (Geography, UBC, 1985). The growing visibility and tolerance of gay identity was also evident during the Gay Games, held in Vancouver in 1990: Stan Persky, 'Tolerant Gay Games Coverage was Right On,' *Vancouver Sun* (18 August 1990). There is also a lesbian community in Vancouver in the Commercial Drive area, a former Italian district, though the institutional profile of this community is not as visible as is that of gay men in the West End. The public display of affection by lesbian couples in a coffee bar on the Drive, and opposition to it, has become a cause célèbre: Brian Truscott, 'Three Arrested after Joe Calls in Police,' *Vancouver Courier* (16 December 1990).

We are indebted to Caroline Mills for our discussion of Fairview Slopes. Her PhD thesis, 'Interpreting Gentrification: Postindustrial, Postpatriarchal, Postmodern?'(Geography, UBC, 1989), includes interviews with residents of Fairview Slopes. We draw on two PhD theses for our discussion of Vancouver suburbs: Isabel Dyck's study of women residents in Coquitlam and Geraldine Pratt's research on residents of Surrey. Dyck's thesis, 'Towards a Geography of Motherhood,' was completed in the Geography Department, Simon Fraser University, in 1988. She has also published excerpts from this work in the *Canadian Geographer* 33 (1989):329-41 and *Society and Space* 8 (1990):459-83. Pratt's thesis, 'An Appraisal of the Incorporation Thesis: Housing Tenure and Political Values in Urban Canada,' was completed in 1984 in the Geography Department at UBC. Statistics for household patterns in Coquitlam, Port Coquitlam, and Port Moody in 1987 are drawn from a report by Share Social Services, *The TriCities:*

A Profile of Our Community (1989). Helen Potrebenko's poem, 'Witness to a Murder,' appears in the collection of her poems entitled *Life, Love, and Unions* (Vancouver: Lazara 1987).

The issue of gang violence in Vancouver has drawn extensive media coverage. Much of this has been sensationalist; see, for example, Terry Gould, 'Leader of the Pack,' in *Vancouver* magazine (September 1990) and the suite of essays under the general heading 'Terror in the Streets,' in *Maclean's* (25 May 1991). There are also frequent stories on gangs in the *Vancouver Sun* and other community newspapers; see, for example: L. Johnson, 'Youth Gangs Lure Young Victims as Activity Spreads to Richmond,' *Richmond Review* (10 April 1990) and M. Becker, 'North Shore Mean Streets,' *North Shore News* (21 September 1990). The cost of providing services for new immigrants is discussed in Gerry Bellett, 'Paying the Price: Federal Immigration is Costing a Fortune; Why Should Vancouver be Stuck with the Tab?' *Vancouver Sun* (21 March 1991), A9. As the title of the above article suggests, politicians in British Columbia are reluctant to increase funding for social services for immigrants, asserting that increasing immigration levels are the result of federal policies, and that the required services should therefore be paid by the federal government. Predictably, federal politicians are unwilling to meddle in areas of provincial jurisdiction. While the argument continues, the need for additional services only grows.

General surveys of Canadian immigration policy, and especially its redefinition in the 1960s, are: Freda Hawkins, *Canada and Immigration: Public Policy and Public Concern*, 2nd ed. (Kingston and Montreal: McGill-Queen's University Press 1988); Jean Leonard Elliott and Augie Fleras, 'Immigration and the Canadian Ethnic Mosaic,' in Peter S. Li (ed.), *Race and Ethnic Relations in Canada* (Toronto: Oxford University Press 1990), 51-76; and Victor Malarek, *Haven's Gate: Canada's Immigration Fiasco* (Toronto: Macmillan 1987). Immigration statistics reported in the chapter are drawn from quarterly and annual reports of Employment and Immigration Canada.

Vancouver's growing Chinese community has received much attention. Margaret Cannon provides an enthusiastic portrayal, though it is marred by constant references to the 'innate' business acumen of Chinese people, in *China Tide: The Hong Kong Exodus to Canada* (Toronto: Harper Collins 1989), while Donald Gutstein,

The New Landlords (Victoria: Press Porcépic 1990), is more sceptical of the value of Asian investment in Canadian cities. Also see Gordon Keast, 'The World's Longest Commute,' in *Equity* magazine (March 1988) and John Mercer, 'New Faces on the Block: Asian Canadians,' in the *Canadian Geographer* 32 (1988):360-2. The development of Vancouver's Chinatown is described in David Chuenyan Lai, *Chinatowns: Towns Within Cities in Canada* (Vancouver: UBC Press 1988) and Kay Anderson's PhD thesis, '"East" as "West": Place, State and the Institutionalization of Myth in Vancouver's Chinatown, 1880-1980,' completed in the Department of Geography, UBC, in 1986; excerpts have been published in the *Annals of the Association of American Geographers* 77 (1987):580-98 and *Society and Space* 6 (1988):127-49. Anderson also provides an important discussion of racism against Chinese-Canadians. Relations between home and work among Chinese women garment workers are investigated in Charles Mather's MA thesis, 'Flexible Manufacturing in Vancouver's Clothing Industry,' completed in the Geography Department at UBC in 1988.

Sources are meagre on Vancouver's South Asian community, and we have relied largely on newspaper accounts and the census for this discussion. We have also found the following accounts tangentially useful: Adrian C. Mayer, 'A Report on the East Indian Community in Vancouver,' working paper (UBC University Institute of Social and Economic Research 1959); James Gaylord Chadney, 'The Vancouver Sikhs: An Ethnic Community in Canada' (PhD thesis, Michigan State University, 1976); and Baljeet Dhaliwall, 'Sikhs in the Vancouver Region: A Descriptive Study of Certain Sikhs' Views of Education Since 1904' (MA thesis, Education, Simon Fraser University, 1985). Sikh settlement in the Scott Road area is discussed in Harold Munro, 'Thriving Surrey Bazaar Caters to Indo-Canadian Community,' *Vancouver Sun* (22 May 1990).

Much has been written on the issues of real estate prices and neighbourhood change in Vancouver. The Laurier Institute, speaking for the real estate industry, has published two reports: the first, by G. Schwann (*When Did You Move to Vancouver? An Analysis of Migration and Migrants into Metropolitan Vancouver*, 1989) is deemed by critics to be superficial, while the argument of the second is belied by its title:

The Housing Crisis: The Effects of Local Government Regulation, (by W.T. Stanbury and J.D. Todd, 1990). Our account of 'monster houses' in Kerrisdale relies heavily on Niall Charles Majury, 'Identity, Place, Power and the "Text": Kerry's Dale and the "Monster" House' (MA thesis, Department of Geography, UBC, 1990). The protest letter written by a resident of Churchill Street is quoted on page 74 of this thesis. Also see Kim Alexander Read, 'Continuity with Change: An Investigation of the "Monster" House Issue in Vancouver's Westside Single-Family Neighbourhoods' (MA thesis, School of Community and Regional Planning, UBC, 1989).

Urban politics and development issues up to 1980 are covered by Ley, 'Liberal Ideology and the Post-Industrial City,' in *Annals of the Association of American Geographers* 70 (1980):238-58. For another perspective, see Donald Gutstein, 'Vancouver,' in Warren Magnussen and Andrew Sancton (eds.), *City Politics in Canada* (Toronto: University of Toronto Press 1983). For some earlier insights into development issues, see Walter Hardwick, *Vancouver* (Toronto: Collier Macmillan 1974). The long view of civic politics is provided by Paul Tennant, 'Vancouver Civic Politics, 1929-80,' in *BC Studies* 46 (1980):3-27.

The freeway debate is discussed in detail in Setty Pendakur, *Cities, Citizens and Freeways* (Vancouver: 1972). The urban renewal struggle is told in H. Kim and N. Lai, 'Chinese Community Resistance to Urban Renewal: The Case of Strathcona in Vancouver,' *Journal of Ethnic Studies* 10 (1982):67-81. The freeway and renewal conflicts are covered in the chapter, 'Chinatown-Strathcona: Gaining an Entitlement,' in Hasson and Ley, *Neighbourhood Organizations and the Welfare State*. For a broader community study of Chinatown, see Kay Anderson, 'Cultural Hegemony and the Race-Definition Process in Chinatown, Vancouver, 1880-1980,' *Society and Space* 6 (1988):127-49. The preservation of Gastown is discussed in Mark Denhez, *Heritage Fights Back* (Toronto: Fitzhenry and Whiteside 1978).

The importance of Ray Spaxman's lengthy tenure as city planning director has scarcely been appreciated. For a start, see Ray Spaxman, 'Mr. Spaxman's Neighbourhoods,' in *Vancouver* magazine (October 1989). Redevelopment in Kerrisdale is discusssed in Danny Ho's MA thesis, 'An Investigation of the Reasons for, and Impact of, Rental Apartment Demolitions in Vancouver's Kerrisdale Neighbourhood' (School of Community and Regional Planning, UBC,

1989). Events surrounding the Expo evictions on the Downtown Eastside are recounted in Kris Olds, 'Planning for the Housing Impacts of a Hallmark Event: A Case Study of Expo 86' (MA thesis, School of Community and Regional Planning, UBC, 1988). The beginnings of gentrification in Grandview-Woodlands were described by Bradley Jackson, 'Social Worlds in Transition: Neighbourhood Change in Grandview-Woodlands' (MA thesis, Department of Geography, UBC, 1984).

A key document in attempts at regional planning in the metropolitan area was *The Livable Region 1976-1986*, a far-reaching document prepared by the Greater Vancouver Regional District in 1975. The report was updated and broadened in *Choosing our Future* (GVRD 1990). For its main points, see Daphne Bramham, 'GVRD Wants Improvements,' *Vancouver Sun* (23 July 1990). Different stages in the Terra Nova land conflict are covered in 'Terra Nova Land Defenders Elated Top Court To Hear Case' (*Sun*, 7 December 1989); 'Arson Suspected at Controversial Development,' by Peter Trask (*Sun*, 19 November 1990); and 'Farmland Votes Oust Mayors,' by Daphne Bramham and Harold Munro (*Sun*, 19 November 1990). The anti-growth sentiment in the suburbs is reviewed by Bramham, 'Voters Put Breaks on Growth' (*Sun*, 19 November 1990).

Acknowledgments: The authors acknowledge the research assistance provided by Debbie Leslie and Bruce Willems-Braun.

Chapter 9: The biophysical environment today

The GVRD prepares and issues both quarterly and annual reports of air quality which summarize ambient air quality as sensed by their network of monitors. Data from a decade of this monitoring formed the basis of the MSc research undertaken by Scott Robeson in his thesis, 'Time Series Analysis of Surface Layer Ozone in the Lower Fraser Valley of British Columbia' (MSc thesis, Department of Geography, UBC, 1987) which investigated the temporal variability of near-surface ozone at both diurnal and annual time scales. He went on to develop a set of statistically based models for the prediction of ozone concentrations. The history of air quality in the GVRD and a comprehensive statement of the current (1989) state of the environment was prepared by the consortium B.H. Levelton and Associates and Concord Scientific Corporation as the *GVRD Air Management Plan – Stage 1: Assessment of Current and Future Air Quality* (GVRD 1989). A rather more detailed analysis of meteorological conditions surrounding a particularly severe ozone episode is to be found in D.G. Steyn, A.C. Roberge, and C. Jackson's report, *Anatomy of an Extended Air Pollution Episode in British Columbia's Lower Fraser Valley* (BC Ministry of Environment and Parks 1990).

The continuing loss of wetland habitat was researched by M.E.A. North and J. Teversham for the *Environmental Impact Assessment of the Vancouver International Airport Expansion* in 1976. The figures given are from the 'Summary Report' dated January 1976. In the same year, Hubbard and Bell prepared a report for Environment Canada, Pacific Forest Research, entitled *Preliminary Description, Mapping and Interpretation of the Vegetation of the Greater Vancouver Regional District*, that mapped the remaining natural, including altered, habitats of the GVRD. A detailed comparison of their map (at 1:50,000 scale) with the 1989 BC air photographs (flight line A27396:1-274, 1:25,000 scale) was made by Maria Cavezza, working with Challenge 1990 funding in the summer of 1990. The figure 24.3 per cent loss of habitat since 1976 was the result of eight weeks of careful mapping and measurement using a dot planimeter. The detailed assessment of habitat in Richmond was prepared by AIM Ecological Consultants for the township of Richmond's 1984 Official Community Plan and was published as Background Paper No. 4, *The Environment: Natural Resources*. Similar research was done in Surrey by Susan Abs. The GVRD's response to the loss of habitat is briefly stated in a summary document entitled *Greater Vancouver's Green Zone: Draft Concept and Identification Guidelines* (GVRD, 24 May 1991). The problems of managing just one of the many protected parks' habitats is discussed in the draft version of the *Pacific Spirit Regional Park Management* Plan (GVRD, July 1991).

One of the most readable introductions to the complex field of water quality management in the Lower Fraser River is Anthony Dorcey's *The Uncertain Future of the Lower Fraser* (UBC Westwater Research Centre 1976). This work is now somewhat dated with regard to pollutant sources and levels in that the data were collected prior to the completion of the Annacis Island

sewage treatment plant. However, a number of federal-provincial reports have since been published under the auspices of the Fraser River Estuary Study (FRES 1980). Individual volumes in this series deal with industrial and municipal effluents, stormwater discharges, the impact of landfills, synthetic organic compounds, and a variety of other water quality themes. In 1985, a new federal-provincial study group, the Fraser River Estuary Management Program (FREMP), was established. The seven volumes in this series published between September 1988 and January 1991 draw upon and update the information in the earlier FRES reports. The keynote volume is *The Fraser River Estuary: An Overview of Changing Conditions* (FREMP 1988). The most recent FREMP overview of changing water quality conditions in the period since the FRES reports is the *Status Report on Water Quality in the Fraser River Estuary* (FREMP 1990).

A good introductory overview of landslide types and processes is G.H. Eisbacher's paper, 'First-Order Regionalization of Landslide Characteristics in the Canadian Cordillera,' in *Geoscience Canada* 6(1979):69-79. This paper includes a section on debris flows in the coastal belt of British Columbia. More detailed treatments of debris flows are given by G.H. Eisbacher and J.J. Clague, 'Urban Landslides in the Vicinity of Vancouver, British Columbia, with Special Reference to the December 1979 Rainstorm,' *Canadian Geotechnical Journal* 18(1981):205-16 and by D.F. VanDine, 'Debris Flows and Debris Torrents in the Southern Canadian Cordillera,' *Canadian Geotechnical Journal* 22(1985):44-68. Geotechnical studies of debris flows include the paper by O. Hungr, G.C. Morgan, D.F. VanDine, and D.R. Lister, 'Debris Flow Defenses in British Columbia,' *Reviews in Engineering Geology* 7(1984):201-22 and by M.J. Bovis and B.R. Dagg, 'A Model for Debris Accumulation and Mobilization in Steep Mountain Streams,' *Hydrological Sciences Journal* 33(1988):589-604. Seismic liquefaction of Fraser Delta sediments is discussed by P.M. Byrne, 'An Evaluation of the Liquefaction Potential of the Fraser Delta,' *Canadian Geotechnical Journal* 15(1978)205-16 and by J.L. Luternauer and W.D.L. Finn, 'Stability of the Fraser River Delta Front,' *Canadian Geotechnical Journal* 20(1983):603-16.

The hydrological background to flooding along the Lower Fraser River, together with engineering analyses of the need for upstream storage reservoirs, are discussed in several publications of the federal-provincial Fraser River Board, including the *Preliminary Report on Flood Control and Hydro-electric Power in the Fraser River Basin* (1958), *Final Report on Flood Control and Hydro-electric Power in the Fraser River Basin* (1963), and *Fraser River Upstream Storage Review Report* (1976). The hydrologic conditions producing floods in smaller basins draining the Coast Mountains near Vancouver are discussed in A.M. Melone, 'Flood Producing Mechanisms in Coastal British Columbia,' *Canadian Water Resources Journal* 10(1985):46-64 and in M. Church and M. Miles, 'Meteorological Antecedents to Debris Flow in Southwestern British Columbia: Some Case Studies,' *Reviews in Engineering Geology* 7(1987):63-80. A detailed account of the impacts of the 1948 flood along the Lower Fraser is given by W.R.D. Sewell, *Water Management and Floods in the Fraser River Basin*, Department of Geography Research Paper No. 100, University of Chicago (1965).

Acknowledgments: The authors acknowledge the research assistance of Christine Jackson.

Epilogue

Denis Cosgrove's *Social Formation and Symbolic Landscape* (London: Croom Helm 1984) provides a helpful historical account of the genealogy of landscape as a way of seeing. The discussion of the 'strategies of visual consumption' that promote selective readings of the landscape rests on Sharon Zukin's *Landscapes of Power: From Detroit to Disneyworld* (Berkeley: University of California Press 1991), which is indispensable for anyone who wants to reflect on the significance of landscape in North America. So too is Alexander Wilson, *The Culture of Nature: North American Landscape from Disney to the Exxon Valdez* (Toronto: Between the Lines 1991). Jan Morris's essay on Vancouver is in her *From City to City* (Toronto: MacFarlane Walter and Ross 1990). David Stoddart's *On Geography and Its History* was published in Oxford by Basil Blackwell in 1986. Paul Carter, *The Road to Botany Bay: An Essay in Spatial History* (Oxford: Oxford University Press 1987) offers a penetrating account of the processes and implications of 'naming' by explorers. The notion of a 'structure of feeling' is taken from Raymond Williams, *Marxism and Literature* (Oxford: Oxford University Press 1977). For an attempt to set time-space compression and the emergent global geographies of modernity (and post-modernity) within a politico-economic framework, see David Harvey, *The Condition of Postmodernity* (Oxford: Basil Blackwell 1989). The cultural dimensions of those changes are probably better explored in the company of James Clifford, *The Predicament of Culture: Twentieth-Century Ethnography, Literature and Art* (Cambridge, MA: Harvard University Press 1988) and Richard Fox (ed.), *Recapturing Anthropology: Working in the Present* (Santa Fe: School of American Research Press 1991) – which includes Appadurai's essay on ethnoscapes.

Credits

Abbreviations

BCARS British Columbia Archives and Records Service
CRA City of Richmond Archives
CVA City of Vancouver Archives
VPL Vancouver Public Library

Chapter 2: The primordial environment

Figure 1 courtesy Jim McKenzie
2 courtesy Canada Centre for Remote Sensing (E-1043-18364)
5 photograph by Olav Slaymaker
6 courtesy BC Ministry of Crown Lands, Surveys and Resource Mapping Branch
7 photograph by Olav Slaymaker
10 *Vancouver Sun* (1981)
11 photographs by Olav Slaymaker and Hans Schreier
15 photograph by Olav Slaymaker
16 photograph by Olav Slaymaker

Chapter 3: The Lower Mainland

Figure 2 Homer Barnett, Field Notes (Special Collections, UBC)
4 courtesy Special Collections, UBC
5 courtesy BCARS (Doc. 616.8 [57], R888 new)
6 top: courtesy BCARS (PDP1769)
bottom: courtesy Special Collections, UBC
7 simplified from 'Map of New Westminster, B.C., 1876' (Special Collections, UBC)
8 derived by Donna Cook from F.W. Laing, 'Colonial Farm Settlers on the Mainland of British Columbia, 1858-71' (BCARS)
9 James Lenihan census, May-June 1877 (DIA, RG10, reel C-10114, vol. 3650, file 8302); George Blenkinsop census, Dec. 1876 (RG88, vol. 494)
10 Annual Reports of the Ministers of Public Works (Department of Marine and Fisheries), *Sessional Papers,* Parliament of Canada
11 Census of Canada, 1881 (NAC, reel C-13285)

Chapter 4: The rise of Vancouver

Figure 1 Airphoto Collection, Department of Geography, UBC
2 Harland Bartholomew and Associates, *Plan for the City of Vancouver*
3 adapted from R.H. Meyer, 'The Evolution of Roads in the Lower Fraser Valley' (BA thesis, Geography, 1967), 18, 24, 31, 37
4 derived from G.I. Howell-Jones, 'A Century of Settlement Change. A Study of the Evolution of Settlement Patterns in British Columbia' (MA thesis, UBC, 1966)
5 courtesy CVA (Dist.P.156 N. 133 #1)
6 courtesy CVA (OUT.P.676 N. 213)
7 M.E.A. North, M.W. Dunn, and J. Teversham, *Vegetation of the Southwestern Fraser Lowland, 1858-1880* (Ottawa: Environment Canada 1979) and A. Ulmer, 'A Comparison of Land-Use Changes in Richmond British Columbia' (MA thesis, Geography, 1964)
8 Airphoto Collection, Department of Geography, UBC

9 courtesy CRA (P77-01-53-1964)

10 courtesy CRA (P77-01-16-1953

11 R.M Bone, 'A Land-Use Study of the Steveston District' (BA essay, Geography, 1955) and CRA (P77-16-02-nd)

12 based on H.I. Ivanisko, 'Changing Patterns of Residential Land Use in the Municipality of Maple Ridge, 1930-1960' (MA thesis, Geography, 1964) and P.H. Connolly, 'A Geographical Analysis of Historical Events in the Maple Ridge District' (BA essay, Geography, 1953)

13 republished by CVA

14 D. McCririck and G. Wynn, '"Building Self-Respect and Hopefulness,"' in G. Wynn (ed.), *People, Places, Patterns, Processes* (Toronto: Copp Clark Pitman 1990)

15 courtesy VPL (#2999)

16 courtesy CVA (7-79)

18 courtesy VPL (#10998)

21 sketches by George Yu

22 sketches by Tom Zuber

23 appeared along with similar advertisements in the *Vancouver Sun* through much of 1990 and 1991 (see especially 5 April, 16 and 23 August, and 14 September 1991)

24 Fraser Valley Centennial Edition, *British Columbian*, New Westminster, BC (27 November 1912)

25 based on P.D. McGovern, 'Industrial Development in the Vancouver Area,' *Economic Geography* 37(1961):189-206

26 Thomas Mawson, 'Vancouver. A City of Optimists,' *Town Planning Review* 4(1913):7-12

27 ibid.

28 Thomas Adams, 'Vancouver Civic Centre Competition,' *Town Planning Review* 6 (1915), 34

29 Harland Bartholomew and Associates, *Plan for the City of Vancouver,* 186

30 ibid., 174

31 ibid., 224

32 ibid., plates 14, 38, and 52

33 ibid., plate 50

34 courtesy VPL (#5266)

35 courtesy VPL (#25095)

36 courtesy VPL (#12786)

37 courtesy CVA (STR.p.351 N.383#1)

38 courtesy VPL (#5238)

39 courtesy CVA (492-43)

40 courtesy VPL (#23945)

Chapter 5: Primordial to prim order

Figure 1 courtesy Jim McKenzie (top) and Allen Aerial Photos (bottom)

3 courtesy CVA (STR.P.276 N.235)

4 CVA (DIST.P.132 N.111)

5 constructed from data supplied by the GVRD

6 J.J. Clague, J.L Luternauer, and R.J. Hebda, 'Sedimentary Environments and Post-Glacial History of the Fraser Delta and Lower Fraser Valley,' *Canadian Journal of Earth Sciences* 20 (1983):1,314-1,326

7 courtesy GVRD Library

10 courtesy CVA (DIST.P.35 N.24)

11 courtesy CVA (VAN.SC.P.147 N.82 and VAN.SC.P.25 N.29)

13 courtesy CVA (260-572)

14 courtesy Special Collections, UBC

15 R. Murray Schafer, Vancouver Soundscape Project, Simon Fraser University, Document No. 5

Chapter 6: Vancouver, the province, and the Pacific Rim

Figure 1 photograph by A.H. Siemens

2 *Financial and Economic Review,* 1989

3 photograph by A.H. Siemens

4 photograph by A.H. Siemens

5 compiled from data provided by Vancouver Port Corporation, Transport Canada
6 photograph by A.H. Siemens
7 photograph by D.W. Edgington
8 photograph by D.W. Edgington
9 photograph by D.W. Edgington

Chapter 7: Vancouver since the Second World War

Figure 1 based on *Greater Vancouver Key Facts: A Statistical Profile of Greater Vancouver, Canada* (Burnaby: GVRD Development Services 1991), 52; Population Census Subdivisions (Historical), Catalogue No. 92-702, *1971 Census of Canada* (Ottawa: Statistics Canada), vol. 1, part 1, table 2, 114-15
2 based on Census Tracts, Profiles: Vancouver (Part 2), *1986 Census of Canada* (Ottawa: Statistics Canada), endpaper maps
3 based on *Industry in Vancouver: Background Report* (Vancouver: City of Vancouver Planning Department 1989), 4, 6; Greater Vancouver industrial areas map (Vancouver: Royal LePage 1990); Greater Vancouver industrial map (Vancouver: Colliers Macaulay Nicolls 1990)
5 based on Census Tracts, Profiles: Vancouver, *1986 Census of Canada* (Ottawa: Statistics Canada); Census Tracts, Selected Social and Economic Characteristics, Vancouver, *1981 Census of Canada* (Ottawa: Statistics Canada); Economic Characteristics, Labour Force, Occupations, *1971 Census of Canada* (Ottawa: Statistics Canada), vol. 3, part 2; Labour Force, Occupations by Sex, Series 3.1, *1961 Census of Canada* (Ottawa: Dominion Bureau of Statistics), passim
6 courtesy Allen Aerial Photos
7 *Industry in Vancouver: Background Report* (Vancouver: City of Vancouver Planning Department 1989), 9, 21
8 courtesy Allen Aerial Photos
9 photograph by A.H. Siemens
10 based on Fraser River Harbour Commission, *Annual Reports,* 1954-90 (New Westminster: Fraser River Harbour Commission); and data from the Harbour Commission
11 photograph by A.H. Siemens
12 photograph by A.H. Siemens
13 photograph by A.H. Siemens
14 based on *Megatrends* (prior to 1984: *Real Estate Trends in Metropolitan Vancouver*) (Vancouver: Real Estate Board of Greater Vancouver 1980-90)
15 ibid. (1976-90)
16 photograph by A.H. Siemens
17 photograph by A.H. Siemens
18 photograph by A.H. Siemens
19 based on *Real Estate Trends in Metropolitan Vancouver* (Vancouver: Real Estate Board of Greater Vancouver, 1970); *Land Prices in Industrial Areas* (Vancouver: Royal LePage, July 1991)
20 based on data from Vanouver Port Corporation; *Annual Report: Harbour of Vancouver* (Ottawa: National Harbours Board 1950-75)
21 North Fraser Harbour Commission

Chapter 8: From urban village to world city

Figure 1 photograph by D. Hiebert
2 photograph by D. Hiebert
3 photograph by K. Olds
4 photograph by D. Hiebert
6 Lloyd Friesen, 'Out of the Closets and into the Street? The Increasing Visibility and Openness of Gay Business in the Urban Landscape' (undergraduate paper, UBC, 1986)
7 photograph by C. Mills
8 photograph by C. Mills
12 photograph by D. Hiebert
13 photograph by D. Hiebert
14 photograph by D. Hiebert
15 photograph by D. Hiebert

Chapter 9: The biophysical environment today

Figure 1   based on D.G. Steyn, A. Roberge, and C. Jackson, 'Anatomy of a Severe Ozone Episode in the Lower Fraser Valley, British Columbia' (prepared for the BC Ministry of the Environment, 1990)

2   from D.G. Steyn and D.A. Faulkner, 'The Climatology of Sea Breezes in the Lower Fraser Valley,' *BC Climatologic Bulletin* 20 (1986):21-39

4   Federal Air Photo A 37597:153

5   from Greater Vancouver *Liquid Waste Management Plan* (Vancouver: GVRD 1989)

6   from M.J. Bovis and B.R. Dagg, 'Mechanisms of Debris Supply to Steep Channels along Howe Sound, Southwest British Columbia,' in *Erosion and Sedimentation in the Pacific Rim,* International Association of Hydrological Sciences, No. 165 (1987), 191-200

7   from G.H. Eisbacher and J.J. Clague, 'Urban Landslides in the Vicinity of Vancouver, British Columbia, with Special Reference to the December 1979 Rainstorm,' *Canadian Geotechnical Journal* 18 (1981):205-16

8   from an appendix to an unpublished report to the Board of Governors, UBC, Cliff Erosion Task Force (4 December 1979)

9   from W.R.D. Sewell, *Water Management and Floods in the Fraser River Basin,* Department of Geography Research Paper No. 100, University of Chicago (1965)

Contributors

Trevor J. Barnes obtained a BSc degree in economics from University College London and MA and PhD degrees in geography from the University of Minnesota. An associate professor, his research interests are in the field of economic geography, broadly conceived.

M. Bovis, associate professor, obtained his BA and M.Phil. degrees from University College London and a PhD from the University of Colorado. His research focuses on the geomorphology of the Western Cordillera, with particular attention to mass movement processes and natural hazards. He teaches introductory physical geography, hillslope geomorphology, and field methods.

Kenneth G. Denike, assistant professor, has an MSc from the University of British Columbia and a PhD from Pennsylvania. He teaches urban and economic geography and has served on the Vancouver School Board.

David W. Edgington, associate professor, has a PhD from Monash University and studies and teaches the geography of Japan.

Derek Gregory obtained his MA and PhD from the University of Cambridge, where he taught from 1973 until coming to the University of British Columbia as professor of geography in 1989. His teaching and research interests focus on the relations between social theory and human geography and on the historical geography of modernity.

Walter G. Hardwick, professor, has an MA from the University of British Columbia and a PhD from Minnesota. His 1974 book *Vancouver*, and his roles as alderman (1969-74), director of the Greater Vancouver Regional District (1972-4), and chair of the Choosing our Future process in 1990-1, reflect his long-continued commitment to providing leadership for the region. His most recent publication, *Shaping a Liveable Vancouver Region* (1991), deals with polls and public responses to key urban issues, attitudes, and challenges.

Cole Harris has a BA from the University of British Columbia and a PhD from Wisconsin. He has been a Guggenheim Fellow and a Canada Council Killam Fellow, and he is also a Fellow of the Royal Society. He teaches the historical geography of Canada, and his current research interests deal with the historical geography of British Columbia, focusing particularly on changing patterns of settlement along the Fraser River in the nineteenth century.

Daniel Hiebert obtained his PhD in geography at the University of Toronto. He teaches the geography of Canada, urban history, and economic geography. His research interests include immigration policy, the formation of ethnic groups in Canada, and the entry of immigrants into housing and labour markets.

David Ley, professor, received his BA from Oxford and graduate degrees from Pennsylvania State University. His research and teaching focus on urban and social geography, with an emphasis on issues in Canadian inner cities. In 1988 he received the Award for Scholarly Distinction from the Canadian Association of Geographers.

Terry G. McGee is director of the Institute of Asian Research and professor of geography. A graduate of Victoria University of Wellington, he has taught at the universities of Malaya, Wellington, and Hong Kong, and at the Australian National University. A past president of the Canadian Association of Geographers and a Fellow of the Academy of Social Science of Australia, he has spent the last thirty years researching urbanization in Southeast Asia.

Robert N. North, associate professor, has his MA from Cambridge and his PhD from the University of British Columbia. A specialist on the USSR and its successor states, and in particular the transport system of the area, he has become interested in the economic development of Vancouver through teaching introductory economic geography, and is currently investigating the fate of firms displaced by rezoning since 1968.

M. North, senior instructor, completed her MA at Kansas and is interested in vegetation change in the Lower Fraser Valley. She teaches environmental studies and physical geography.

T.R. Oke, professor and head of the Department of Geography, has his BSc from Bristol and his MA and PhD from McMaster University. His research focuses on urban climates and the energy and water balance of cities. He teaches climatology and the physical environments of cities. He is a Fellow of the Royal Society of Canada and the Royal Canadian Geographical Society, and is the recipient of a Guggenheim Fellowship, the CAG Award for Scholarly Distinction, and the President's Prize of the Canadian Meteorological Society.

Geraldine Pratt, associate professor, has her BSc from the University of Toronto and her MA and PhD from the University of British Columbia. Her current research and teaching interests are in feminist and social geography and in the gender-, class-, and race-based segmentation of urban labour markets in Worcester, MA and Vancouver, BC.

J. Ryder, adjunct professor, has her BSc from Sheffield, her MA from McMaster University, and her PhD from the University of British Columbia. With research and consulting interests in quaternary and glacial geomorphology, quaternary and environmental geology, terrain analysis, and air-photo interpretation, Dr. Ryder teaches quaternary and glacial geomorphology and the physical environment of British Columbia.

O. Slaymaker, professor of geography and associate vice-president Research, has his BA from Cambridge, MA from Harvard, and his PhD from Cambridge. His research interests focus on the geomorphology and the sustainability of mountain environments, especially in British Columbia. He teaches geomorphology and hydrology, environmental science, society-environment interactions, and the history and philosophy of geography. President of the Canadian Association of Geographers in 1991-2, he has been awarded a certificate of merit from the Japanese Geomorphological Union, and is a member of the Norwegian Academy of Science and Letters.

Alfred Siemens, professor, has an MA from the University of British Columbia and a PhD from Wisconsin. His research focuses on pre-Hispanic wetland agriculture in meso-America. He teaches courses on Latin America, landscape interpretation, and cultural geography.

D.G. Steyn, an associate professor, obtained his BA and MA in physics from the University of Cape Town and his PhD in physical geography from the University of British Columbia. He teaches air pollution meteorology, boundary layer meteorology, field methods, meteorological instrumentation, and research methods in physical geography. His research involves modelling and measurement studies of ozone air pollution in the Lower Fraser Valley.

Graeme Wynn, professor of geography and associate dean of Arts, has his BA from Sheffield, and his MA and PhD from Toronto. An historical geographer with research interests in the development of Canada and other new-world societies, he has published most extensively on the Canadian Maritime provinces. He teaches the historical geography of Canada, cities, and environmental attitudes as well as the cultural geography of North America. He held a Canada Council Killam Research Fellowship in 1988-90.

Index

Note: bold page numbers indicate that the entry occurs in a figure caption.

A

Abbotsford, 27, 73, 232
Abbott Street, **143**
Adams, James, 91
Adams, Thomas, 121
Aerospace industries, 172
Aesthetic use of land. *See* Land
Agriculture, 84, 203, 264, 265, 267
Agrodome, 167
Air cargo. *See* Freight handling; Vancouver International Airport
Air pollution, 160-1, 164, 231, 267, 268-72, **269**, 291
Air quality monitoring, **269**
Airport. *See* Vancouver International Airport
Alaska, 18, 228
Alberni Canal, 63
Albert Creek, 31
Alcan, 178, 286
Alder trees, 23, 24, 34, 288
Alex Fraser Bridge, **216**
Allman, W.H., 101
Alouette River, 22, **132**
Alouette Valley, 86
Alpine Tundra zone, 32, 34-5
ALRT. *See* Light Rapid Transit
Altitudinal zonation, and biogeoclimatic regions, **36**
Aluminium, 178
Ambleside (neighbourhood), 220
Anglo-British Canning Company, 107
Angus, Henry, 140
Animals, non-Native, xiv, 45, 152, 158, 159
Annacis Island, 206, 211, **216**, 276, 277
Annance, F.-N., 41
Appadurai, Arjun, 292
Aquarium (Stanley Park), 167
Arbutus trees, 22, 33
Archibald, Harry, 129
Architecture, 50-2, 88-9, 101-5
Arrow Transfer Company, 220
Arthur Laing Bridge, 211
'Arts and Crafts' movement, 102
Asbestos, 178
Asia Pacific Foundation, 195
Asia-Pacific region. *See* Pacific Rim countries
Asian studies, 195
Asiatic Exclusion League, 134
Athabasca River, 197

Australia, 192; as trading partner, 190
Authority, of Chief Trader regarding Natives, 46, 47
Automobile industry, 172
Automobiles: and air pollution, 270, 272; industry, 172, 192, 197, **216**, 216; and transportation system, 73

B

Ballantyne Pier, 108-9
Bank of British Columbia, 91
Banks. *See* Financial services; *names of specific banks*
Barnet, **9**
Barnett, Homer, 39, 43, 46, 64
Barnston Island, **11**
Bartholomew, Harland, 259
Bartholomew and Associates (town planners), 123-9
Bathymetry (of Vancouver region), **19**
Baynes, E.G., 101
BC. *See* British Columbia
BC Magazine, 103
BC Place, **238**, 239
BC Saturday Sunset, 132
Beaches, 21, 117, 155, 252-3, 267
Beaufort Range, 24
Beaver (animal), 48
Beaver (boat), 48
Bellingham, 233
Bennett, W.A.C., 177
Bentall Centre, 208
Biogeoclimatic regions, **36**
Biophysical environment, 149-70, 290
Birds, 159, 267
Black Ball Ferries, 211
Board of Trade. *See* Chinese Board of Trade; Vancouver Board of Trade
Boardinghouses. *See* Rooming houses
Boundary Bay, 22, 287
'Boutiquing' (of retail areas), 182
Bowen Island, **2**, 32-3, 232, 289
Brascan, 184
Brentwood Mall, 217
Bretton Woods Currency Agreement, 172
Bridge River, 286
Bridgepoint Market, 223
Bridgeport (neighbourhood), 79, 82
Bridgeport Road, 224
Bridges, 73, 148, 201, 206, 210-11
Brighouse (neighbourhood), 79, 82
Brighouse Industrial Park, 211

British, presence in Vancouver region, 45, 52, 249, 250, 257
British Columbia, history of, 38, 49
British Columbia Canning Company, 107
British Columbia Electric Company, 206
British Columbia Electric Railway (BCER), 73, 95, 96
British Columbia Forest Products, 206
British Columbia Forest Service, 226
British Columbia Homes, 104
British Columbia Hydro, 286
British Columbia Land and Investment Agency, 91
British Columbia Mills and Timber Company, 102
British Columbia Packers, 107, 206, 223
British Columbia Railway, 177, 214
British Columbia Resources Corporation (BCRIC), 183
British Columbia Sugar Refinery, **8**
British Columbia Telephone Company, 212
British Properties (neighbourhood), 137, 252, 254
Broadway Corridor, 221
Brockton Point, 166
Bronfman family, 184
Brunette River, 29, **32**, 156-7
Bryn Mawr (neighbourhood), 89
Buck, F.E., 129
Buckley, Alfred, 129
'Builders' Special' (1920s), 104-5, **105**
Building materials, local, 155-6
Bungalows, California, 103, **103**, 104, 144
Burnaby, 28, 69, 117-18, 156, 157, 202, 206, 207, 247, 252, 263
Burnaby Lake, 73, 157
Burnaby Mountain, **3**, 18, 26, 35
Burnaby Street, **131**
Burnham, Daniel, 120
Burns Bog, 37, 149, 169
Burrard Bridge, **220**
Burrard Inlet, 22, 41, 58, 63-5, 73, 117-18, 135, 206, 213, **216**, 223; aerial view, **2**, **9**; pollution of, 154-5; seaports, 214, 226
Burrard Street, 121, 142, 206
Burrard Street Bridge, 162
Burrard Thermal Generating Plant, 161
Burrard Uplands, 168
Buses, 73, 212
Businesses, small. *See* Small business

C

C. Itoh and Company, 193
Cabin (style of house), 101-2, **102**, 144
Cable, trans-Atlantic, 53
California, as trading partner, 190, 233
California bungalow, 103, **103**, 104, 144
Cambie Street Bridge, **7**, 162
Cambridge (England), 292, 293
Camosun Bog, 168-9, 288
Campbell, Gordon, 265
Campbell, Tom, 259, 260
Campbell River, 232
Campus Canyon, 281
Canada Place, **4**, **5**, **194**, **222**
Canadian Club, 118, 132, 133
Canadian Fishing Company, 206
Canadian Forest Products, 201
Canadian National Railway (CNR), **8**, 213
Canadian Northern Pacific Railway, 108
Canadian Pacific Railway (CPR), **5**, **6**, 69, 85, 87, 104, 162, 201, **210**, 213, 223, 260
Canadian Paperworkers Union, 176
Canadianoxy Chemical Ltd., **215**
Canneries, 50, 52, 60-2, **61**, 63, 107, 111, 201
Canton Alley, 136, 137
Cantonese language, 63
Canyons, river, 18
Capilano Reservoir, 117-18, 156
Capilano River, 21, 25, 30, **32**, **33**
Capilano River Valley, 25, 35
Capilano watershed, 289
Capitol Hill, 26, 35
Carrall Street, 136
Cascade Mountains, 232
Cassiar mine, 178
Cedar trees, 24, 33, 63, 166, 289
Cemp/Eaton, 208
Centerm, 214
Charles Creek, 31
Cheakamus River, 18
Chemainus, 184-5, 232
Chemicals: industries, 216; and water pollution, 275-7
Cherry trees, 152
Chickens, 46
Children: as captives of Natives, 43; child-care services, 244-5
Chile, as trading partner, 190
Chilliwack, 73, 286
Chilliwack River, 47
Chimakuan (linguistic family), **40**
China. *See* People's Republic of China
Chinatown: New Westminster, 52; Vancouver, 134-142, **138**, 250, 252, 260, 262; Victoria, 136
Chinese, presence in Vancouver region, 50, 52, 56, 61, 64, 65, 66, 84, 130, 135-42, 250, **251**, 252, **253**, 253-4, 254, 255-6, 258, 295

Chinese Benevolent Association, 141
Chinese Board of Trade, 140
Chinese Immigration Act, 137
Chinook language, 45, 63, 64
Chlorination, of water supply, 118
'Chuck-a-luk,' 140
Church of England, 57
Cirques, glacial, 21
Citizens' League, 114
Citizenship, exclusion of Natives and Chinese, 50
'City Beautiful' movement, 120
City planning, 118-29, 198, 208, 212, 224, 231-2, 234-5, 261. *See also* Urban development
Civic Improvement League, 122
Clallum people, 41
Clark Drive, 162
Class differentiation, in Vancouver neighbourhoods, 89, 102, 104-5, 130
Cliff failures, 281-2
Clifford, James, 293
Climate, 22-3, 24-32, **27**, 153, 159
Clothing, Native, 43
Coal, as export product, 48, 106, 107, 180, 191, 196, 197, 213, 214, 232
Coal Harbour, 119-20, 121, 263
Coast hotels, 194
Coast Mountains, 18-19, 20, **21**, **23**, 24, 267, 285, 286, 288
Coast Salish, xvi, **40**, 41, 48
Coastal Alpine Tundra subzone, 35
Coastal Douglas Fir zone, 32-3, 35, 288
Coastal Western Hemlock zone, 23, 32, 35, 288, 289
Coastline (of Vancouver region), 19-22, 153-5, **154**
Coliform count, 276
Coliseum, 167
Collieries, 48
Colombia, as trading partner, 190
Colonization, British Columbia. *See* British Columbia
Colorado, xiv
Columbia Cellulose, 183
Columbia River, 41, 177, 286
Cominco, 178
Commercial use of land. *See* Land
Commission of Conservation (federal), 123
Committee of Progressive Electors (COPE), 261, 265
Communism, 140
Community Arts Council, 260
Community involvement, in city planning, 237, 260, 262
Community Resources Boards, 262
Company towns. *See* Single-resource communities
Competitions, civic, 121, 122
Computers, 184, 231

Concord Pacific development, **6**, 238
Condition of Postmodernity, The, 296
Condominiums, 237, 243, 262
Conifer trees, 24, 32, 37, 288
Conservatism, political, 182, 188-9
Conserver land, 289
Conserver use of land. *See* Land
Construction, 221. *See also* Building materials
Consumer goods, markets for, 172, 173
Container terminals, 180, 214, 226, 228
Contracting out, 189
Cook, James, 120, 294
COPE. *See* Committee of Progressive Electors (COPE)
Copper, as export product, 178
Coquihalla Highway, 189, 221, 232
Coquitlam, 32, 69, 156, 244, 245, 247, 263
Coquitlam Lake, 117
Coquitlam River, 20, **32**
Coquitlam River Valley, 20, 35
Coquitlam Water Company, 117
Coquitlam watershed, 149, 169, 289
Cordillera Mountains, 17, 18, 20
Corporations, multinational, 171, 172, 178, 230
Cost of living (Vancouver region), 113
Counter-culture, 259, 260, 261
Cowichan people, 41, 42, 46-7, 48, 58-9, 60, 67
CPR. *See* Canadian Pacific Railroad (CPR)
Crab fishery, effect of pollution on, 287
'Craftsman' movement, 102
Crestwood Industrial Park, 211
Crime, 231, 237, 247, **248**, 249
Crofton, 216, 232
Crofton House (school), 133, 134
Crown Colony. *See* British Columbia
Crown Heights (neighbourhood), 89
Crown Zellerbach, 175, 183, 184
Cruise ships, **5**, 221, **222**, **227**, 228
Cuestas, 18, 35
Culverts, 156
Cumyow, Won Alexander, 138

D

Dairy farms, **13**, 79
Daishowa Paper Manufacturing Company, 197
Daiwa Securities Company, 197
Dams, 118, 286
Davis, Chuck, 267
Dawson, George M., 63
Dawson Creek, 177
Dazzled, 237
de Champlain, Samuel, 49
Deas Island Tunnel, 211
Deas Slough, 276
Debris flows, 29, 31-2, 277-8
Decentralization (of Vancouver region), 263, 264-5

Deciduous trees, 34, 37, 288
Deer Lake, 157, 161
Deglaciation, of Vancouver region, 21
Delta, municipality of, 69, 153, 252, 264, 266, 274, 286
Delta formation, after Fraser Glaciation, 21
Demographics. *See names of cities and towns;* Population
Density, of residential areas, 219-20, 224
Dentallium shells, 43
Denver, xiv
Deposition, by glacial activity, 19
Depression (1915), 113, 121. *See also* Recession (1980)
DERA. *See* Downtown Eastside Residents' Association (DERA)
Deserts, 159
Design Vancouver, 234
Development. *See* Urban development
Devine, Harry, 93
Discharge, maximum daily (of rivers), **33**
Discrimination, 134-40, 255, 256. *See also* Racism
Distribution of goods, Vancouver as centre of, **180**, **181**, 189, 199
District Lot 139 (Dunbar), 96, 97, 100. *See also* Dunbar (neighbourhood)
District Lot 301 (False Creek), 90, 91. *See also* False Creek
Diversity, biological, 268
Divine Light Mission, 237
Divorce, 245
DL 139. *See* District Lot 139 (Dunbar)
DL 301. *See* District Lot 301 (False Creek)
Dogwood trees, 151
Dominion of Canada, establishment of, 38
Dominion Trust Building, 88
Douglas, James (Governor), 48, 54, 59, 63, 67
Douglas College, **220**, 223
Douglas fir (*Pseudotsuga menziesii*), **14**, 22, 23, 24, 63, 157, 166, 288, 289
Dow Chemicals, 213
Downtown Eastside (neighbourhood), 221, 237, **238**, 239, 262, 265
Downtown Eastside Residents' Association (DERA), 237, **238**, 262, 265
Downtown Vancouver. *See* Vancouver, city of
Dr. Sun Yat-sen Classical Chinese Garden, **196**
Dragon-boat races, 195
Drainage, **14**, **31**, 151
Dredging, 153-4, 162, **163**, 214
Drier Maritime Coastal Western Hemlock subzone, 35
Drought, 28
Dumping, of waste materials, 154-5, 156, 164

Dunbar (neighbourhood), 96-101, 98, **99**, **100**, **101**
Dunbar, Charles Trott, 96
'Dunbar Heights.' *See* Dunbar (neighbourhood)
Dunbar Street, 95, **97**
Dupont Street, 135, 139
Durrant and Durrant, 101
Dutton, Thomas, 157-8
Dyck, Isabel, 244
Dykes, 36, 78, 148, 153, 154, 285, 286, 287

E
Earthquakes, 18, 268, 282-3, 290
East Vancouver (neighbourhood), 89, 137, 249, 252
Eaton's, 208
Eburne, 73, 201
Ecological Reserve Act (1971), 289
Economic Advisory Commission (of City of Vancouver), 198
Economies of scale, and 'Fordism,' 173
Economy: British Columbia, 183-90, 192, 208; Canadian, 172, 203, 206; in 19th-century Lower Mainland, 56; Vancouver region, 106-16, 229, 236
Edmonds, 156
Edmonds, Henry V., 91, 95
Education: Asian studies, 195; foreign students, 229, 255; higher education, 209; and immigrant children, 258
Efficiency, of land use, 122-9
Effluents, industrial. *See* Water pollution
Elderly, 245, 256, 265
Electors' Action Movement, The (TEAM), 208-9, 261
Electronics industry, 172, 230
Elite class, 52, 130-5, 250. *See also* Class differentiation; Middle class
Elite Directory, 130
Empire Stadium, 167
Employment. *See* Labour force; Unemployment
'Empty nesters,' 220, 223
Enchantment Lake, 19
English as a Second Language (ESL) classes, 249
English Bay, **2**, **3**, 22, 41, 117, 155, **222**, **273**
English Bay Cannery, 95
Entrepreneurs, 194, 219, 254
Environment: as election issue, 265; Native relationship with, 43-4; protection groups, 236, 261; transformation by human action, xiv-xv, xvi, 290-1; and urban development, 147-70, 231
Environment, Ministry of (BC), 269, 272
Epic Data, 219
Erickson, Arthur, 234, 235

Erosion, by glaciers, 19, 21
Ethnic groups, 134-45, 236, 249-58
Europe, and investment capital, 196
European Economic Community, 185
Europeans, presence in Vancouver region, xiii, xiv, 38, 130, 250, 295
Exchange students, Asian, 195
Exhibition Park, 167, 168
Experience of Place, The, xvi
Explorers, 38
Expo '86, **6**, 165, 189, 213, 221, 223, 235, 237, **238**, 239, 255, 256
Export products, 63, 106, 109, 110, 173, **175**, 179, 213-14
Export Zones (Pacific Rim), 172
Eyak Athapaskan (linguistic family), **40**

F
Fairview (neighbourhood), 89, 96, 137, 187, 243, **243**
Fairview Slopes. *See* Fairview (neighbourhood)
False Creek: as former site of University of British Columbia, 114-15; as industrial area, 87, 90-2, **164**, 206, **210**, 213, 217, 230; pollution of, 155, 162-5, 259; public market, 223; redevelopment of, 165, 170, 219-20, **220**, 223, 238-9, 263; residential areas, 209, 223, **238**; as site of Native villages, 58; views, **7**, **162**, **163**, **210**
Family, changes in, 236, 237, 239-45
Fantasy Gardens, 266
Farmers, 56-7, 59-60, 67-8, 84-5
Farmland, 84, 114, 231, 261, 264, 265
Farms, 50, 52, 56-7, 78-81, 85-7, 86
Feasts, Native (potlatch), 43
Feng shui, 258
Ferries, 73, 211, 232
Fertilizers, 110, 152
Fifth Avenue, 94
Filipinos, 193
Films. *See* Motion picture industry
Financial services, 182, 187, 195-6, 197, 208
Finger lakes, 21
Fir trees. *See* Conifer trees; Douglas fir (*Pseudotsuga menziesii*)
Firearms, sale to Natives, 46-7, 48
Fires, and destruction of forests, 37, 64
Fish: impact of water pollution, 157, 274, 275; meal, as export product, 110; and Native economy, 41-3, 47, 60, 62
Fishing industry, 62-3, 203, 206
Fjords, 21-2
Floods: causes of, 30, 153, 268, 283-5; and dyke construction, 78; and gravel bed rivers, 31
Fog, 28, **161**
Food: Asian, 195; early settlers, 45; food banks, 245; industries, 110; Native, 43

'Fordism,' 171, 172, 173-82
Forest products industry: diversification of, 206; and export products, 173-4; and 'Fordism,' 175-6; on Fraser River, 202, 203, 214; and government concessions, 177; and Japanese investment, 193, 196, 197; labour force, 60, 113; profits, 177, 184; relocation to small towns, 232; restructuring (early 1980s), 183, 184; technological innovations, 184-5, 221. *See also* Lumber; *names of specific companies*; Pulp; Sawmills
Forests: aerial view, **12**; disappearance of, 71, 78; impact of removal, 149, 151, 156; original state of, 24, 35, 37, **150**, 159, 288
Fort George, 73
Fort Hope, 47, 49
Fort Kamloops, 47
Fort Langley, 38, 45-9, **47**. *See also* Langley
Fort St. John, 177
Fort Vancouver, 45
Fort Victoria, 48
Fort Yale, 47
Forty-third Avenue, 95
Fotheringham, Allan, 235
Fourth Avenue, 94, 95, **162**, **220**, 237, 259
Fraser, Simon, 44, 48
Fraser Canyon, 41, 42
Fraser Delta, 22, 35, 41, 169, 213, 276, 282-3, 286
Fraser Glaciation, 19-22, 37
Fraser Lowland, **16**, 20, 21, 29, 35, 267, 286, 287, 288
Fraser Mills, 202
Fraser River: aerial view, **17**; canyons, 18; dam proposals, 285, 286; dredging of, 153-4; and early inland transportation, 53; floods, 36, 286; flow characteristics, **32**, 272; gold rush, 38; Main Arm, 11, 272; as Native fishing site, 41-2; North Arm, 39, 201, 202, 214, 226, 272, **273**, 274; pollution of, 154-5, 272, 287; size, 29-30; South Arm, 214, 216, 287
Fraser River Board, 285
Fraser Street, 201
Fraser-Surrey docks, 214
Fraser Valley: air pollution, 231, 269-70; draining of marshes, 114; flood areas, **284**; fog, **29**; geology, 35; temperature, 26, 28; urbanization of, 231; vegetation, 36-7; wind patterns, 26
Fraserport, 211, **211**, 226
Free Trade Agreement, 173, 193, 233, 236
Free Trade Zones (Pacific Rim), 172
'Freeway debate,' 260
Freeways. *See* Highways

Freight handling industries, 213
Freight traffic, **211**, 73, **227**, **228**
French, presence in Vancouver region, 130
French Canadians, presence in Vancouver region, 45, 57, 65, 67
'Frenchtown,' 65
Friends of Stanley Park, 167
Fruit, as export product, 107
Frye, Northrop, xvi
Fur trade, 44, 47
'Future Growth. Future Shock,' xii

G

Gabriola Island, 232
Gambling, 140
Gangs, 237, 249, 258
Garden City movement, 116, 129, 144
'Garden Competition,' 121
Garden parties, 132
Gardens: private, 152, 159; public, 195, **196**. *See also* Parks
Garibaldi Park, 18, 19, 21
Gastown, 208, 260, 292
Gays, in West End, 240, 242
Gender relations, changes in, 239-45
Gentrification, of neighbourhoods, 182, 187, 262-3
Geography, science of, 294
George Massey Tunnel, 211
Georgia-Puget ice lobe, 20
Georgia Straight (newspaper), 237, 261
'Georgia Strait Region' (planning region), 232
Georgia Street, 119, **119**, 206
Geothermal springs, 18
Germans, presence in Vancouver region, 130
Gifts, at Native feasts (potlatches), 40, 43, 46
Gill nets, 62
Glaciers (Vancouver region), **17**, 18-24, 168
'Glen Brae' (mansion), 133
Glynn-Ward, Hilda, 134
Goat Mountain, 19
Gold, as export product, 106
Gold rush: Fraser River, 38, 49, 56, 135; Klondike, 107
'Golden Horseshoe,' 174
Golf courses, 264, 287
Gore Street, 137
'Gothic Revival' (house style), 102, **102**
Grain: and early settlers, 45, 47; as export product, 200; handling, 109, 213, 232
Grandview (neighbourhood), 90, 92, 93, 137, 142, 263
Grant Hill, 18
Granville (village), 58, 59
Granville Island, **162**, 209, **210**, **220**, 223, 292

Granville Street, 87, 89, 130, **150**, **162**, 208, 257
Granville Street Bridge, 162
Grasslands, 36, 37, 287
Gravel bed rivers, 31
Gravel deposition, 20
Gray, John, 237
Great Britain, 109, 192, 196; and colonization of Lower Mainland, 38, 49; as trading partner, 106. *See also* British
'Great freeway debate,' 260
Greater Vancouver Regional District (GVRD): boundaries, **202**; economic development of, 232; and Livable Region Plan, 263, 264-5; manufacturing industries in, 226; successor to Lower Mainland Regional Planning Board, 212; surveys, 147
Greater Vancouver Sewerage and Drainage District, 117
Greater Vancouver Water District, 118, 207
Green, Jim, 262, 265
'Green belt.' *See* 'Green zone'
'Green chain,' **175**, 185
Green Party, 236
'Green zone,' 231, 261, 268, 288, 289
Greenpeace, 236, 237, 261
Grouse Mountain, 25
Guildford Mall, 217
Gulf Islands, 28, 41, 211
Gulf of Georgia, 18, 267
Gutstein, Donald, 243

H

H.A. Simons, 229
Haiqua (dentallium shells), 43
Halkomelem language, **40**, 41, 63
Hall, J.Z., 94, 96
Hammond Lumber Company, 86
Haney, 23, 24, 73, 85
Harbour. *See* Port of Vancouver
Harcourt, Michael, 235
Hardwick, Walter, 147, 234, 235
Harland Bartholomew and Associates (town planners), 123-9, 151
Harmac, 216
Harris, Cole, xiii
Harrison Hot Springs, 18
Harrison Lake, 26
Harrison River, 47
Harvey, David, 295-6
Hastings Mill, 63, 64, 214
Hastings Park, 167-8, 170
Hastings Street, **4**, 136
Hawaiians, 45, 64, 65
Head tax (on Chinese immigrants), 135-6
Hemlock trees, 23, 24, 32-5, 157, 166. *See also* Coastal Western Hemlock zone; Mountain Hemlock zone

Herbicides, effect of, 152
Heritage preservation, 236, 260-1, 262
High-rise buildings, 208, 262
Highland Valley Mine, 178
Highways, 208, 211, 260
Hillcrest. *See* False Creek (neighbourhood)
Hindus, 130, 253
Hippies, 237, 259-60
Hiss, Tony, xvi
Hobbes, Thomas, 67
Hobby farms. *See* Farms
Hodograph, Vancouver International Airport, **271**
Hollyburn Mountain, precipitation, 25, **26**
Holy Trinity Anglican Church (New Westminster), 50, **51**
Home ownership, 105, 129, 244-5
Homeless, 115
Homer Street, 121
Homosexual groups. *See* Gays
Honda Corporation, 192
Hong Kong: immigrants from, 194, 250, 252, 254; as source of investment capital, 255; as trading partner, 172, 190
Hong Kong and Shanghai Banking Corporation, 196
Hong Kong Canada Business Association, 195
Honshu Paper Company, 197
Horne, J.W., 91
Horse-raising, 86
Horseshoe Bay, 73, 212
Hotel Georgia, 115
Hotel Vancouver, 89
Hotels, 51, 219, 221
House groups, Native, 39-40
Houses: company houses (Steveston), **83**; designs, **101**, 101-5, 102, **103**, 104, 144, 257, **257**; materials, 152; Native, **40**. *See also* Real estate
Housing: costs, **246**; multifamily, 245
Howard, Ebenezer, 116
Howe Sound: air temperature, 26; debris flows, 277-80, **279**; as fjord, 21, **22**; geology of, 18, 20; as Native habitat, 41
Hudson's Bay Company, 45, 49, 106, 115
Hui, Terry, 256
Human geography, of Vancouver region, xv, 38-68, 252
Hydroelectric power, 177, 286
Hydrology (of Lower Mainland), 29-32, 156-7

Ice ages, 19
'Ideal house' (style), 104, **104**
Ideal Lumber Company, 100
Igneous rocks, 18

Immigrants: country of origin, 236, 250, 252; discrimination against, 255, 256; early settlers, 49, 68; policy towards, 250; and social problems, 249
Immigration, early settlers, 57
Imports, Japanese, 192, 193
Income, polarization of, 245, 247
India, immigrants from, 250
Indian Affairs, Department of, 53
Indian Arm, **9**, 21, 149
Indian potato (*Sagitarria latifolia*), 42
Indian River Valley, 35
Indians. *See* Natives; Southeast Asians
Indonesia, as trading partner, 190
Industrial age (in BC), 49
Industrial areas, 84, 123, 202-3, 206, 211, 213, 223, 224, 230
Industries, 107, 110, 111, 174, 202. *See also names of specific industries*
Information industry, xvii, 182, 184, 187, 195, 231
Inner Harbour, 214, 223
Innocent Traveler, The, 129
Insecticides, effect of, 152
Instant Towns Act (1965), 178
Interior Salish, linguistic family, **40**
International Finance Centre, 187, 195-6, 197
International Woodworkers of America (IWA), 176-7
Inversions, 18, 270
Investment: by American corporations, 171; by Asian corporations, 96-198, 194; by immigrants, 254, 255
Iona Island, 276, 287
Iran, immigrants from, 254
Irish, presence in Vancouver region, 57
Iroquois people, 45
Irrigation, effect of, 152, 153
Italy, immigrants from, 130, 250
IWA. *See* International Woodworkers of America (IWA)

'J' bar, 185
Jackson Street, 142
Jacobs Creek, **32**
Jacob's River, 31
Japan: as market for BC products, 185, 189, 190, 191; as source of investment capital, 196; as trading partner, 109, 191, 192-3
Japanese: discrimination against, 85, 142, 255, 295; as immigrants, 130, 135, 250; in Steveston, 83
Jericho Beach, 155
Jews, 130, 142
Jim Pattison Group, 193
Joint Sewerage and Drainage Boards, 207
Juan de Fuca plate, 18

K
Kaiser Resources, 178, 180, 196
Kamloops, 232, 286
Kanaka Creek, 31, **32**
'Kanaka Road,' 65
Kaon project, 230
Kee, Sam, 137
Keefer Street, 136
Kennedy Lake, 19
Kentucky, xiv
Kerrisdale (neighbourhood), 89, 117, 220, 252, 254, 256, **257**, 257-8, 262
Kettle Valley Railway, 108
'Killarney' (mansion), 94
King Edward Avenue, 144
King George Highway, 217
Kingsway Street, 156, 217
Kirk, Mrs. L.M., 94
Kitimat, 178
Kitsilano (neighbourhood), 89, 94, 96, 134, 142, 187, 219, 237, **237**, 259, 261
Kitsilano Beach, 155
Kitsilano Indian Reserve, 108
Klondike Gold Rush, 107
Kluckner, Michael, xii
Knight Street Bridge, 211
Kuwait, immigrants from, 254
Kwantlen people, 39, 41, 44, 46, 58

L
Labour Code, changes to, 189
Labour force: earnings, 113, 180; foreign workers, 136, 137, 193-4; growth of, 107, 180, **188**; and international division of labour, xvii, 172; and labour market, 52, 111-12, 188-9, 217; unionization of, 173. *See also* Occupations
Labour Temple, 112
Ladies' Aid, 133
Ladies' Minerva Club, 133
Ladner, 73
Ladner Slough, 276
Lake City (industrial park), 206
Lakes, drainage of, 78
Land: and Native suppression, 55, 67; pre-emptions, 53-6, **55**; surveys, 54, **54**, 55, 122, 147; use, xii, 123, 235, 236, 286
Land Registry Act, 286
LANDSAT image of Vancouver, **17**
Landscape, concept of, 291
Landslides, 156, 268, 277-82
Langley, **12**, **13**, 245, 247, 274. *See also* Fort Langley
Larch Street, 133
Laurier Institute, 256
Law, British, and control of Lower Mainland, 48, 49, 52, 67
Law firms, 187
Le Corbusier, 235
Lead, as export product, 177, 178

Lebanon, 254
Legislative buildings (Victoria), 106-7
Lekwiltok people, 41, 44, 46-7, 48
Lexington, xiv
Li, Victor, 256
Li Ka-shing, 239
Light industries, 216-17, **218**
Light Rapid Transit, 189
Lighthouse Park, 159, 288
Lighthouses, **2**
Lillooet people, 41
Linguistic families, Native, **40**
Lions, The (mountains), 19, 20, 25
Lions Bay, 32
Lions Gate Bridge, 73, 148, 167, 169, 201
Liquefaction, of soil during earthquakes,
    282-3
Little Ice Age, 23
Little Mountain: geology of, 26;
    reservoir, 118, 156; rock quarry, 156
'Little Tokyo,' 250
Livable Region Plan, 212, 263, 264-5,
    288
Lodgepole pine, 23
Logging, in 19th century, 50, 65, 166,
    167
London, 53
Lonsdale Avenue, 73
Lonsdale Quay, 220, 223
Lord Kitchener School, 98
Lost Lagoon, **165**
Lougheed Highway, 85
Lougheed Mall, 217
Low gradient rivers, 29
Lower Fraser Valley. *See* Fraser Valley
Lower Mainland, history of, 38-68
Lower Mainland Regional Planning
    Board, 87, 207, 212, 232
Luck, Hip Tuck, 137
Lulu Island, 41, 78, 79, 84-5, 108, **109**,
    211, 276, 277
Lumber. *See* Forest products industry
Lumber, as export product, 63, 106, 107,
    109-10, 174, 200
Lushootseed people, 41
Lynn Canyon Park, 159, 169
Lynn Creek, 30, **30**, 63
Lytton, Edward Bulwer, 67

M

Macau, 194, 254
MacDonald Dettwiler, 219
Machinery industry, 180
MacKenzie (town), 178
MacMillan Bloedel, 167, 175, 184, 206,
    230
MacMillan Export Company, 184
Main Street, **7**, 108, 136, 142
Main Street Bridge, 162
Main Street Market, 252
Mainland Street, 206

Majury, Niall, 258
Malaysia, 190, 194, 254
Male (Musqueam village), 39
Manning Park, 23
Mansions (West End), 130-1, **131**
Manufacturing industries, 110-11, 171,
    173-82, 185, 187, 200, 213, 226
Maple Ridge, 85-7, **86**
Maple Street, 158
Maple trees, 151
Maps. *See names of cities, rivers, towns, etc.*
Marathon Realty, **5**, 221
Marine Drive, 223
Marion Lake, 23
Maritime Forested Mountain Hemlock
    subzone, 35
Markets, public, 223, 238, 252
Marpole, 201
Marriage, between white men and Native
    women, 45-6
Marshes. *See* Tidal marshes
Massey Tunnel, 211
Mather, Charles, 254
Matsqui, 232
Matthews, J.S., 157-8
Mawson, Thomas, 118-20, 121, 129
McCleery, Fitzgerald, 57
McGeer, Gerry, 115
McRoberts home (Richmond), **79**
McSpadden, George, 93, 96
Meadows, subalpine, 35, **35**
Meager Mountain, 18
Medical service industries, 219
Medium gradient rivers, 29
Melville Street, 206
Metal industries, 206
Metals: as export product, 110; and
    water pollution, 275, 276
Methodists, 51
Métis, 45
Metropolitan Club, 132
Metrotown, 218, 223, 263
Microclimate, 153, 159. *See also* Climate
Middle class, and community planning,
    261
Migration, of Natives, 41-3
Military force, and control of Lower
    Mainland, 67
Miller Creek, **23**
Mills, Caroline, 243
Millstream, **26**
Mineral Royalties Act (1974), 183
Minerals, as export product, 174
Mining industry, 111, 171, 177-8, **179**,
    183, 191, 196, 197, 206
Minority groups, 261
Miss Gordon's (school), 133
Mist, 26. *See also* Fog
Mitchell Island, 201
Mitsubishi Bank, 197
Mitsubishi Canada, 193

Mitsubishi Trading Company, 196, 197
Mitsui Bank of Canada, 197
Mitsui Corporation, 193
Mobile Data International, 219
Molluscs, 21
Molson Brewery, **220**
Molybdenum, as export product, 178
Monkey puzzle trees, 158
Monoliths, 18
'Monster houses,' 257, 258
Montane Wetter Maritime subzone, 35
Montreal, 229
Moody, S.P. (Colonel), 49, 51, 52, 63, 64
Moodyville, 58, 59, 65
Moral Reform Association, 140
Moran Canyon, 285
Morris, Jan, 290
Morris, William, 102
Mortuary sites, 40
Mosquito Creek, 31
Mosses, 24
Motion picture industry, 230, 231, 291
Motion pictures, Asian, 195
Mount Baker, **11**, 18, 267, 295
Mount Brunswick, 20
Mount Cayley, 18
Mount Garibaldi, 18
Mount Pleasant (neighbourhood), **7**, 89,
    92, 137, 219, 262
Mount Waddington, 23
Mountain creeks, 29, 31-2
Mountain Hemlock (*Tsuga mertensiana*),
    23, 24, 32-4
Mountain Hemlock zone, 32-4
Mountains, 21; geology of, 18
Multinational corporations. *See* Corpora-
    tions, multinational
Municipal Act, 286
Munitions industry, 200
Museum of Anthropology, 282
Musqueam Band, 226, 288
Musqueam people, xiv, 39, 41, 44, 58, 59
Mynah, Crested, 158

N

Nanaimo, 48, 216, 232
Nanaimo people, 41
Nanika Dam, 178
Nannies, 193
National Harbours Board, 214
Natives: and Christianity, 68;
    discrimination against, 52-3;
    displacement by Europeans, xvi, 295;
    early contact with Europeans, 44; and
    environment, 43; exclusion from
    citizenship, 50; feasts, 43; as fishermen,
    41, 59, 60, 62, 63; foods, 42, 43; impact
    of colonialism on, xiii, xiv, 67;
    inter-tribal wars, 43; as labourers, 47,
    52-3, 56, 59-60, 63, 64-5; loss of land,
    38, 55, 67-8, 78; mythology, 44;

population, 39, 57-8, **58**, 66; and rapid social change, 38, 46, 68; and reserves, 38, 78; seasonal migration patterns, 41-3; socioeconomic relationships, 42, 43; and traders, 47; use of fire in shrub management, 37; vaccination against smallpox, 57-8. *See also names of specific groups;* Women, Native

'Nature,' attitude towards, 290, 291

Nawittee (trading post), 44

NDP. *See* New Democratic Party (NDP)

Nechako River, 286

Neighbourhoods: adult-oriented, 243-4; and class differentiation, 89, 102, 104-5, 130; and community planning, 261-2; demographics, 240, 253-4; effects of redevelopment, 231, 236-9, 259; preservation of, 265. *See also names of specific neighbourhoods*

'Neighbours' (gay bar), 242, **242**

Neptune Terminals, **215**

New Democratic Party (NDP), 183, 213, 261, 262, 265

New Industrial Economies, 190

New International Division of Labour (Pacific Rim), 172

New Westminster: aerial view (1948), **70**; as capital of British Columbia, 52; construction by Royal Engineers, 50-1, 54; industrial areas, 202, 206; and interurban transportation, 73; in 19th century, 49, 50-3, 57; original Native residents, 39, 58, 59; original plan of, 50; population, 52, 69, **76-7**; port, **51**, 214; as regional town centre, 263; socioeconomics, 245, 247; water supply, 117; waterfront redevelopment, **220**, 223, 238

New Westminster Quay, 220, 223

New York, 221

New Zealand, 190

Newly Industrialized Economies, 172

Newsprint, as export product, 174

Nexus Engineering, 219

Nicomekl River, 286

Nicomekl River Valley, 35, 36

Nineteenth Avenue (West), 100, 101

Nobbs, Percy, 116

Noise measurements (Stanley Park), **166**

Non-Partisan Association (NPA), 259, 260, 262, 265

Nooksack people, 41

Noranda, 178, 184, 206, 230

North Alouette River, 31

North East Coal, 189, 196, 221

North Shore, 28, 156, 207. *See also* North Vancouver; West Vancouver

North Shore Mountains: aerial view, **15**, **16**; debris flows, 29, 31-2, 278-81; difficulty of land development, 169; geology of, 18, 19, 37; gravel bed rivers,

29; precipitation, 25; protected areas, 288

North Vancouver, 32, 202, 214, **215**, 245, 247, 249

NPA. *See* Non-Partisan Association (NPA)

O

Oak Street, **151**

Oak Street Bridge, 206, 210-11

Oakridge (neighbourhood), 254

Oakridge Centre, 206

Oblates, 52, 53

Occupations, 52, 187, 188, 200-1, **205**, 219, 229, 231, 237, 243, **253**, 253-4, 260

Ocean, effect on temperature, 28

Ocean Falls, 178, 183

Office buildings, 88, 181, 200-1, 206, 217-18, **218**, 263

Oil prices, effect of increase, 172, 183, 196, 229

Oji Paper Company, 197

Okanagan region, 107

Oliver, John, 114

Olympia, 20

Omni-Max Theatre, **8**

On Geography and its History, 294

Ontario, 57, 171, 174

Oolachan, 41, 62

OPEC. *See* Organization of Petroleum Exporting Countries (OPEC)

Opium dens, 140

Oppenheimer, David, 167

Orcadians, 57

Oregon, as trading partner, 190

Organic wastes, and water pollution, 275, 276

Organization of Petroleum Exporting Countries (OPEC), 172, 183

Oxen, use in logging, 65

Oxygen, dissolved, effect on fish, 274, 276

Oyster fishery, effect of pollution on, 287

Ozone, 161, 268-72, **271**

P

Pacific Centre Mall, 218

Pacific Great Eastern Railway, 73, 114, 177, 214

Pacific High Pressure System, 270

Pacific Rim countries, 172, 190-9, 250, 254

Pacific Spirit Park, 168, 288

Pan Pacific Hotel, **5**, 194, **194**

Panama Canal, 53, 110, 229

Park Royal (shopping centre), 206, **207**

Parks, 129, 166-9, 288; public, 123

Passage Island, **2**

Passenger traffic, 73

Pattern books (house styles), 102, 103

Pattison, Jim, 193

Pattullo Bridge, 73, 201

Pavement, effect on ecosystem, 152, 153

Paving Paradise, xii

Peace River, 177, 197, 286

Peat bogs, 79, 84, 289

Peat moss, 203

Pender Street, 121, **136**, 206

Pension plans, 115

People's Republic of China, 109, 191, 194

Peru, as trading partner, 190

Petroleum industry, 202

Philippines, immigrants from, 250, 254

Phillips, Art, 235

Phosphate, 214

Physical geography, urban, 146-70

Physiography (Vancouver region), **17**

Piedmont ice lobes, 20, 21

Pine trees, 24

Pitt Lake, 22

Pitt Polder, 289

Pitt River, 22, 31, **32**, 41, 42, 43

'Plan for Chicago,' 120

Planning. *See* City planning

Plants. *See* Vegetation

Plants, non-Native, 152, 158

Playland, 167

Pleistocene Epoch, 18-24, 280

Plum trees, 152

Podzolic soils, 24

Poets, Romantic, influence on Royal Engineers, 51

Point Atkinson, **2**, 26, **26**, 288

Point Grey (neighbourhood), 89, 101, 117, 118, 122; development of, 96, 142; early settlement, 133; ethnic mix, 130; exclusion of Chinese, 137; house styles, 104

Point Grey peninsula, 20, 21, 25, 26, 115, 273

Point Grey Towing Company, 226

Point Roberts, 39, 41, 48

Poland, 250

Pollution, 148-9, 268-77. *See also* Air pollution; Ozone; Soil pollution; Water pollution

Population, 39, 41, 66, 66, 76-7. *See also names of places or ethnic groups*

Port Coquitlam, **10**, 212, 213, 245, 247

Port Hammond, 85

Port Haney, 85

Port Mann, 213

Port Mann Bridge, 211

Port Mellon, 213, 232

Port Moody, 73, 156, 202, 245, 247, 250

Port of Vancouver: bulk-loading facilities, 180, 213; climate advantages, 26; container terminals, **4**, **8**, 213; cruise ship traffic, **227**, 228; expansion after the Second World War, 190-2; freight traffic, 226-7, **227**; history of, 108-10; Inner Harbour, 214; and job

creation, 226; redevelopment of, 238; truck barge dock, **5**
Portland, 233
Portugal, 250
Potash, as export product, 191, 214
Potatoes, 45, 46
Potlatch feasts, 43, 46
Potrebenko, Helen, 249
Poverty, 115, 236-7, 245, **247**
Powell River, 232
Powell Street, 142
Prairies, as importers of BC products, 107-8
Precipitation, 24-5, **25**, 274, 278, 285
Predicament of Culture, The, 293
Prefabricated cottages, 102, **102**
Presbyterians, 51, 57
Prince George, 286
Prince Rupert, 183, 189, 232
Private clubs, 132. *See also names of specific clubs*
Private enterprise, and political conservatism, 182
Private schools, 133
Privatization, of government services, 189
Processing industries, 276
Productive land, 286
Project 200, 260
Property, private, 53-6, 59, 67
Property values, 123, 256-7. *See also* Land
Prospect Point, 18
Prostitution, 139
Public markets. *See* Markets, public
Public transit system, 212
Puget glacial lobe, 20
Puget Lowland, 20, 289
Puget Sound, 295
Pulp: as export product, 174, 176, 180, 191, 197; mills, **179**, 183. *See also* Forest products industry
Punjab, 252
Punjabi Bazaar, 252

Q
'Quality of life,' 231
Quarries, 156
Queen Anne (house style), 102
Queen Charlotte Channel, **2**
Queen Elizabeth Park, 18, 156
Queen Victoria, 49, 52
Queensborough Bridge, 211
Quiney, James L., 94, 95, 96

R
Racism, 52-2, 68, 134, 258, 295-6
Railroads, 52, 73, 87, 106, 108, 177
Rainfall. *See* Precipitation
Rainforests, and vegetation, **35**
Rattenbury, Francis Mawson, 89
Ravines, and debris flows, 280-1

Raw materials, as export products, 171
Rayonier, 184, 206
Reagan, Ronald, 189
Real estate: commercial, 194, 197, 198, **225**, 229, 236, 239; residential, 54, 90-6, 123, 225, 256-7. *See also* Land
Recession (1980), 183
Recreation facilities, 123, **124**, **125**
Red River frame (fort), 45
'Regatta, The,' (condominium), 256
Regional districts, 212
Regional Taxation Centre (federal), 212
Regional town centres, 223
Relief camps, 115
Research industry, 230
Reserves, 38
Reserves, Native, 59
Reservoirs, water, 117-18, 156
Residential areas: density, 219-20, 224; planning, **126**; and rezoning, 209; safety concerns, 231. *See also names of specific neighbourhoods*; Neighbourhoods
Resource industries, 49, 52, 177, 178, 199, 230, 236
Resource procurement sites, 40, 42; Native, 41-3, 59
Restaurants, Asian, **194**, 195
'Restraint' (economic policy), 188-9
Retail areas. *See names of specific malls*; Shopping areas
Revelstoke, 108
Richmond, 69, 78-85, **80-1**, **82**, **83**, 84, 108, **109**, 148, 153, 180, 206, 207, 224, 238, 247, 249, 252, 263, 264, 265, 267, 274, 286
Richmond Square, 217
Richmond Town Centre, 223
Rio-Algam, 178
Rituals, Native, 40
Rivers: effect of urban development on, 156; flow characteristics, **32**; types, 29. *See also names of specific rivers and creeks*
Roads, 73, 166, 177, 277-8
Robberies (break and entry), **248**
Roberts Bank, 180, 196, 214, 226
Robson Street, 121, 208, 217
Roche Point, **9**
Rock quarries. *See* Quarries
Rock slides. *See* Debris flows; Landslides
Rocky Mountains, 17
Rogers, B.T., 130
Roman Catholics, 57, 68
Rooming houses, 51, 136, 243, **243**
Roosevelt, Theodore, xiv
Ross Street, 252
Rossland, 112
Roundhouse (CPR), **6**
Royal Commission, on Native land question, 59

Royal Engineers, 50-1, 54, 59, 67
Royal Mail Stage, **150**
Royal Navy, 67
Royal Vancouver Yacht Club, 132
Rubinowitz & Company, **143**
Runoff, **33**, 156, 157, 274, 285
Ruskin, John, 51
Russians, as trading partners, 47

S
Saanich people, 41, 42
Sage, Walter, 145
Salmon: canning industry, 60-3; as export product, 60, 106, 110; and flood control measures on Fraser River, 285; as food of settlers, 45; Fraser River run, 30; impact of urban development on streams, 157-8; in local rivers, 29
Salmon River, 29, **30**, **32**
Salmon River Valley, 35
'Salmonopolis' (Steveston), 79
San Francisco, 53, 106, 112, 194, 221
Sand, 20, 155
Sandwell-Swan Wooster, 229
Sanitation services. *See* Sewage disposal
Sanyo Kokusaku Pulp Company, 197
Sapperton, 50, 51, 52
Saskatchewan Wheat Pool grain elevator, **215**
Save Richmond Farmland Society, 264
Sawmills, 50, 52, 63-5, 85, **176**, 201, 202, 206, 214, 230. *See also* Forest products industry
Scandinavians, presence in Vancouver region, 130
Schools, 68, 249
Science World, **8**
Scots, presence in Vancouver region, 45, 57
Scott Road, 252, 253
Sea cargo. *See* Freight handling; Port of Vancouver
Sea Island, 78, 79, 211, 216, 226, **273**. *See also* Vancouver International Airport
Sea level, 20, 21
Seaboard Terminal, **215**
Seabus, **5**, 223. *See also* Port of Vancouver
Seaports, 226
Seattle, xiv, 20, 112, 194, 228, 233
Sechelt people, 41, 65
Second Narrows Bridge, 148, 201, 202
Sedge (*Carex lyngbyei*), 287
Sediment, in Fraser River, 153-4, 273, **273**
Sedimentary rocks, North Shore Mountains, 18
Seismic activity. *See* Earthquakes
Sequoia trees, 258
Serpentine River, 286
Serpentine River Valley, 35, 36

Service industries, 173, 180, 181, 185, 187-8, 189, 208-9, 219, 229, 231, 236, 237
Sewage: disposal, 116-21, **275**; treatment, 274-5, 277
Sewers, 156, 157
Sexual assault, 247, **248**, 249
Seymour Mountain, 25
Seymour River, 30, **32**, 118
Seymour River Valley, 25, 35
Seymour watershed, 149, 289
Shanghai Alley, 136
Sharp and Thompson (architects), 121
Shaughnessy (neighbourhood), 89, 104, 133, **135**, 137, 144, **151**, 152, 169, 252, 254, 262
Shaughnessy Heights. *See* Shaughnessy (neighbourhood)
Shelburn Refinery, **9**
Shellfish, as Native food, 43
Sher-a-Punjab (company), 252
Shipbuilding, 110, 172, 200, 202, 206
Shipping. *See* Freight traffic; Port of Vancouver; Railroads
Shopping areas, 201, 206-7, 208, 217-18
Shrubs, 24
Sidewalks, 90, 151
Sikhs, 142, **143**, 252, 253
Simon Fraser University, **3**
Singapore, as trading partner, 172, 190
Single parents, 243, 245
Single-resource communities, 178
Site, definition of, 171
Situation, definition of, 171
Siwash Rock, 18
Skeena River, 41
Skid roads, 65
Skid row (Downtown Eastside), 237
Skytrain, **220**, 223, 263. *See also* Light Rapid Transit
Slaves, kept by Natives, 39-40, 44, 48
Small business, 107, 123, 198-9, 201, 217, 230, 240, 242
Smallpox, 44, 57-8
Smelting industry, 171, 177-8
Snauq, 161
Snow, 25, 30, 31, 283-5
Snowmelt floods, 30, 31
Snukal, Sherman, 237
Social Credit Party, 177, 183, 189, 213
Social geography, of Vancouver region, xv, 38-68, 252
Social housing, 262
Social problems. *See* Crime; Discrimination; Poverty; Racism
Social services. *See* Welfare services
Socialists, 129
Society Promoting Environmental Conservation (SPEC), 237
Socioeconomics. *See names of cities and towns*

Soft Rock Café, 237
Software industry, 197
Sōgō shōsha. *See* Trading companies, Japanese (sōgō shōsha)
Soil, changes due to development, 151, 153
Soil pollution, 164
Soils, podzolic, 24
Solidarity Coalition, 239
Songs, Native, 40
Soup kitchens, 115
South Asians, presence in Vancouver region, **251**, 252, 253, 254
South Granville (neighbourhood), 220
South Korea, 172, 190, 191
South Vancouver (neighbourhood), 28, 90, 95, 117, 118, 122, 130, 137, 142
Southeasterly winds, 26
S.P. Moody and Company, 63
Spanish Banks, **2**, **3**, 20, 95
Spanish explorers, 38
Sparwood, 196
Spaxman, Ray, 234, 235, 262
SPEC. *See* Society Promoting Environmental Conservation (SPEC)
Speculation, land, 78, 92-4, 100-1
Spetifore lands, 264
Sphagnum moss, 168
Spokane, xiv, 108
Sproat, Gilbert Malcolm, 59, 64, 65
Spruce (*Picea*), 23
Squamish, 18, 214, 232
Squamish people, 41, 42, 64
Squamish River, **284**
Squirrels, 158-9
Stamp, Edward, 63, 64
Standard of living, 115
Stanley Park, 18, 65, 87, 119, 130, 134, 144, 151, 160, **160**, **166**, 166-7, 169, 288
Staple goods, 189
Starlings, 158
Stawamus Chief (monolith), 18
Stawamus River, 30, **32**
Steam power, 62, 63
Steamboats, 51, 53
Stearman, W.C., 133
Steel, as export product, 172, 196
Stephen, L.C., 101
Steveston, 73, **79**, 79, 83, **83**, 250
Stick style (house), 102
Stickley, Gustave, 102
Still Creek, 29, **32**, 157
Still Creek Valley, 35
Stock market crash (1929), 115
Stoddart, David, 294
Strait of Georgia, **16**, 20, 28, 37, 153-4, 206, 287
Straits people, 41
Strathcona (neighbourhood), **4**, 89-90, 96, 208, 254, 261, 265

Strathcona Property Owners and Tenants Association (SPOTA), 260
Streams, effect of urban development on, 156
Streetcars, 72-3, 88, 96, 97, 206
Streets, planning of, 87, 123, 128-9
Strikes, 112-13, 114, 115
Stucco cottages, 104, **104**
Stumpage fees, 177
Sturgeon, 41, 45, 62
Sturgeon Bank, 276
Suburbs, growth of, 209, 211, 213, 240, 244-5, 264
Sulley, Mrs., 132
Sulphur, as export product, 191, 213, 214
Sumas Lake, draining of, 114
Sumitomo Canada, 193
Sunshine Coast, 232
Surrey, 69, 78, **192**, 212, 247, 249, 252, 253-4, 263, 264, 274
Surveys. *See* Land
Sustainable development, xii
Swamps, drainage of, 78
System E, 285

T
Tait, W.L., 133
Taiwan, 172, 190, 191, 194, 252, 254
Talking Dirty, 237
Tantalus Range, 19, **21**
Tasu Mine, 178
Taylor, Carole, 234
TEAM. *See* (The) Electors' Action Movement (TEAM)
Tectonic plates, and earthquakes, **18**
Ted Creek, **30**, 31
Teit, James, 41
Telecommunications, 172, 182, 195, 231
Telegraph lines, 53
Temperature (in Vancouver region), 26-9, 28, 159-61, **160**
Terminal City, 130, 134
Terminal City Club, 132
Terra Nova farmlands, 264, 265, 287
Thackeray, William, 116
Thatcher, Margaret, 189
Thetis (boat), 48
Thompson, David, 44
Thompson people, 41
Thompson River, 285
Through the Heart of Canada, 71
Thunderstorms, 27
Thurlow Street, 133
Tidal marshes, 36-7, 114, 154-5, 164, 169, 287
Tilbury Island, 213
Timber leases, 64, 65
Tokyu Corporation, **194**
Tools, Native, 43
Topography (of Vancouver region), 17, **19**

Torrent tracks. *See* Debris flows
Tourism industry, 185, 189, 194, 197, 221, 229
Tourist attractions, 140-1
Tourist industry, 188, 194
Town Planning Act, 122
Town Planning Commission, 128
Town Planning Institute of Canada, 122, 129
Town Planning Institute of Great Britain, 122
Townhouses, 243, **243**
Towns, in early British Columbia, 50
Toyota Motor Corporation, **192**, 192, 197, 230
Trade, with Natives, 41, 46, 47
Traders, American, 45, 48
Trading companies, Japanese (sōgō shōsha), 191-3
Trail, 178
Trails (roads), 53
Trains. *See* Railroads
Trans-Canada Highway, 15, 211, 212
Transportation, and urban development, 53, **72**, **74-5**, 84, 87-8, 123, 148, 180, 203, **204**, 209, 211, 221-2, 259. *See also* Automobiles; Ferries; Freight traffic; Highways; Railroads; Roads; Seabus; Skytrain; Vancouver International Airport
Trapping, 48
Treaties, trade, 172
Trees, 28, 151-2, 153, 289. *See also* names of specific trees
TRIUMF project (UBC), 198, 230
Trucks, 73
Trudeau, Pierre Elliott, 261
Truman, Harry, 171
Tsawwassen, 25, 211
Tsawwassen people, 39, 41, 58
Tsunamis, and earthquakes, 283
Tudor Revival (house style), 103, 104, **104**
Tumbler Ridge, 189, 196, 221
Twigg Island, 226

U

Uncertain Future of the Lower Fraser, 272
Unemployment, 111, 112, 115, 183
Union Pacific Railroad, 53
Unions, xvii, 113, 174, 176-7, 178, 189, 228. *See also* Strikes
United States, 192, 193, 196, 250, 293
United States Navy, 118
University Endowment Lands, 137, 288, 289
University of British Columbia, 114-15, 120-1, 198, 230, 260, 281-2
Urban development: biophysical impacts, 147-70; and city politics, 259-66. *See also* City planning; Gentrification

Urban geography, physical, 146-70
Urban-industrial land class, 286
Urban theory, and composition of neighbourhoods, 240
Utah Mining Company, office building, 206

V

Values, of settlers, 49, 50, 52
Van Horne, W.C., 165
Van Horne Industrial Park, 211
Vancouver, city of: aerial view, **3**, **4**, **6**, **7**, **17**, **70**, **88**, **146**, **186**, **222**; aesthetic qualities, xi, 267, 290; as business and financial centre, 181, 199, 208, 229; construction boom, 221; as core-ring city, 212; cost of living, 113; crime rate, 247; demographics, 245, 292; downtown, 3, **4**, 181, 182, 206-7, 221; ethnic mix, 249-58; expansion, 69-145, 72, 200-33; geology of, 17-24; and government of BC, 181; labour market, 208-9; in original state, **16**; population, xi, xii-xiii, 69, 70, 73, **76-7**, 200-1, **201**, 234, **241**, 255, 259; regional policy and planning, 207; socioeconomics, xiii; transportation routes, **72**, **74**, 87-8; and Victoria, 106; view of skyline, **6**, **15**; water supply, 117-18; waterfront redevelopment, 238. *See also* Vancouver City Council
Vancouver, George, xi, xiii, xvi, 120, 294
Vancouver and District Waterfront Workers' Association, 116
Vancouver and Districts Joint Sewerage and Drainage Board, 117
Vancouver Board of Trade, 122, 132, 195
Vancouver City Beautiful Association, 121
Vancouver City Council, 198, 208, 259, 265. *See also* Vancouver, city of
Vancouver Club, 132, 134
Vancouver Community College, 249
Vancouver Cricket Club, 132
Vancouver General Hospital, 133
Vancouver Harbour Commission, 108, 226
Vancouver International Airport, **27**, 84, **161**, 169, 180, 191, 194, 211, 216, 226, **271**
Vancouver Island, 17, 24, 26, 28, 211
Vancouver Island mountains, 267
Vancouver Lawn Tennis Club, 132, 134
Vancouver Lowlands, 19-21
Vancouver Planning and Beautifying Association, 121
Vancouver Police Force, 140
Vancouver Port Corporation, 214
Vancouver Riding Club, 132
Vancouver Rowing Club, 132
Vancouver Social and Club Register, 133

Vancouver Stock Exchange, 181, 206, 219
Vancouver Water Supply Area, 289
Vancouver Water Works Company, 117
Vander Zalm, William, 265-6
VanTerm, **5**, 214
Vegetables, 45
Vegetation: biogeoclimatic zones, 32; and ecosystem, 28, **35**, 151, 270; Native, 32-7, 157, 287-8; non-Native, xiv, 152, 158
Venereal diseases, at Fort Langley, 46
Veterans' Land Act, 82
Victoria: as capital of British Columbia, 52, 53; decline as trade centre, 107; early importance as port, 53, 106; population, 106; and raw sewage discharge, 277; rivalry with Vancouver, 106-7
Victoria Drive, 93
Victoria Order of Nurses, 133
Victory Square, march by relief camp workers (1935), 115
Vietnam, immigrants from, 250
Views of water, as status symbol, 131
'Village' mentality, 235, 236
Volcanic activity, in Vancouver region, 18, **18**
Voluntary organizations, 133
Voting rights, 140
Voysey, C.F.A., 104
'Voyseyesque' cottage, 104, **105**, 144

W

Wages, 112, 113, 115
Wagon roads, 53
Wakashan, as linguistic family, **40**, 41
Wallerstein, Emmanuel, xvi
Wapato (*Sagitarria latifolia*), 42
Warehouses, 180, 206, 213, 216-17, **218**
Wars, inter-tribal, 43
Washington, 190
Water Commission, 156
Water District and Joint Sewerage and Drainage Board, 118
Water pollution, 157, 164, 267, 268, 272-7
Water quality, monitoring of, 276-7. *See also* Water pollution
Water Quality Objectives, 276, 277
Water shortages. *See* Drought
Water Street, 206
Water supply, 116, 117-18, 267, 289
Waterfront, redevelopment of, 223, 224, 238
Webber, Ellen, 44
Weber, Adna Ferrin, xiv
Welfare services, reduction in, 189
Welfare state, 236
Wesleyans, 57
'West Coast' lifestyle, 292

West End (neighbourhood), **3**, **4**, 102, 130, 133, **134**, 144, 151, 208, 217-18, 219, 240, 242
West Germany, 250, 254
West side (of city of Vancouver), 89
West Vancouver (neighbourhood), 25, 32, 202, 206, 245
Westerly winds, 24, 26
Western Hemlock (*Tsuga heterophylla*), 23, 157
Western Red Cedar (*Thuja plicata*), 23, 33, 166, 289
Westin Bayshore Hotel, 194
Westminster and Vancouver Tramway Company, 91
Weyerhaeuser, 175
Wheat, as export product, 109, 191
Whistler (town of), 221
White Paper on Immigration (1966), 250
White Rock (town of), 245
Widgeon River Valley, 35
Wildlife, 157, 158, 162, 267
Williams, Raymond, 290
Willow trees, 23, 24
Wilson, Alexander, 291
Wilson, Ethel, 71, 102, 118-29, 133
Winch, R.V., 130
Wind patterns, 24, 26-8
Winning of the West, The, xiv
Winnipeg, 108
Witness to a Murder, 249
Women: captured by Natives for ransom, 43; as foreign service workers, 193-4; in the labour force, 237, 244; marital status, **246**; occupations, 243; as percentage of population, **241**; as purchasers of condominiums, 243
Women, Chinese, 136
Women, Native, 45, 46, 47, 57, 60, 61, 62, 63, 65
Women's Auxiliary, 133
Women's Council of the YMCA, 133
Work, John, 44
Work camps, 49, 60, 111
Workers. *See* Labour force
Workers Compensation Act, 115
'World System,' xvi
World Tower, 88
World War One, 108-9, 121
World War Two, 110, 200-3
World's Fair (Chicago 1892), 120
Writing on the Wall, The, 134
Wybourn, Sidney T., 95-6

Y

Yale, James Murray, 45-6
Yaletown, **6**
Yorkshire Guarantee and Securities Corporation, 91

Yorkson Creek, **32**
Young Men's Christian Association, Women's Council, 133
Youth gangs. *See* Gangs
Yuen, Gim Lee, 137
'Yuppies,' 220, 237

Z

Zinc, as export product, 177, 178
Zoning, 101, **127**, 128, **128**, **209**, 219, 224, 262
Zoo (Stanley Park), 166, 167